LIFE, RE-SCALED

Life, Re-Scaled

The Biological Imagination in Twenty-First-Century Literature and Performance

*Edited by Liliane Campos
and Pierre-Louis Patoine*

© 2022 Liliane Campos and Pierre-Louis Patoine. Copyright of individual chapters is maintained by the chapter's authors.

This book was published with the support of the Institut Universitaire de France, the Sorbonne Nouvelle University, and the PRISMES – EA 4398 research laboratory.

This work is licensed under a Creative Commons Attribution-NonCommercial 4.0 International (CC BY-NC 4.0). This license allows you to share, copy, distribute and transmit the text; to adapt the text for non-commercial purposes of the text providing attribution is made to the authors (but not in any way that suggests that they endorse you or your use of the work). Attribution should include the following information:

Liliane Campos and Pierre-Louis Patoine (eds), *Life, Re-Scaled: The Biological Imagination in Twenty-First-Century Literature and Performance*. Cambridge, UK: Open Book Publishers, 2022, https://doi.org/10.11647/OBP.0303

Further details about Creative Commons licenses are available at https://creativecommons.org/licenses

All external links were active at the time of publication unless otherwise stated and have been archived via the Internet Archive Wayback Machine at https://archive.org/web

Updated digital material and resources associated with this volume are available at https://doi.org/10.11647/OBP.0303#resources

Every effort has been made to identify and contact copyright holders and any omission or error will be corrected if notification is made to the publisher.

ISBN Paperback: 9781800647497
ISBN Hardback: 9781800647503
ISBN Digital (PDF): 9781800647510
ISBN Digital ebook (EPUB): 9781800647527
ISBN Digital ebook (AZW3): 9781800647534
ISBN XML: 9781800647541
ISBN HTML: 9781800647558
DOI: 10.11647/OBP.0303

Cover: 'Life Along the Nile', image by Earth Resources Observation and Science (EROS) Center (2014), https://www.usgs.gov/media/images/life-along-nile. Public domain. Cover design by Katy Saunders.

Contents

Acknowledgements	ix
Notes on Contributors	xi
1. Introduction	1
Liliane Campos and Pierre-Louis Patoine	
Imagination, Science and Power	5
Questions of Scale	14
Aesthetic Trends	19
Chapter Presentation	23
Works Cited	33
I. Invisible Scales: Cells, Microbes and Mycelium	**39**
2. Human Environmental Aesthetics: The Molecular Sublime and the Molecular Grotesque	41
Paul Hamann-Rose	
The Molecular Sublime	45
Imagining Microbes: From the Molecular Sublime to the Molecular Grotesque	54
Molecular Landscapes: New Ways of Reading the Anthropocene	60
Conclusion: The Big Moment of the Very Small	65
Works Cited	66
3. Still Life and Vital Matter in Gillian Clarke's Poetry	69
Sophie Laniel-Musitelli	
The Poetry of Stone	71
Playing with Scale	76
Images of Metamorphosis and Development	80
Sounding the Flesh	85
Science in the Landscape	87
Works Cited	91

4. Mycoaesthetics: Weird Fungi and Jeff VanderMeer's *Annihilation*	93
Derek Woods	
Weird Ecology, Weird Fiction	98
Wood Wide Web as Ecological Genome	103
The Fungal Kingdom	108
Works Cited	116

II. Neuro-Medical Imaging and Diagnosis   121

5. To Be or Not to Be a Patient: Challenging Biomedical Categories in Joshua Ferris's *The Unnamed*	123
Pascale Antolin	
Challenging Medical Knowledge and Classifications	127
Challenging Neurological Reduction	132
Challenging Social and Literary Categories	135
Works Cited	143
6. Neurocomics and Neuroimaging: David B.'s *Epileptic* and Matteo Farinella and Hana Roš's *Neurocomic*	147
Jason Tougaw	
The Tools of Comics	153
The Tools of Neuroimaging	158
A Person Surrounds This Brain	164
Works Cited	179

III. Pandemic Imaginaries   181

7. The Fiction of the Empty Pandemic City: Race and Diaspora in Ling Ma's *Severance*	183
Rishi Goyal	
Works Cited	201
8. Dead Gods and Geontopower: An Ecocritical Reading of Jeff Lemire's *Sweet Tooth*	203
Kristin M. Ferebee	
Works Cited	226

9. Depopulating the Novel: Post-Catastrophe Fiction, Scale,
   and the Population Unconscious   229
Pieter Vermeulen
   The Population Unconscious   229
   Cosy Catastrophe   234
   Population between Science and Speculation in Science Fiction   238
   Survival at Scale in Post-Catastrophe Science Fiction   242
   Utopian and Realist Fictions   245
   Conclusion: Downscaling Survival   249
   Works Cited   256

IV. Ecological Scales   **259**

10. The Everyday Pluriverse: Ecosystem Modelling in *Reservoir 13*   261
Ben De Bruyn
   Introduction: The Rural Mesocosm   261
   Noticing Nonhuman Narratives   268
   Visualising Coexistence, Part I   274
   Modelling Interspecies Assemblages   280
   Visualising Coexistence, Part II   287
   Conclusion: Scale and Stoicism in the Everyday Anthropocene   291
   Works Cited   295

11. The Narrative and Aesthetic Strategies of Climate Change
    Comics   299
Susan M. Squier
   Making the Global Threat Personal   300
   Anthropomorphic Figures   304
   Biography and Autobiography   308
   Scientific Distance Versus Intimate Experience   312
   Works Cited   321

12. Displacing the Human: Representing Ecological Crisis
    on Stage   323
Kirsten E. Shepherd-Barr and Hannah Simpson
   'It's Actually Not About Us': The Paradox of Human-Centric Ecological
   Drama   326
   Shifting the Boundaries: The Spatial, the Temporal, and the Sensory   332

'Fragments, Shards, Whispers': Imagining the Impossible Other	342
Conclusion	347
Works Cited	349

13. Staging Larger Scales and Deep Entanglements: The Choice of Immersion in Four Ecological Performances — 353
Eliane Beaufils

Intermingling Life Forms and Scales	355
Forms of Displacement by Immersion	363
Reading Signs	369
The Place of the Spectator	371
A Diplomatic Theatre	374
Works Cited	376

List of Illustrations	379
Index	385

Acknowledgements

The work presented in this book was carried out with the support of the Institut Universitaire de France and the PRISMES laboratory at the Sorbonne Nouvelle University in Paris. We would like to thank all our contributors for their dedication and enthusiasm over the three years we spent on this project, making this book a truly collaborative work. We are also grateful to the other researchers who contributed to our preparatory workshops and made them so wonderfully stimulating: Frédérique Aït-Touati, Eric Bapteste, Sarah Bouttier, Pierre Cassou-Noguès, Josie Gill, Catherine Larose, Marc Porée, Dan Rebellato, Kerry-Jane Wallart, Apolline Weibel and Marion Clanet.

We would also like to thank Yale University Press, who were kind enough to allow us to publish an adapted version of Jason Tougaw's chapter, 'Neurocomics and Neuroimaging'. And our warmest thanks go to the editorial team at Open Book Publishers, for their constant support and excellent work on this project.

Notes on Contributors

Pascale Antolin is Professor of American Literature at Bordeaux Montaigne University, and head of the research group CLIMAS. A specialist of American modernism and naturalism, she has published books and articles on F. Scott Fitzgerald, Nathanael West, Frank Norris and Stephen Crane. For several years now, she has focused her research on the representation of illness in literature and published numerous articles on the subject in French, European and American journals. Recently, she has developed a special interest in the neuronovel, and published several articles, a special issue of *EJAS*, and a Wiley encyclopedia entry on brain fiction.

Eliane Beaufils is Habilitated Assistant Professor in Theatre Studies at the University Paris 8 and Vice-Head of the Doctoral School of Arts EDESTA. She studied German literature and civilization at Sorbonne University, as well as politics and sociology at the Institut d'Etudes Politiques de Paris. Her research fields comprise European contemporary theatre and performance, mainly in relation to violence, war, poetry, resonance, and the meaning and means of criticality today. Her most recent publications are with Eva Holling, *Being With in Contemporary Performing Arts* (2018), with Alix de Morant, *Scènes en partage. L'Être-ensemble dans les arts performatifs* (2018) and *Toucher par la pensée. Théâtralités critiques et résonances poétiques* (2021). In 2018 Eliane began a pluriannual research project on 'Theatre with regard to Climate Future'.

Ben De Bruyn teaches English Literature at UCLouvain, Belgium. He is the author of *The Novel and the Multispecies Soundscape* (2020) and *Wolfgang Iser: A Companion* (2012) as well as the co-editor of *Planetary Memory in Contemporary American Fiction* (with Lucy Bond and Jessica Rapson, 2018) and *Literature Now: Key Terms and Methods for Literary*

History (with Sascha Bru and Michel Delville, 2016). He is currently working on a new monograph provisionally entitled *Beyond Cli-Fi*.

Liliane Campos is a Lecturer in English and Theatre Studies at the Sorbonne Nouvelle and a research fellow of the Institut Universitaire de France. Her research explores ways in which contemporary novelists, poets, and performers engage with the images and discourse of biology. She is the author of *The Dialogue of Art and Science in Tom Stoppard's* Arcadia (2011) and *Sciences en scène dans le théâtre britannique contemporain* (2012). She has co-edited special issues of *Epistémocritique, Alternatives théâtrales,* and *Sillages Critiques,* the collective volume *Lectures de Tom Stoppard:* Arcadia (2011), and conference proceedings, *Living Matter / Literary Forms* (20th-21st centuries) on fabula.org. Her essays on contemporary fiction have appeared in *Textual Practice, Etudes Britanniques Contemporaines* and *Modern Fiction Studies*.

K.M. Ferebee is a Postdoctoral Research Fellow with the Narrating the Mesh Project in the Department of Literary Studies of Ghent University. Her interdisciplinary research interests are located in the environmental humanities, particularly in the intersections of postcolonial theory and critical posthumanism. Her current work includes a re-evaluation of subjectivity in the Anthropocene and the growing area of eco-deconstruction.

Rishi Goyal is Associate Professor of Emergency Medicine at the Columbia University Medical Center (in Medical Humanities and Ethics and the Institute for Comparative Literature and Society) and Director of the medical humanities major at ICLS. He co-edits the health humanities journal *Synapsis,* https://medicalhealthhumanities.com/.

Paul Hamann-Rose is Assistant Professor in the Department of English Literature and Culture at the University of Passau, Germany. Before joining the English Department at the University of Passau, he has held positions at the Goethe-University Frankfurt, the University of Siegen and the University of Hamburg. He studied at the University of London Institute in Paris and at the University of Hamburg, where he received his PhD. While pursuing an ongoing research focus on the history of literary engagements with genetics and proto-genetics, he is currently finalising a monograph on the genetic renegotiation of life itself in the contemporary novel. He has published articles in *Medical Humanities* and the *Journal of Literature and Science*, as well as book chapters on cultural representations of genetic science in contexts from postcolonialism to privacy. He has recently spent two extended research stays as visiting scholar at Vanderbilt University, USA, working on the NIH-funded transdisciplinary GetPreCiSe project on genetic privacy.

Sophie Laniel-Musitelli is Associate Professor at the University of Lille and a Junior Fellow at the Institut Universitaire de France. Her research focuses on the interactions among literature, the sciences, and philosophy in the Romantic era. She is the author of *"The Harmony of Truth": Sciences et poésie dans l'œuvre de P. B. Shelley* (2012), and of several articles and chapters on the works of Erasmus Darwin, William Blake, William Wordsworth, Percy B. Shelley, and Thomas De Quincey. She has edited *Sciences et poésie de Wordsworth à Hopkins* (2011), co-authored *Muses et ptérodactyles: La poésie de la science de Chénier à Rimbaud* (2013) and co-edited *Romanticism and the Philosophical Tradition* (2015), *Romanticism and Philosophy* (2015), *Inconstances Romantiques* (2019) and *Romanticism and Time* (2021).

Pierre-Louis Patoine is Assistant Professor of American Literature at Sorbonne Nouvelle University, co-director of the Science/Literature research group (litorg.hypotheses.org) and co-editor of the journal *Epistemocritique*. He has published a monograph on the role of the empathic, physiological body in the experience of reading (2015), co-edited collections on David Foster Wallace (2017) and Ursula K. Le Guin (2021), and articles exploring biosemiotic, ecocritical and neuroaesthetic approaches to immersion and altered states of consciousness, virality, planetary life, postmodern speciation and anthropocenic acceleration in the work of Ursula K. Le Guin, William S. Burroughs, J. G. Ballard, Kim S. Robinson and Frank Herbert.

Kirsten E. Shepherd-Barr is Professor of English and Theatre Studies at the University of Oxford. Books include: *The Cambridge Companion to Theatre and Science; Science on Stage: From Dr Faustus to Copenhagen; Theatre and Evolution from Ibsen to Beckett; Twentieth-Century Approaches to Literature: Late Victorian into Modern,* co-edited with Laura Marcus and Michele Mendelssohn; and *Modern Drama: A Very Short Introduction*.

Dr Hannah Simpson is Lecturer in Drama and Performance in the Department of English Literature at the University of Edinburgh. Her research is broadly interested in the staging of the human body, and the political, ethical, and affective charges that attend it. She is the author of *Samuel Beckett and the Theatre of Witness: Pain in Post-War Francophone Drama* (2022) and *Samuel Beckett and Disability Performance* (2022). Her current research project explores the forgotten stage plays of modernist novelists.

Susan Squier is Brill Professor Emerita of Women's, Gender, and Sexuality Studies and English at Pennsylvania State University. From 2017-2021 she was Einstein Visiting Fellow, Freie Universität, Berlin. She is the co-author of *Graphic Medicine Manifesto* (2015) and co-editor of *PathoGraphics: Narrative, Aesthetic, Contention, Community* (2020). Among her other works, she is the author of *Epigenetic Landscapes: Drawing as Metaphor* (2017), *Poultry Science, Chicken Culture: A Partial Alphabet* (2011), *Liminal Lives: Imagining the Human at the Frontier of Biomedicine* (2004), and *Babies in Bottles: Twentieth-Century Visions of Reproductive Technology* (1994). Squier is co-editor of the Graphic medicine book series at Penn State University Press, and past president of the Graphic Medicine International Collective, whose mission is to guide and support the uses of comics in health. She is currently collaborating on a book on Comics and One Health.

Jason Tougaw is the author of *The Elusive Brain: Literary Experiments in the Age of Neuroscience* and *The One You Get: Portrait of a Family Organism*. His essays are published in *Modern Fiction Studies, Literature and Medicine, Electric Literature, Literary Hub*, and *Out* magazine. He teaches literature and creative writing at Queens College, The City University of New York.

Pieter Vermeulen is an Associate Professor of American and Comparative Literature at the University of Leuven, Belgium. He is the author of *Romanticism After the Holocaust* (2010), *Contemporary Literature and the End of the Novel: Creature, Affect, Form* (2015), and *Literature and the Anthropocene* (2020), and a co-editor of, most recently, *Institutions of World Literature: Writing, Translation, Markets* (2015), *Memory Unbound: Tracing the Dynamics of Memory Studies* (2017), and a double special issue of *LIT* on "Contemporary Literature and/as Archive" (2019–2020). His current writing project studies the notion of world literary value.

Derek Woods is Assistant Professor of Communication Studies and Media Arts at McMaster University. He has held positions at the University of British Columbia and in the Society of Fellows at Dartmouth College. He writes about ecology, technology, scale, and modern narrative in relation to the history of science. With Karen Pinkus, he recently edited an issue of *diacritics* (47.3, 2019) on the topic of terraforming. With Joshua Schuster, he published *Calamity Theory: Three Critiques of Existential Risk* (2021). His book in progress is called 'Terrarium: A Media Theory of the Ecosystem'.

1. Introduction

Liliane Campos and Pierre-Louis Patoine

Looking back at the twentieth century, evolutionary biology and genetics stand out as two immensely influential sciences, whose images and discourses shaped the imagination of life across cultural forms. While Darwin's 'plots', as Gillian Beer noted in 2000, continued to generate productive and conflicting narratives, Darwin's presence 'in argument and in popular imagination' was strengthened by the progressive rise of genetic research.[1] The 'modern synthesis'[2] of the theory of evolution with that of Mendelian heredity, and the subsequent discovery of DNA and mapping of genes, prolonged and renewed many of the controversies surrounding the first reception of *The Origin of Species*. Human exceptionalism was questioned once again by newly discovered proximities between humans and other life forms, all equally reduced to 'elaborate contraptions [...] constructed and controlled by genes'.[3] Determinism seemed to return in a new biological guise. Just as the keywords of nineteenth-century evolutionary theory—words such as *struggle, nature, family*—drew what Beer calls their 'story-generating' strength from the variety of meanings they evoked,[4] certain words related to DNA—particularly its description as a *code* or *programme*—presented it as the 'logos of life' through a terminology whose very

---

1 Gillian Beer, 'Preface to the Second Edition', in *Darwin's Plots: Evolutionary Narrative in Darwin, George Eliot and Nineteenth-Century Fiction* (Cambridge: Cambridge University Press, 1983, 2000), pp. xvii–xxxii (p. xxiii), https://doi.org/10.1017/CBO9780511770401
2 Julian Huxley, *Evolution: The Modern Synthesis* (London: Allan & Unwin, 1942).
3 Evan Thompson, *Mind in Life: Biology, Phenomenology and the Sciences of Mind* (Cambridge MA: Harvard University Press, 2007), p. 173.
4 Beer, 'Preface', p. xxv.

familiarity was rich with narrative potential.⁵ The language of biological theory, as it percolates through the different media of its time, is always already rife with images and ambiguities. The growing field of 'science and literature studies' has demonstrated that novelists, poets and playwrights were quick to reveal and respond to such figurative and narrative potentialities.⁶

If natural selection and the genetic code were defining paradigms for the imagination of the nineteenth and twentieth centuries, what might be the equivalent for the twenty-first century? Twenty years into the promised 'century of biology',⁷ this book begins to answer that question, by investigating engagements with life sciences and biopolitical questions in recent European and North American fiction, poetry, graphic novels and performance. Though it is clear that images drawn from evolutionary theory and genetics, particularly those that move beyond genocentrism, continue to inform representations of organic life, the essays gathered here turn to some of the fields that have attracted most attention in recent years.⁸ They ask how artistic work integrates

---

5   Claire Hanson, *Genetics and the Literary Imagination* (Oxford: Oxford University Press, 2020), p. 1, https://doi.org/10.1093/oso/9780198813286.001.0001

6   Recent titles include Kirsten E. Shepherd-Barr's *Theatre and Evolution from Ibsen to Beckett* (New York: Columbia University Press, 2015); Natania Meeker and Antónia Szabari's *Radical Botany: Plants and Speculative Fiction* (New York: Fordham University Press, 2019), https://doi.org/10.5422/fordham/9780823286638.001.0001; Josie Gill, *Biofictions: Race, Genetics and the Contemporary Novel* (London: Bloomsbury, 2020), https://doi.org/10.5040/9781350099869; Tom Idema, *Stages of Transmutation: Science Fiction, Biology, and Environmental Posthumanism* (New York: Routledge, 2019); Clelia Falleti, Gabriele Sofia, and Victor Jacono, eds, *Theatre and Cognitive Neuroscience* (London: Bloomsbury, 2016); Jason Tougaw, *The Elusive Brain: Literary Experiments in the Age of Neuroscience* (New Haven: Yale University Press, 2018); Sam Solnick, *Poetry and the Anthropocene: Ecology, Biology and Technology in Contemporary British and Irish Poetry* (New York: Routledge, 2016), https://doi.org/10.4324/9781315673578; Lejla Kucukalic, *Biofictions: Literary and Visual Imagination in the Age of Biotechnology* (New York: Routledge 2021), https://doi.org/10.4324/9781003132325; Lara Choksey, *Narrative in the Age of the Genome: Genetic Worlds* (London: Bloomsbury Academic, 2021).

7   Craig Venter and Daniel Cohen, 'The Century of Biology', *New Perspectives Quarterly*, 21 (2004), 73–77, https://doi.org/10.1111/npqu.11423. See also W. John Kress and Gary W. Barrett, eds, *A New Century of Biology* (London: Penguin Random House, 2016).

8   The essays gathered in this volume were written following a series of workshops supported by the Institut Universitaire de France (IUF) and organized at the Sorbonne Nouvelle in 2019 and 2020. The workshops invited scholars working in the fields of literature and science, medical humanities, ecocriticism and microbiology

new knowledge about neurons, microbes or fungi, how it echoes new perspectives brought by epidemiology, ecosystem modelling, or Earth System science, and how it contributes to critical thought concerned with those fields. As they tease out the key images and ideas informing representations of human and other-than-human life, the studies in this volume demonstrate contemporary Western culture's fascination for the life sciences—from microbiology to ecology. The fields under scrutiny also include demographics, climate science and geology. Indeed, firm boundaries between 'life' sciences and physical sciences seem increasingly artificial when life is considered at the scale of an 'Earth System'—a system, moreover, whose balance is rapidly shifting.

It will be no surprise, therefore, that this volume highlights intricate connections and overlaps between different scientific imaginaries in contemporary literature and performance. Susan M. Squier, in her analysis of climate change comics, demonstrates that the imagination of biodiversity loss is linked with fraught questions of climate science communication. Through her study of Welsh poet Gillian Clarke, Sophie Musitelli argues that contemporary biological images are infused with the geological awareness of Anthropocene writing. Her investigation of fossilized life is echoed by Kristin Ferebee's analysis of the graphic novel *Sweet Tooth*, which contrasts the 'fossilized death' driving oil-based economies with the connection, preserved in indigenous ontologies, between life and fossil strata. In their critical readings of post-apocalyptic fiction, Pieter Vermeulen and Rishi Goyal ask what unspoken biopolitical ideologies, fantasies of biopower and population control may lurk behind pandemic imaginaries. Contemporary performance, explored here by Kirsten E. Shepherd-Barr, Hannah Simpson, and Eliane Beaufils, encounters challenging yet generative paradoxes when it engages with the non-anthropocentric perspectives of ecology and climate science.

These essays prove it is worth asking, once again, the questions formulated by Gillian Beer: 'What new tales are being unleased from scientific work now? And what new forms for storytelling?'[9] And, we might add, what new scales? For the scale at which we imagine life is a recurrent, central concern in the artistic work examined here. The

---

to explore new scales and images derived from biology in twenty-first-century literature and performance.
9   Beer, 'Preface', p. xxviii.

writers and performers presented in this volume not only question our relation to microscopic or macroscopic scales, from cellular biology to climatology, but also experiment with aesthetics that connect disparate scales, including alternating focalizations, 'pluriverse' perspectives, 'multiscale narration', utopian microcosms, and 'neo-sublime' or grotesque aesthetics. In the neuronarratives examined by Jason Tougaw and Pascale Antolin, the key question is the 'explanatory gap' between neurobiological phenomena and human experience, in a culture suffused with misleadingly transparent images of the brain. For the ecocritical studies in this volume, the gaps are rather those that separate everyday perception from the realization that human life depends on microbiological and macroecological phenomena. From those perspectives, it becomes clear that relations between different scales of life is a crucial, perhaps *the* crucial question that emerges from our contemporary biological imagination.

The idea of forms being 'unleashed from scientific work' may seem outdated, suggesting that science is a source or resource to be tapped. Rather than viewing either science, literature, or performance as the primary source of biological imaginaries, we follow N. Katherine Hayles' advice to beware of the idea that 'influence' flows from one field to another. The 'cross-currents', as Hayles points out, 'are considerably more complex than a one-way model of influence would allow', since 'culture circulates through science no less than science circulates through culture'.[10] These complex currents appear clearly when we pay attention to how popular science—the forms science takes when engaging general audiences—interacts with contemporary imaginaries. In their contributions to this volume, Paul Hamann-Rose and Derek Woods are attentive to the evolution of popular biology texts, and of their figurative and affective strategies, over time. As they attempt to pinpoint the defining traits of contemporary 'mycoaesthetics', or the 'molecular sublime', they draw our attention to the circulation of images and narrative structures between fiction and popular science. Cross-currents are also mapped out by Pieter Vermeulen and Rishi Goyal in their respective discussion of population politics and aesthetics, moving from economist Thomas Malthus' 1798 *Essay on the Principle of Population*, to

---

10 N. Katherine Hayles, *How We Became Posthuman: Virtual Bodies in Cybernetics, Literature, and Informatics* (Chicago: University of Chicago Press, 1999), pp. 21–22.

biologists Paul and Anne Ehrlich's 1968 best-seller *The Population Bomb*, to the depopulated utopias of post-catastrophe fiction, and the unequal treatment of infectious diseases according to a prejudiced demographic imagination. Gillian Beer's phrasing, however, remains productive in the way it allows content to slip into form, in a conscious echo of Propp's morphological approach to narrative. Contemporary fiction, as Ben De Bruyn's essay on Jon McGregor demonstrates, experiments with forms comparable to ecosystem monitoring and modelling. Beer, moreover, does not satisfy herself with the angle of influence. 'Are there stories', she asks in the same paragraph, 'to be told from places and organisms until now unrecognised?'[11] Contemporary writers and performers, we find throughout this book, try to tell tales from places and organisms that, although perhaps not 'unrecognised', are unexpected—the anthropomorphized perspective of an ailing 'Gaia', the 'invasive' species of destabilized habitats, the 'weird' bodies of fungi, or the microbial 'landscapes' of the human body.

Bringing together scholars in literature and performance studies, our enquiry does not claim to be in itself interdisciplinary, instead it explores interdiscursivity and the cross-fertilizing of imaginaries between contemporary artistic work, popularizations of the life sciences, and philosophy. The resulting collection outlines literature and performance's encounter with, and creation of, an emergent and multifaceted biological imagination.

Imagination, Science and Power

Imagination is not the remit of artistic work. This volume explores the visual and figurative strategies of literature and performance, but also how they engage with the imaginative dimensions of contemporary ecological, biomedical and biopolitical discourses. Such engagements unfold against the background of rising interest, in both the arts and the humanities, for life forms and perspectives that challenge earlier modes of representation: composite, relational entities such as symbionts or holobionts; previously neglected forms of consciousness, captured by terms such as 'neurodiversity' or the ecological 'pluriverse', including

---

11   Beer, 'Preface', p. xxviii.

animated, vibrant matter; newly studied types of communication, within fungal networks or microbial communities; and phenomena whose fluidity and complexity defy easy representation, many of which tend to be viewed, in the humanities, through the prism of Timothy Morton's catchword 'hyperobject'.[12] The poems, novels and performances examined in this volume seeks out those representational challenges: they attempt to visualize, narrate, perform, perhaps even model such entities.[13]

In his influential work on the idea of a 'scientific imagination', Gerald Holton proposes to differentiate between visual imagination, analogical imagination, and thematic imagination. Holton defends the key role of scientific imagination in the forging of theories: how Galileo's knowledge of Euclidean geometry, for instance, may have helped him visualize and understand the shapes of the moon; or how analogies such as Darwin's 'tangled bank' or the military metaphors of medical discourse (like an 'invading virus', or 'losing the battle with cancer') shaped emerging narratives of life. He distinguishes from visualization and metaphor a third type of imagination which he identifies as *thematic*: *themata*, an example of which would be 'discontinuity' in the early formulations of quantum physics, are 'the often unconfessed or even unconscious basic presuppositions, preferences, and preconceptions that scientists may choose to adopt, even if not led to do so by the data or current theory'.[14] Although Holton is mostly interested in these three types of imagination as tools that can 'energize the initial phases of research', visualization, analogy and *themata* are also productive entries for our enquiry into the cross-currents between scientific and artistic imaginaries.

---

12 Timothy Morton, *Hyperobjects: Philosophy and Ecology After the End of the World* (Minneapolis: University of Minnesota Press, 2013). See Kristin Ferebee, Susan M. Squier, Kirsten E. Shepherd-Barr and Hannah Simpson's chapters in this volume (chapters 8, 11 and 12).

13 Caroline Levine defends 'model thinking' as a way of reading literature: because they move across scales and media, models 'sharpen or set in motion our knowledge of a reality that is not available to direct perception' and 'allow us to understand forms at work. That is, by detaching shapes, orders, and arrangements from particular contexts, they allow us to play out the affordances of forms, especially in their interactions with other forms' ('Model Thinking: Generalization, Political Form, and the Common Good', *New Literary History*, 48.4 (2017), 633–53, p. 644, p. 643, https://doi.org/10.1353/nlh.2017.0033).

14 Gerald Holton, 'On the Art of Scientific Imagination', *Daedalus*, 125.2 (Spring 1996), 183–208 (p. 201).

Undeniably, the way organic life is imagined in the twenty-first century is shaped by scientific visualizations, whether in the form of medical imaging (chapter 6), ecosystem modelling (chapter 10), demographic graphs (chapter 9), satellite images (chapter 11) or other kinds of data visualization which are popularized and sometimes misconstrued in journalistic and social media. The problematic 'reality effect' of certain visualizations is a well-studied phenomenon:[15] in our volume this issue arises particularly around brain imaging technology—a particularly striking case of the mainstream media's tendency to hide the complexity of data visualization behind easily 'readable' images, which give the illusion of direct access, in this case to human consciousness. Such images remain techniques of what Foucault refers to as *savoir*, forms of knowledge that subsume power relations and hierarchizations beneath apparent transparency.[16] Aesthetic forms that engage with these images provide opportunities for critical, sometimes satirical, distance. Beyond the specific case of medical imaging, each modelling choice carries epistemological orientations that have simultaneously aesthetic and political ramifications: a key question for contemporary evolutionary biologists, for example, is the benefit of moving away from arborescent models towards network representations,[17] a move that, by changing how we perceive evolution, changes how we conceive of progress and of the power hierarchies historically established in its name. Epistemic forms thus often carry within them ethical options and figurative choices, which operate as metaphors beyond the literal visualizations they inform, so that Holton's distinction between visual, analogical, and thematic imagination reveals itself to be porous, identifying three angles of enquiry rather than three separate domains.

The analogies and metaphors that exist in scientific discourse—but also, perhaps more importantly for our enquiry, in popular science—play a key role in artistic attempts to grapple with new conceptions of organic life. One of the most interesting aspects of such figurative language is shifts in metaphors: how microbiology is gradually moving

---

15 For a summary of scholarship discussing this issue in neuroscience, see Jason Tougaw, *The Elusive Brain*.
16 Michel Foucault, *Histoire de la sexualité I : La volonté de savoir* (Paris: Gallimard, 1976).
17 See for instance Julie Beauregard-Racine et al., 'Of woods and webs: possible alternatives to the tree of life for studying genomic fluidity in *E. coli*', *Biology Direct*, 6.39 (2011), https://doi.org/10.1186/1745-6150-6-39.

away from military lexical fields towards ecological similes;[18] or how the individualistic imaginary of the 'selfish gene', associated to molecular Neo-Darwinism, has been undermined by a post-genomic approach to the genome as a system reacting to internal and external environments.[19] Here too, evolving representations in biology carry consequences for ethics and politics. The foregrounding of recursive relations between gene, organism, and environment—where before we saw 'selfish' genes and mostly linear causation—may, for example, help to challenge the political models that used evolutionary terms such as 'survival of the fittest' to naturalize individualism or liberalism.

Ecological theory has also undergone crucial figurative shifts over the past century. Ben De Bruyn, in his contribution to this volume, highlights the distrust expressed towards the image of the ecosystem by researchers in the environmental humanities, such as Vinciane Despret, who links it to the 'machine analogy' of the 'balance of nature', or Elizabeth Deloughrey, who has studied the intertwined history of island ecosystem ecology and military agendas.[20] Stepping back from the term 'ecosystem' to find the figures that tend to represent it, Derek Woods argues that:

> In influential works of ecological science writing, two tropes are prevalent: scala, which I define elsewhere as the substitution of "one object for another across at least a degree of magnitude," usually starting with an object perceivable by our senses; and technomorphism, which substitutes a technological object for a natural one (63). The figures of the chain, the wheel, the terrarium or aquarium, and the computer or digital network are all examples of at least one of these tropes.[21]

---

18 Eric Bapteste et al., 'The Epistemic Revolution Induced by Microbiome Studies: An Interdisciplinary View', *Biology*, 10.7 (2021), 651, https://doi.org/10.3390/biology10070651.

19 Hanson, *Genetics*, p. 2.

20 Vinciane Despret and Michel Meuret, 'Cosmoecological Sheep and the Arts of Living on a Damaged Planet', *Environmental Humanities*, 8.1 (2016), 24–36 (p. 26), https://doi.org/10.1215/22011919-3527704; Elizabeth DeLoughrey, 'The Myth of Isolates: Ecosystem Ecologies in the Nuclear Pacific', *Cultural Geographies*, 20.2 (2013), 167–84, https://doi.org/10.1177/1474474012463664.

21 Derek Woods, 'Scale in Ecological Science Writing', *Routledge Handbook of Ecocriticism and Environmental Communication*, ed. by Scott Slovic, Swarnalatha Rangarajan, and Vidya Sarveswaran (Oxon: Routledge, 2020), pp. 118–28 (p. 120), https://doi.org/10.4324/9781315167343-11.

Such tropes, according to Woods, allowed writers to move beyond the image of the superorganism inherited from early twentieth-century ecological science, and which later formulations and figures of the ecosystem tried to leave behind. Nevertheless, we can wonder whether images like the 'supercomputer' are fundamentally different from the superorganism: '[d]espite the critiques and alternatives, writers keep reinventing the wheel of the larger-scale organism, pulled in by the gravitational force of synecdoche'.[22] Against this force, which implies the possibility of substitution between parts and a larger whole, Woods examines the work of scala and technomorphism, emphasizing the capacity of certain tropes, the network in particular, to avoid substitution across scales.

While Derek Woods has questioned the cultural work of synecdoche, Ursula Heise has emphasized the key role of allegory in the images of planetary life premised on 'synthesis, holism and connectedness', that dominated the 1960s and 70s. Images such as Lovelock and Margulis' Gaia reflected a conception of 'global ecology as harmonious, balanced, and self-regenerating'.[23] While Heise notes a decline of such allegories in the last decades of the twentieth century, they have more recently been set to work by philosophers such as Bruno Latour, who tries, through the figure of Gaia, to articulate the epistemic earthquake required to understand the Earth System as reacting to human action. The contemporary artworks studied by our contributors seem to confirm Latour's assertion that such a system can only be viewed from the 'inside'.[24] Wary of the distanced view, they zoom in. Seasonal migrations in McGregor's *Reservoir 13*, variations in fish population in Kurlansky's *The Story of Kram and Ailat*, coal formation in Gillian Clarke's poetry, or

---

22   Ibid., p. 127.
23   According to Heise, 'from McLuhan's "global village", Fuller's "Spaceship Earth", and Lovelock's "Gaia" to visual portrayals of Planet Earth as a precious, marble-like jewel exposed in its fragility and limits against the undefined blackness of outer space, these representations relied on summarizing the abstract complexity of global systems in relatively simple and concrete images that foregrounded synthesis, holism and connectedness' (Ursula K. Heise, *Sense of Place and Sense of Planet: The Environmental Imagination of the Global*, Oxford University Press, 2008, p. 63, https://doi.org/10.1093/acprof:oso/9780195335637.001.0001).
24   Bruno Latour, *Facing Gaia*, trans. by Catherine Porter (Cambridge: Polity Press, 2017). See also Frédérique Aït-Touati and Bruno Latour's 2017 performance lecture *Inside* (http://www.bruno-latour.fr/node/755.html).

the 'rewilding' of a specific urban garden in Tobias Rausch's *Planttheater* are limited, situated entry points that open toward more abstract visions of living systems. Such privileging of the 'inside view' implies a partial turning away from a Modern episteme predicated on a distanced view, supposed to guarantee objectivity and universality—but that in effect served to impose forms of imperial and colonial power from afar, both on human populations and on ecosystems.[25]

The image of the network, which pervades contemporary representations of ecological interdependence, could be considered an example of what Horton calls *thematic* imagination. The portrayal of fungi-forest symbioses as a 'wood wide web' is in itself a technological image which, as Woods analyses in his contribution to this volume, often hides the mycelium behind the tree: he agrees in this with Jedediah Purdy, who has satirized the inevitability with which, 'after centuries of viewing forests as kingdoms, then as factories (and, along the way, as cathedrals for Romantic sentiment), the 21st century would discover a networked information system under the leaves and humus'.[26] As a dominant image through which life forms are imagined today, the web in itself is an organic image turned technological. The 'web', like the 'virus', may have started out as organic images for information technology, but they return, somewhat uncannily, as technological entries into biological realities, when the now familiar world of computers and internet connectivity helps us to imagine strange biology.

This volume contains many timely reminders that the imagination of life is unavoidably biopolitical. Michel Foucault's conception of biopower, as a type of power 'situated and exercised at the level of life, the species, the race, and the large-scale phenomena of population',[27] remains an indispensable hinge around which to think about the relations constructed by biomedicine, health policies and environmental discourse. The power exercised by medical representations, and the racial and social dimensions of public health, emerge clearly from the sections focused on neuronarratives and pandemic fictions. As Rishi

---

25 Aníbal Quijano, 'Coloniality and Modernity/Rationality', *Cultural Studies*, 21.2–3 (2007), 168–78, https://doi.org/10.1080/09502380601164353.

26 Jedediah Purdy, 'Thinking Like a Mountain', *N+1*, 29 (2017), https://nplusonemag.com/issue-29/reviews/thinking-like-a-mountain/.

27 Michel Foucault, *The History of Sexuality, Volume 1: An Introduction*, trans. by Robert Hurley (New York: Vintage, 1990), p. 137.

Goyal reminds us, following Foucault, 'racism is the 'indispensable precondition' that authorizes the biopolitical state's right to kill' (chapter 7). In the context of a pandemic management that reactivated a history of treating racialized 'flesh'[28] as impersonal, manageable matter, pandemic fictions might serve to destabilise what Sylvia Wynter has identified as the Modern reduction of the human to 'Man'—a patriarchal, white version of the human rooted in the racism that accompanied the birth of modernity in the fifteenth and sixteenth centuries. For Wynter, the human is not only a racialized conception, but also a biologized one, since Western modernity employed Darwinian thought to contrast the Western Bourgeois 'human' to those subhumans considered to have been 'dysselected' by evolution. As a result, what Wynter calls the 'biocentric ethnoclass genre of the human' uses both 'biogenetic and economic notions of freedom' to justify the continuing sacrifice of 'the peoples of African hereditary descent and the peoples who comprise the damned archipelagoes of the Poor, the jobless, the homeless, the "underdeveloped"'. Injustice is thus justified by biology, in a mode of thought that conveniently attributes our contemporary order to 'the imagined agency of Evolution and Natural Selection'.[29]

The pandemic imaginaries presented in this book highlight the biopolitical exclusion or invisibilization of racialized and economically marginalized others. The racialized, falsely 'empty' cities analysed by Rishi Goyal in chapter 7 resonate with what Pieter Vermeulen calls the 'population unconscious' of certain pandemic novels, fictions which assuage unspoken fears of overpopulation by emptying the world of the undesirable many, so that the happy few may thrive (chapter 9). Imagined catastrophes thus place a spotlight on existing political hierarchies, including the extractive practices of 'settler-capitalist world ecology' analysed by Kristin Ferebee in chapter 8. The Foucauldian biopolitical angle can be broadened by exploring the relations between human, nonhuman, and even posthuman species in the era Haraway has named Chthulucene. Such multispecies poetics are examined here

---

28 Hortense Spillers, 'Mama's Baby, Papa's Maybe: An American Grammar Book', *Diacritics*, 17.2 (1987), 64–81, https://doi.org/10.2307/464747.
29 Sylvia Wynter, 'Unsettling the Coloniality of Being/Power/Truth/Freedom: Towards the Human, After Man, Its Overrepresentation—An Argument', *The New Centennial Review*, 3.3 (Fall 2003), 257–337 (p. 317), https://doi.org/10.1353/ncr.2004.0015.

with the help of philosophical work that opens up perspectives for biopolitical imaginaries beyond the human, including Emanuele Coccia's investigations into the life of plants.[30] The limitations of biopower, as a concept, are also probed by more recent formulations that attempt to escape life/non-life binaries and boundaries, notably Elizabeth Povinelli's work on 'geontologies', presented here by Kristin Ferebee in her study of Jeff Lemire's graphic fiction.[31] As it gradually reveals a long history of transgression, in which both lands and bodies are invaded by Western science, Lemire's *Sweet Tooth* recalls the key points raised by Sherryl Vint in her analysis of contemporary biopolitics: the commodification of life by biotechnology, the erosion of the boundary between the organic and the inert, and the renewal of inherited patterns of dispossession. Lemire's speculative fiction thus explores 'epivitality', this 'life becoming thing' that Sherryl Vint locates in the current flux of biotechnological capitalism,[32] but anchors it in the longer colonial history of North America. This colonial context leads Ferebee to read the novel through Povinelli's concept of 'geontopower', asking what ecocritical insight can be gained from this alternative framework, where questions of governance and power are shifted from *bios* to *geos*.

The acknowledgement of interdependencies between human and nonhuman life is so fundamental to contemporary art and philosophy that it might be more accurate to refer to current biopolitics as 'eco-biopolitics'. Many of the works studied in this volume engage with the dominant aesthetic forms of contemporary eco-biopolitical discourse: for instance, what Latour calls the 'pornography of catastrophism' in ecological rhetoric; or the tragic or elegiac modes that shape much discourse concerned with climate disaster and extinction.[33] As Heather

---

30   Emanuele Coccia, *The Life of Plants: A Metaphysics of Mixture* (Hoboken: Wiley, 2018).
31   Donna J. Haraway, *Staying with the Trouble: Making Kin in the Chthulucene* (Durham: Duke University Press, 2016), https://doi.org/10.1215/9780822373780; Elizabeth Povinelli, *Geontologies: A Requiem to Late Liberalism* (Durham: Duke University Press, 2016), https://doi.org/10.1215/9780822373810.
32   Sherryl Vint, 'Neoliberalism and the Reinvention of Life', in *Biopolitical Futures in Twenty-First-Century Speculative Fiction* (Cambridge: Cambridge University Press, 2021), 1–24, https://doi.org/10.1017/9781108979382.
33   Bruno Latour, quoted in 'Décor as Protagonist: Bruno Latour and Frédérique Aït-Touati on Theatre and the New Climate Regime', Sébastien Hendrickx and Kristof van Baarle, *The Theatre Times* (18 February 2019), https://thetheatretimes.com/decor-is-not-decor-anymore-bruno-latour-and-frederique-ait-touati-on-

Houser has argued, data itself is aestheticized and integrated into twenty-first-century art in reaction to the *infowhelm* produced by overwhelming amounts of information about ecological emergencies.[34] On the other hand, absent images are just as important as those that assail us every day: the dominance of white male experts in the media reporting scientific discourse contributes, for instance, to the racial dimension of pandemic management, and to the gendering of relations between scientific subjects and objects, be they human or Gaian. Anthropocentrism leads biodiversity conservation discourse, and many fictional eco-narratives, to focus on forms of life that may become 'characters': as Ursula Heise has argued, 'charismatic megafauna' tend to dominate in stories of extinction.[35] Jeff VanderMeer's *Southern Reach* trilogy, which gives centre stage to the fungal kingdom, may therefore be read in contrast to works like Richard Powers' *Overstory* or even Tobias Rausch's *Planttheater*, which foreground the more easily perceptible and character-like plant kingdom. Anthropomorphism, however, need not be rejected too hastily: Vinciane Despret argues that it provides precious entry points into perspectives beyond that of the human. Some contemporary work, like Deke Weaver's *Unreliable Bestiary* performances, unabashedly embraces the shaping of representations by human imagination (chapter 12).

Bio-medical and ecological imaginaries are thus fundamentally intertwined in twenty-first-century representations. *Silent Spring*, Rachel Carson's now-classic indictment of pesticide-based agriculture, used health as a narrative angle on environmental degradation, alerting readers to their inseparable nature.[36] Such overlaps are common in

---

theatre-and-the-new-climate-regime/. For examples of tragic rhetoric, see the Deep Adaptation movement and Jem Bendell's 'Deep Adaptation: A Map for Navigating Climate Tragedy' (2018, 2020), http://lifeworth.com/deepadaptation.pdf. Ecological elegy is discussed by Timothy Morton, 'The Dark Ecology of Elegy', *The Oxford Handbook of Elegy*, ed. by Karen Weisman (Oxford: Oxford University Press, 2010), https://doi.org/10.1093/oxfordhb/9780199228133.013.0015.

34 Heather Houser, *Infowhelm: Environmental Art and Literature in an Age of Data* (New York: Columbia University Press, 2020). See also Victoria Vesna, ed., *Database Aesthetics: Art in the Age of Information Overflow* (Minneapolis: University of Minnesota Press, 2007).

35 Ursula K. Heise, *Imagining Extinction: The Cultural Meanings of Endangered Species* (Chicago: University of Chicago Press, 2016), pp. 23–25, https://doi.org/10.7208/chicago/9780226358338.001.0001.

36 Rachel Carson, *Silent Spring* (Boston: Houghton Mifflin, 1962).

contemporary representations of the 'sick' earth, and more generally in what Anne Hawkins calls *ecopathography*, a narrative mode which 'links a personal experience of illness with larger environmental, political, or cultural problems', suggesting that illness is the 'product of a toxic environment'.[37] In the anticipatory fictions studied here, imagined eco-biological futures oscillate between politically questionable emptiness—the city in Ling Ma's *Severance*, or the post-apocalyptic, depopulated worlds of *Station Eleven* or *Sweet Tooth*—and excess, where viral or fungal contamination, or simply the sheer size of human population, produce potentially horrific, uncontrollable proliferation. In the works focused on the present, *solastalgia*—Glenn Albrecht's term for a proleptic homesickness linked to awareness of environmental change[38]—is present (chapter 13), but coexists with other affects, including sublime awe (chapter 2), ecostoicism (chapter 10), intimate strangeness (chapter 12), and the soothing lull of disturbingly 'cosy' catastrophes (chapter 9).

Questions of Scale

Across the essays presented here, the most striking aspect of these twenty-first-century imaginaries is their engagement with scale. On the one hand, writers and performers are reacting to the worsening climate crisis and to accelerating, anthropogenic ecological damage with a rich array of aesthetic strategies that make the larger scale of planet or ecosystems perceptible and relatable. On the other hand, contemporary artworks are increasingly exploring microscopic life, the invisible scales of microbial or neuro-chemical phenomena that resist everyday human perception, yet shape every moment of our existence: creative practice is responding to the new areas of knowledge opened up by microbiome studies, and to new technical developments such as brain imaging practices. The intimate connection between those two epistemic gestures, scaling *up* and scaling *down*, has been forcibly impressed on us by the COVID-19 pandemic. But the 'double zoom' of

---

37 Anne H. Hawkins, 'Pathography: patient narratives of illness', *Western Journal of Medicine*, 171.2 (1999), 127–29.
38 Glenn Albrecht, 'Solastalgia: the distress caused by environmental change', *Australasian Psychiatry*, 15 (2007), 95–98, https://doi.org/10.1080/10398560701701288.

this lens-shift, towards the very small and the very large, is not limited to epidemiological perspectives: it is a recurrent trait of contemporary ecological discourse. Our working hypothesis, therefore, is that the contemporary biological imagination foregrounds and problematizes relations between the meso-scale of everyday human experience and the micro or macro scales of our increased biomedical and ecological awareness.

It is through questions of scale, then, that creative work engages with the ethical, philosophical, and political issues raised by this century's shifting views of life. The influence of invisible, microscopic entities on everyday life is explored through aesthetics that emphasize defamiliarization: the sublime and the grotesque, in Hamann-Rose's analysis of molecular and microbial representations (chapter 2); but also the weird, in Woods' reading of VanderMeer's trilogy (chapter 4), where disturbing fungal and genetic images embody the 'otherness of microscopic scale and its putative ability to control what happens at the scales of human senses and social systems'. Although strange otherness also pervades contemporary representations of the brain, Antolin and Tougaw show us that contemporary novels and graphic novels expose and question the reduction of meso-scale experience to invisible, neurochemical determinism, so that 'control' of either scale by the other is always an effect of representational strategies. Relations are thus questioned between different scales of life, but also between different scales of perception. The 'local' view, or the close zoom, are used not only as synecdoches for larger ecosystems, but as sites through which to question the possibility of zooming out or in. As Shepherd-Barr and Simpson argue in their analysis of site-specific performances by climate activists, the scattered, simultaneous 'guerrilla theatre' orchestrated by Extinction Rebellion can be viewed as replicating the properties of hyperobjects, so that they are, in fact, '*nonlocal*' in Morton's terminology, just like the 'massively distributed in time and space' ecological hyperobjects that can only be partially perceived (chapter 12).[39] Such planetary ambitions are not the only form that scalar awareness may take: Ben De Bruyn's narratological analysis reveals how a focus on apparently normal, everyday life, such as we find in McGregor's fiction,

---

39   Morton, *Hyperobjects*, p. 1 (emphasis in the original), p. 48.

may combine scalar 'modesty' with scalar 'flexibility', where attention paid to the minute forms of life and to day-to-day bio-logging places the reader in the position of ecological detective or citizen scientist.

As scale has rapidly become a keyword for much ecocritical research, recent publications in the humanities have outlined several important distinctions between different uses of the term. Following Andrew Herod's distinction between scale as a 'mental device' and as a 'material social product', Dürbeck and Hüpkes contrast a materialist perspective that stabilizes 'scales as socially constructed yet ontologically fixed entities' with an epistemological perspective that 'allows us to conceives of scales as contingent ways of framing space and time'.[40] Pushing distinctions further, Zach Horton identifies four 'disciplinary models of scale': *scale as relational ratio*, derived from cartography; *scale as absolute size domain*, derived from physics; *scale as compositional structure of parts to whole*, a model dominant, according to Horton, in engineering and biology; and scale as a *homologous scaling operation* in mathematics.[41] All these models are relevant to representations of life in the current context of planetary ecological crisis, where global perspectives and Anthropocenic views are both necessary and problematic, and the smooth scalability of certain economic models—namely, those of global capitalism—are increasingly contrasted with non-scalable structures. This volume, however, is mainly concerned with literature and performance's interest in the micro and macro scales of life, the connections they explore between those other scales and the meso-scale of human experience, and the representational strategies they invent to play with scale as an epistemological framework and a visualizing tool. The definitions most relevant to this book are therefore Dürbeck and Hüpkes' second (epistemological) perspective, and Horton's first and second models (*scale as relational ratio* and *scale as absolute size domain*). Our collection also benefits from the trend identified by Derek Woods as 'scale critique', research in environmental humanities that is attentive to

---

40 Gabriele Dürbeck and Philip Hüpkes, 'The Anthropocene as an Age of Scalar Complexity', in *Narratives of Scale in the Anthropocene: Imagining Human Responsibility*, ed. by Gabriele Dürbeck and Philip Hüpkes (New York: Routledge, 2022), p. 5, https://doi.org/10.4324/9781003136989.

41 Zachary Horton, *The Cosmic Zoom: Scale, Knowledge, and Mediation* (Chicago: University of Chicago Press, 2021), pp. 14–22, https://doi.org/10.7208/chicago/9780226742588.001.0001.

the effects of scale on the framing of issues. Anna Tsing's work on non-scalability, Bruno Latour's anti-zoom, and Zachary Horton's critique of the cosmic zoom all provide an inspiring context in which to improve our 'scalar literacy'.[42]

Two main areas of political debate frame our contributors' exploration of trans-scalar relations. As climate change and environmental degradation become ever more pressing topics, this volume reacts against accusations of a 'failure' of environmental imagination, arguing that contemporary work is answering appeals for imaginative action, such as Martin Puchner's search for 'stories for the future', Bruno Latour's call for new 'Gaiagraphies', and the now widely shared assertion, within ecocriticism, that one role that art can play is to perform a 'scalar translation' of more-than-human processes.[43] Although increasingly difficult to distinguish from such environmental questions, the second area of debate concerns the biopolitics of health and its transformation by recent advances in microbiological science, including microbiomics, virology, cellular biology and neurology. Here too, scalar translation is a key issue, between invisible, microscopic agency, everyday perception, and the large-scale politics of pandemic management. The imagination of viral contagion is intimately linked to the social and racial politics of public health, examined in this collection through pandemic fictions where haunting realities, such as overpopulation or the racialization of disease, are revealed as the dark underside of epidemiological and ecological discourses. Racial politics, which engage the scale of nation or planet, are thus entwined with the microscopic scales of contemporary biomedicine and biotechnology, where the 'politics of life itself', as defined by Nikolas Rose, is concerned with our 'growing capacities

---

42 See Derek Woods, 'Scale Critique for the Anthropocene', *Minnesota Review*, 83 (2014), 133–42 (p. 40), https://doi.org/10.1215/00265667-2782327, and Timothy Clark, *The Value of Ecocriticism* (Cambridge: Cambridge University Press, 2019), pp. 38–56, https://doi.org/10.1017/9781316155073.

43 Martin Puchner, *Literature for a Changing Planet* (Princeton: Princeton University Press, 2022); Alexandra Arènes, Bruno Latour, and Jérôme Gaillardet, 'Giving Depth to the Surface: An Exercise in the Gaia-Graphy of Critical Zones', *The Anthropocene Review*, 5.2 (2018), 120–35, https://doi.org/10.1177/2053019618782257; Timothy Clark, *The Value of Ecocriticism*, p. 49.

to control, manage, engineer, reshape, and modulate the very vital capacities of human beings as living creatures' on a biochemical level.[44]

A tension arises here between seemingly opposed ecological and medical paradigms: whereas environment discourse privileges figures of connectedness and entanglement, Rose asserts that the twenty-first century's economics of vitality is characterized by 'molecularization', a separation of vitality into 'distinct and discrete' objects.[45] Where a socio-economic analysis may reveal what Rose calls the 'dis-embedding' of vitality, a cultural analysis of popular science discourse may conversely highlight, as Woods suggests, 'a shift, in cultures of science, from 'bio' to 'eco': from concern with genomes and DNA to Anthropocene ecosystems, climate change, and weird ecologies' (chapter 4). The opposition may seem somewhat artificial: even if 'molecularization' can be conceived primarily as an effect of the pre-eminence of genetics in the biological imagination of the twentieth century's last decades, the development of environmental genomics in the 2000s and 2010s, thanks to high-throughput sequencing, led to a renewed interest for microorganisms within ecosystems, including the human body with its vast and diverse microbiota. These shifts in microbiology and molecular biology might reconcile 'ecological connectedness' with a form of 'molecularization'. Such overlaps are confirmed by Musitelli's analysis (chapter 3), which connects poetic uses of images drawn from cellular biology to environmental awareness, and Hamann-Rose's reading of representations of the human body as microbial ecosystem (chapter 2). Nevertheless, Nikolas Rose's socio-economic emphasis on separation and discreteness, which he links to biovalue and biocapital, remains illuminating for some representations of medical science in

---

44 Nikolas Rose, *The Politics of Life Itself* (Princeton: Princeton University Press, 2007), p. 3. According to Rose 'The "style of thought" of contemporary biomedicine envisages life at the molecular level, as a set of intelligible vital mechanisms among molecular entities that can be identified, isolated, manipulated, mobilized, recombined, in new practices of intervention, which are no longer constrained by the apparent normativity of a natural order' (p. 6). For an investigation of biotechnological imaginaries, which lie beyond the scope of this volume, see Sherryl Vint, *Biopolitical Futures in Twenty-First-Century Speculative Fiction* (Cambridge: Cambridge University Press, 2021).

45 Rose observes that 'vitality is decomposed into a series of distinct and discrete objects—that can be isolated, delimited, stored, accumulated, mobilized, and exchanged, accorded a discrete value' (*The Politics of Life Itself*, p. 7).

contemporary culture. As Kristin Ferebee demonstrates in her reading of Jeff LeMire (chapter 8), the view of medical research as a process of anatomization, separation, and even as a form of 'extraction' comparable to the extractive practices of fossil fuel economies, is a key part of this century's biological imagination.

Aesthetic Trends

The arguments gathered in this book suggest that epistemological and biopolitical questions are at work within aesthetic form itself. New labels have been proposed by artists and critics, signalling the appropriation of ecological and medical topics in 'climate change art', 'neurofiction', 'syndrome novels', 'pandemic fiction', but also the appearance of new forms of sensibility, such as 'Anthropocene noir', a feeling of disillusion, guilt and disempowerment that shares structural characteristics with the genre of *noir* fiction.[46] The possibility of deriving aesthetic form from unusual biological perspectives is highlighted in this volume by neologisms such as 'mycoaesthetics' (chapter 4) or 'planttheater' (chapter 13). Well-known forms, meanwhile, come under pressure. Familiar modes, such as the pastoral, are distorted by new environmental patterns, leading, for example, to the post-pastoral style of McGregor's *Reservoir 13*. Old heroes, like the 'ecological detective' identified by Sara Crosby in Edgar Allan Poe's writing,[47] return in new guises, in the narrative voices constructed by McGregor or VanderMeer. In Adam Dickinson's poem *Anatomic*, the speaker's own body is the object of his ecological detective work, as he seeks out the biochemicals and microbial life to which he is a host. Those findings inscribe the 'industrial powers and evolutionary pressures' of his time in the poet's body, which becomes 'a spectacular and horrifying crowd'.[48] Tracking the minute traces of all the ways in which a polluted world pollutes the body, Dickinson's collection builds up, according to Hamann-Rose,

---

46   Deborah Bird Rose, 'Anthropocene Noir', *Arena Journal*, 41/42 (2014), 206–19.
47   Sarah Crosby, 'Beyond Ecophilia: Edgar Allan Poe and the American Tradition of Ecohorror', *Interdisciplinary Studies in Literature and Environment*, 21.3 (Summer 2014), 513–25 (p. 515), https://doi.org/10.1093/isle/isu080.
48   Adam Dickinson, *Anatomic* (Toronto: Coach House Books, 2018), p. 9.

an imaginary of 'human environmental aesthetics as transformed by capitalist-industrial molecular writing' (chapter 2).

A preference for the local, immersed or entangled view, and the tendency to destabilise anthropocentrism, are two recurrent features in the texts, images and performances studied here. Human agency is decentred by ecosystemic perspectives, which entangle it with the agencies of both living and non-living matter. The human subject itself is questioned by certain scales, when it becomes the object of neurological study, a partner in symbiotic association with microbial forces, or a site on which the polluted environment writes.[49] Such attempts to change perspectives may clash with inherited artistic forms, as when the cyclicity of biological phenomena resists the linearity of narrative (chapter 10). Though it has the advantage of being able to create physical encounters with other forms and places of life—notably trees, plants, and bacteria in the shows presented in this volume—theatre, like fiction, struggles with the 'anthropocentric bias' of its own medium.[50] As Shepherd-Barr and Simpson point out, realism 'may be in part to blame' for this focus on the human story (chapter 12). Faced with the challenges of microscopic and planetary scales, fantasy, weird fiction or science fiction sometimes seem better suited to the new biological imaginary. Certainly, realism has been the target of much criticism: Patrick Lonergan writes that theatrical 'models of realism might be inhibiting our ecological awareness', while Amitav Ghosh denounces the disjunction between a contemporary world in which the planetary plays an increasingly prominent role, and the conventional form of the Western novel, with its emphasis on the everyday and on individual consciousness.[51] As Susan M. Squier

---

49 Through these pluralities and interdependencies, the destabilized human subjects described in this volume resonate with Rosi Braidotti's definition of posthuman subjectivity as shifting the focus 'from unitary to nomadic subjectivity' within an 'eco-philosophy of multiple belongings' which, crucially, does not remove accountability (*The Posthuman*, Cambridge: Polity, 2013, p. 49).

50 For this idea in relation to narrative, see Monica Fludernik, *Towards a "Natural" Narratology* (Routledge, 1996), p. 9, https://doi.org/10.4324/9780203432501, and Marco Caracciolo, 'Posthuman Narration as a Test Bed for Experientiality: The Case of Kurt Vonnegut's *Galápagos*', *Partial Answers*, 16.2 (2018), 303–14, https://doi.org/10.1353/pan.2018.0021.

51 Patrick Lonergan, 'A Twisted, Looping Form: Staging Dark Ecologies in Ella Hickson's *Oil*', *Performance Research*, 25.2 (2020), 38–44 (p. 41), https://doi.org/10.1080/13528165.2020.1752575; Amitav Ghosh, *The Great Derangement: Climate Change and the Unthinkable* (Chicago: Chicago University Press, 2016).

and Ben De Bruyn's chapters demonstrate, however, even realist and graphic novels still play a key part in the search for relations to other-than-human life forms and imperceptible scales of being.

A number of ecocritics have argued that certain genres may be better suited than others to the representational challenges of ecological complexities. Following Greg Garrard and Susanna Lidström, Timothy Clark suggests that, because of poetry's ability to accommodate multiplicity, uncertainty, and ambiguity, 'the true complexity of environmental issues has been perhaps easier to represent in new or revised forms of poetic practice than in prose forms like the novel, short story or the non-fiction essay, or in theatre'.[52] This volume, however, does not find one genre more suited to multi-scalar aesthetics than another. Rather than which genre, the key question is which forms may best attend to heterogeneous scales of life—individual and species, local ecosystem and planet, virus and host, cell and organism—and their disparate temporal scales, from the milliseconds of neural firing to the eons of geological periods. The narrative challenge of representing the scale of species has led historian Dipesh Chakrabarty to defend the necessity of 'multiple-track narrative',[53] and literary scholar David Herman to search for 'multiscale narration' in his work on 'bionarratology'.[54] Although he focuses on written narratives, many of the strategies that Herman identifies for 'storytelling at species scale' are relevant to other artistic forms. Allegory, for instance, and the multi-layered readings he refers to as 'allegorical laddering'[55]—switching between individuals and species, allowing the reader to glimpse evolution and deep time—are equally vital to the environmental work of poets such as Gillian Clarke or Adam Dickinson, and indeed to the plotting of ecologically minded theatre

---

52 Clark, *The Value of Ecocriticism*, p. 59.

53 According to Chakrabarty, 'we will need to develop multiple-track narratives so that the story of the ontologically-endowed, justice-driven human can be told alongside the other agency that we also are—a species that has now acquired the potency of a geophysical force, and thus is blind, at this level, to its own perennial concerns with justice that otherwise forms the staple of humanist narratives'. (Dipesh Chakrabarty, 'Brute Force', *Eurozine* (2010), https://www.eurozine.com/brute-force/).

54 David Herman, 'Coda: Toward a Bionarratology; or, Storytelling at Species Scale', in *Narratology beyond the Human: Storytelling and Animal Life* (Oxford: Oxford University Press, 2018), pp. 249–94 (p. 252), https://doi.org/10.1093/oso/9780190850401.001.0001.

55 Ibid., p. 263.

performances, even when little is left of 'dramatic' plot notions. Kris Verdonck's *Exote 1*, in which spectators wearing lab coats encountered the invasive species of Belgian landscapes, thus placed them in 'a planetary garden that they themselves have deregulated', as Beaufils points out, while EdgarundAllan's *Beaming Sahara* presented them with a block of soil 'speaking' on behalf of the forest (chapter 13).

Representing heterogeneous scales of being entails shifts, not only in the temporal and spatial dimensions of texts, images, and performances, but also in how we envisage causation, as we face systems whose complexity defy current scientific understanding. Largely distributed, recursive, multivalent causation resists representation. The aesthetic challenge is thus intertwined with the epistemological difficulties, and the political implications, of scale-related choices. In a graphic novel like Jeff Lemire's *Sweet Tooth*, the choice of whether to focus on microbes and virus-resistant DNA, or on the perspective of a young boy who innocently triggers a pandemic, or on a dwindling, guilty human species, will determine key questions of ethical responsibility. The awareness of events happening at imperceptible scales makes critical the question of how 'reality' is constituted: through tenuous, ongoing collaborations, controversies, and negotiations within political, artistic, and scientific communities, and through practices ranging from grant writing to field work, from lab experiments to international conferences. One key political question is what normative, political implications each scale of representation carries with it. The reduced scale of utopian narratives, as Vermeulen argues, may resist scalability and rely implicitly on the necessity of mass death. On the other hand, the focus on unusual scales may argue for a renewed politics of attention, working against the perceptual limitations of the human, the 'unavailability of emergent systemic behavior and multi-scalar patterns to immediate phenomenal experience'.[56]

Artistic work may thus tread a fine line between adding to an ultimately numbing discourse of crisis,[57] or, on the contrary, contributing

---

56   Dürbeck and Hüpkes, 'The Anthropocene', p. 3.
57   Timothy Clark highlights the risk of 'stuplimity' in the face of our 'ongoing biodiversity crisis' (Clark, *The Value of Ecocriticism*, p. 13): this term was first proposed by Sianne Ngai in 'Stuplimity: Shock and Boredom in Twentieth-Century Aesthetics', *Postmodern Culture* 10.2 (2000), https://doi.org/10.1353/pmc.2000.0013.

to what Yves Citton calls an *Ecology of Attention*.[58] The affective dimension of the works studied here is no doubt a key to this possibility of attention. Scale may be mobilized to elicit certain affective responses: in Mary M. Talbot and Brian Talbot's graphic novel *Rain*, discussed here by Squier, the protagonist's distress when faced with a local flood leads her to confront the global reality of climate change. The local similarly becomes infused with feelings of vulnerability and fragility in Beaufils' account of *After ALife Ahead*, an installation where the deep time of geological excavations was juxtaposed with the scientific observation of cancer cells, bacteria, and larger aquatic organisms. Such multi-scalar aesthetics anchor the lives of individuals and communities in both regional terrains and global instability, confronting different timescales in a way that resonates with descriptions of the Anthropocene as an 'age of scalar uncertainty'.[59]

Chapter Presentation

Section I, 'Invisible Scales', explores texts in which a biological imaginary—inspired by developmental biology, mycology or microbiology—connects microscopic, invisible scales of life to environmental questions. In 'Human Environmental Aesthetics: The Molecular Sublime and the Molecular Grotesque' (chapter 2), Paul Hamann-Rose examines the aesthetic categories brought into play by a microbiological imaginary. Drawing examples from popular biology, fiction and poetry, this chapter argues that the microscopic scales of genetics, biochemistry and microbial science tend to be represented through sublime and grotesque aesthetics. After revisiting the history of the concept, Hamann-Rose anchors his analysis in Burke's conception of the sublime which, unlike Kantian formulations, includes the 'wonders of minuteness' provoked by the discovery of life along a 'diminishing scale of existence, in tracing which the imagination is lost as well as the sense'. Burke's sublime can thus serve as a prism through which to approach contemporary visions of molecular and microbial life as

---

58 Yves Citton investigate the 'micropolitics' of attention and proposes to shift from the concept of an 'economy' of attention to that of an 'ecology' of attention, in *The Ecology of Attention*, trans. by Barnaby Norman (Cambridge: Polity, 2017).

59 Dürbeck and Hüpkes, 'The Anthropocene', pp. 1–2.

challenges to human subjectivity and agency. Hamann-Rose examines these aesthetics alongside the equally important 'molecular grotesque', which he conceptualizes in accordance with Mikhail Bakhtin and Susan Stewart's definitions. Although both categories reveal porous boundaries between the human and the nonhuman, and between the body and the environment, the molecular grotesque emphasizes distortion and materiality. Hamann-Rose tests these aesthetic categories in close readings of Simon Mawer's novel *Mendel's Dwarf*, popular science books *I Contain Multitudes* and *Life on Man*, and Adam Dickinson's poetry collection *Anatomic*. His analysis reveals recurrent images, such as the body-as-landscape or the 'Gulliver trope', and a shift, in the twenty-first century, towards an ecological microbial imaginary. While the sublime and the grotesque often coexist, he concludes that the latter combines environmental anxiety with more concrete imaginaries of microscopic materialities.

In chapter 3, 'Still Life and Vital Matter in Gillian Clarke's Poetry', Sophie Musitelli guides us through the intricate relations between geological and biological images in the work of Welsh poet Gillian Clarke. Musitelli chooses a new materialist framework to illuminate the vitality of matter in Clarke's most recent collections. Registering both the life of stone and the minerality of organisms, Clarke's environmental poetry displays an acute awareness of the 'incommensurable yet intersecting scales' of rocks and organisms and of geological and biological time. Musitelli's reading examines the role of medical imaging, cellular biology and epigenetics in these poetics, through key notions such as metamorphosis, fossilization, ontogeny and epigenetic landscapes. The poet's interest in coexisting yet heterogeneous timescales explains why developmental biology has become a key scientific paradigm in her work, this chapter argues. The recent 'environmental turn' taken by her poetry answers the urgency of climate change awareness, and is nourished by scientific images attuned to heterogenous scales of transformation and sedimentation. This overlap between geological and biological imaginations allows Clarke's poems to perform the vitality of both organic and inorganic matter. Musitelli's close reading of her work demonstrates how contemporary environmental poetry may be 'both deeply postromantic, and uncannily close to new materialist thinking'.

The fourth chapter, 'Mycoaesthetics: Weird Fungi and Jeff VanderMeer's *Annihilation*', examines the cultural representation of fungi in the context of global warming and digital modernity. Derek Woods pays close attention to the language through which fungi are described in science communication, and to the figurative role played by fungi themselves both in popular science and in fiction. He argues that the weirdness of fungi tends to be 'captured' by metaphors like the wood wide web, which indicate a new biological scale, but whose plant-centrism and enthusiasm for networks need critical attention. Woods presents mycoaesthetics as a cultural field shaped both by this symbolic role of fungi and by the intrinsic, ontological weirdness of subterranean mycelial life. His analysis examines the weird as a category in literary studies, alongside reflections on ecological weirdness in twenty-first-century environmental humanities, which use fungi to figure 'ecological estrangement'. Using Jeff VanderMeer's *Southern Reach* novels as a case study, the chapter demonstrates that the fungal writing at the heart of VanderMeer's trilogy overlays network metaphors with a genomic imaginary. Woods remains attentive to the cultural origins of such images, noting that the horizontal image of a fungal 'web' reflects the mutual influence between cybernetics and ecology, and the importance of systems in twenty-first-century cultures of science. VanderMeer's fiction, in this analysis, is exemplary of an ambivalent mycoaesthetics that uses fungi as figures of connectedness yet asks that we acknowledge their distinctive ontology. Woods goes on to examine the relatively recent emergence of the fungal kingdom as a third biological category, through which the plant/animal binary 'cracks open to yield a multiplicity—which is not to say an open-ended or unlimited diversity of categories'. A strong reading of mycoaesthetics, he concludes, would read not only human, but also mycelial agency in the production of this emergent field, inviting us to imagine eco-formalist approaches to contemporary literary forms.

Section II, 'Neuro-medical imaging and diagnosis', investigates narrative forms that grapple with a contemporary neurobiological imaginary. In chapter 5, Pascale Antolin reviews recent literary criticism surrounding the emerging 'neuronovel', which she examines as a sub-genre of the 'syndrome novel'. 'To Be or Not to Be a Patient: Challenging Biomedical Categories in Joshua Ferris' *The Unnamed*' focuses on a narrative that introduces tension into the genre by refusing to name the main character's condition. The lack of diagnosis leaves Ferris' character a 'medical orphan', whose sense of self is increasingly disrupted by an irresistible compulsion to walk. Antolin's close reading highlights the overlap of contemporary images—such as brain imaging devices—with older figures such as the body as a machine. Like many neuronovels, Ferris' satirical narrative resists materialist reductions of consciousness to a 'synaptic self' (LeDoux) or a 'neurochemical self' (Rose).[60] Antolin's analysis suggests, however, that *The Unnamed* has a much broader satirical resonance: a syndrome novel without a syndrome, it resists categorization and subverts an American 'corporate pastoral'. Although the main character initially stands for the 'normate', as defined by Garland Thomson,[61] his condition turns him into a paradoxical figure, both over-abled and disabled. Among the many binaries challenged by the novel, the polarities of health and illness interact in unexpected ways with those of ability and disability, social inclusion and exclusion. Through an overview of the novel's critical reception, Antolin's reading also highlights several recent facets of 'Neuro Lit Crit'.

In 'Neurocomics and Neuroimaging in David B.'s *Epileptic* and Farinella and Roš's *Neurocomic*', Jason Tougaw analyses the aesthetics of graphic narratives inspired by neuroscience (chapter 6). The autobiographical memoir *Epileptic* and the popular science book *Neurocomic* both investigate what philosopher Phillip Levine calls *the explanatory gap*, the resistance of immaterial experiences—such as consciousness or subjectivity—to neurobiological explanations. Like Susan M. Squier in chapter 11, Tougaw asks what specific tools graphic narratives can bring to the epistemological challenges they explore. According to Versaci, Chute or DeKoven, the medium foregrounds representation by refusing 'a problematic transparency'.[62] Tougaw accordingly contrasts graphic narratives, which 'make meaning by

---

60 See Joseph LeDoux, *Synaptic Self: How Our Brains Become Who We Are* (New York: Penguin, 2002) and Nikolas Rose, 'Neurochemical Selves', *Society*, 41 (2003), 46–59, https://doi.org/10.1007/BF02688204.
61 Rosemarie Garland Thomson, *Extraordinary Bodies: Figuring Physical Disability in American Culture and Literature* (New York: Columbia University Press, 2017), p. 8.
62 Rocco Versaci, *This Book Contains Graphic Language: Comics as Literature* (New York: Bloomsbury Academic, 2007), p. 6; Hillary Chute and Marianne DeKoven, 'Introduction: Graphic Narrative', *Modern Fiction Studies*, 52.4 (2006), 767–82 (p. 767), https://doi.org/10.1353/mfs.2007.0002.

inviting readers into the representational process', with the tendency, in science journalism and popular science, to present the visual products of brain scanning technologies like PET, SPECT and fMRI as 'seductively transparent'[63] images of neural activity. *Neurocomic* combines textbook images with fantasy, including 'a shrinking man' who wanders through the forest of his own brain, recalling the 'Gulliver trope' noticed by Hamann in popular microbiology (chapter 2). *Epileptic* also mingles realism and fantasy, representing the tests and treatments endured by the author's brother in images that ironically emphasize the brain's resistance to representation. Graphic narratives, Tougaw concludes, demonstrate that meta-representation is one of literature's key contributions to contemporary understandings of the brain.

Section III examines how pandemic imaginaries, deployed in three fictions published between 2009 and 2018 (*Severance*, *Sweet Tooth* and *Station Eleven*), rearrange narratives of daily life within urban, regional, and continental spaces and history. By following their protagonists as they navigate post-catastrophe, depopulated storyworlds, these chapters invite us to question the demographic ideology and positivist ontology that dominated a certain Modern, colonial, extractivist and racist worldview. They thus shed light on how infectious diseases and microbial agents shape contemporary literary representations of life and human community, situating utopia within specific scales of belonging, existence, and population.

During the COVID-19 spring of 2020, quasi pastoral images of streets left to deer and foxes proliferated. But not all human urbanites had been able to flee to greener pastures. In 'The Fiction of the Empty Pandemic City: Race and Diaspora in Ling Ma's *Severance*' (chapter 7), emergency medicine physician and literary scholar Rishi Goyal unveils the biopolitical structuring of such images. He shows how, from Daniel Defoe's fictionalized account of the 1665 great plague of London (*Journal of the Plague Year*, 1722) to Ling Ma's pandemic novel *Severance* (2018)—in which a global fungal infection depopulates the streets of New York, abandoning them to plants, Hispanic cabdrivers and the Chinese-American protagonist—literary depictions of the city in times of spreading disease re-negotiate spatial politics. According to

---

63 N. Katherine Hayles, 'Brain Imaging and the Epistemology of Vision: Daniel Suarez's *Daemon* and *Freedom*', *Modern Fiction Studies*, 61.2 (2015), pp. 320–34 (p. 322).

Goyal, *Severance* can be read as re-activating a twentieth-century history of 'white flights' during which American downtowns saw their white population leave neighbourhoods as racialized communities settled in. The novel's 'empty pandemic city' moreover questions the melancholic role of Asian-Americans in the national narrative, and that of nostalgia in diasporic and immigrant identity formation. Pandemic fiction is revealed as a stage on which we can contemplate the invisibilization of racialized communities, their relegation to the domain of expendable life by an urban cartography that erases social inequalities. By redrawing this very cartography, *Severance* invites a recalibration of how the American community imagines itself.

In Jeff Lemire's graphic novel series *Sweet Tooth* (2009–2013), a pandemic also makes room for a world in which the history of unequal relations between cultures is re-envisaged. The novel depicts how a mysterious plague—brought about by the white European excavation of an indigenous burial ground—destroys civilization and leaves the Earth to human-animal hybrid children, and to Tekkeitsertok and his fellow Inuit gods. In chapter 8, 'Dead Gods and Geontopower: An Ecocritical Reading of Jeff Lemire's *Sweet Tooth*', Kristin Ferebee reads this narrative, through the prism of indigenous epistemology, as a critique of extractive capitalism. She describes how the European colonization of North America has exercised what anthropologist Elizabeth Povinelli has called 'geontopower', a 'settler ontology' that has sidelined indigenous animist knowledge, establishing well-guarded borders between Life and Nonlife, and devitalizing geophysical entities to better exploit them. Disrupting such geontopower, *Sweet Tooth*'s plague—microscopic in its elements, macroscopic in its reach—foregrounds modes of relationality (the co-existence, rather than assimilation of Inuit and Western ontologies) that Ferebee understands through Karen Barad's onto-epistemology. She demonstrates how the comic portrays the life of its hybrid characters as enmeshed with the god-like entities animating the North American landscape and fossil underground, in a 'disaggregated' way that does not erase the difference of the conjoined parts, a compositional 'partedness' that also marks Lemire's visual and narrative choices.

A pandemic similarly plays a crucial, somehow emancipatory role in Emily St. John Mandel's *Station Eleven* (2014), a novel in which a troupe composed of survivors of a flu that killed most of the Earth's population travels across the North American Great Lakes region, performing Shakespeare's plays to the communities slowly recomposing in the ruins of our present civilisation. This representation of a relatively bucolic post-apocalyptic life is, according to Pieter Vermeulen's 'Depopulating the Novel: Post-Catastrophe Fiction, Scale, and the Population Unconscious' (chapter 9), burdened with scalar politics in which the good life is defined against fears of overpopulation, and so hinges on a drastic demographic reduction. Such reduction, Vermeulen argues, is unproblematised in *Station Eleven*, as it is in much post-catastrophe fiction, an increasingly popular genre he situates within the literary genealogies of utopian fiction, science fiction, and the realist novel. Tracing these genealogies, he unearths the genre's Malthusian foundations and shows how the 'cosy catastrophe' subgenre (identified by science fiction writer Brian Aldiss in 1973)[64] responds to fears of environmental exhaustion and planetary overcrowding with fantasies of a world made beautiful again by the removal of excess population. By exploring the normative scales of utopian, realist, and science fiction, Vermeulen invites us to reconsider how literary representations of proliferating life intersect with contemporary population politics.

The fourth and final section of *Life Re-Scaled* turns toward the representational challenges posed by climate change and the planetary ecological crisis. It explores how novels, graphic narratives, theatre, and performance have developed an array of strategies to engage with the spatial and temporal scales of animal, vegetal and microbial life alongside geophysical phenomena. These scales call for a renewal of narrative, visual and theatrical forms, in which the human is regularly decentred or inextricably enmeshed in the thick materiality of ecosystems. The section opens with 'The Everyday Pluriverse: Ecosystem Modelling in *Reservoir 13*', in which Ben De Bruyn explains how Jon McGregor's 2017 novel, which depicts thirteen years in the seasonal life of an English village, succeeds in evoking a rural mesocosm (located between, yet gesturing towards, micro- and macro- ecological scales), by combining the multiple perspectives of vegetal, insect and animal species (including humans), without subsuming them in an anthropocentric narrative. Reviewing existing scholarship on post-pastoral aesthetics, De Bruyn

---

64 Brian Aldiss, *Billion Year Spree: The History of Science Fiction* (London: Weidenfeld and Nicolson, 1973), pp. 315–16.

locates McGregor's novel within emerging strands of realism that train us to pay closer attention to the biological cycles surrounding us, such as reproduction and predation, bird migrations and plant flowering. The renewed attention given to cyclical phenomena manifests itself notably through *Reservoir 13*'s experimentation with repetition and variation, formal strategies that participate in its ecosystem modelling. For De Bruyn, we should then consider the novel as a fully-fledged technique for capturing, logging, and modelling nonhuman lives and environmental processes, comparable to the selected photographs (Stephen Gill's, collected in *Night Procession* and *The Pillar*), maps and diagrams (James Cheshire and Oliver Uberti's *Where the Animals Go*) or citizen science projects (Akiko Busch's *The Incidental Steward*) he discusses. Reflecting on the affordances of mesocosmic storyworlds for ecological storytelling, the chapter engages with concepts such as the 'ecological detective' and 'ecostoicism', and defends the value of narratives that represent familiar phenomena in a more scale-sensitive manner.

In the following chapter, 'The Narrative and Aesthetic Strategies of Climate Change Comics' (chapter 11), Susan M. Squier examines the potential of graphic narrative as a medium to tackle the challenge of representing climate change. Drawing from a range of recent European and American examples, Squier identifies scale-switching as a recurrent narrative structure: these comics alternate between the small scale of individual lives, and the large-scale complexities of climate change. Those complexities are conveyed by scientific discourses and images connecting global warming to biological realities perceivable by human characters, including biodiversity loss, extinction, and medical impacts on health and development. The comics under study in this chapter often choose unexpected perspectives, focusing, for instance, on the less charismatic species threatened by dwindling food chains, rather than on charismatic megafauna usually foregrounded by the discourse of extinction.[65] They also highlight the correspondence between socio-economic vulnerability and exposure to the hazards of a warmer climate, in particular the less well-known medical and epigenetic risks of increasingly toxic environments. Approaching climate change as a hyperobject, Squier demonstrates how the challenge

---

65 See Heise, *Imagining Extinction*.

it poses to representation may be answered through strategies such as anthropomorphism, synecdoche, and the combination of scientific distance with personal experience. The chapter presents the combination of narrative and image specific to comics as a tool able to tackle the collective 'failure of imagination' that threatens to plague twenty-first-century approaches to environmental crisis.

The possibility of countering such a 'failure of imagination' on the theatre stage is investigated by Kirsten E. Shepherd-Barr and Hannah Simpson in the twelfth chapter: 'Displacing the Human: Representing Ecological Crisis on Stage'. According to Shepherd-Barr and Simpson, theatre faces specific challenges when exploring the domain of ecological forces, but it also has powerful tools at its disposal thanks to its capacity to function in a range of spatial, temporal, and sensory dimensions. Reviewing earlier, twentieth-century stagings of creatures such as sentient plants (by Susan Glaspell, Alan Menken and Howard Ashman) and microbes (by George Bernard Shaw), or of posthuman landscapes and depersonalised beings (by Maeterlinck and Beckett), they describe how theatre has striven to depart from human concerns and onstage presence. They then guide us through the formal tools developed by contemporary theatre-makers to engage with the more-than-human scales of climate change and biological extinction. Among these tools, we find the use of miniature ecosystems to represent geological forces, exemplified by a fish tank in Steve Waters' *The Contingency Plan* (2009); the writing of 'unrepresentable' scales into interscenes, such as '*A million newborn babies gasp for breath*', in Hickson's *Oil* (2016); the dispersion, replicability and viral quality of the performances created by Earth Ensemble, the protean troupe associated with activist movement Extinction Rebellion; or the intermixing of media and discourses, factual or false, scientific or mythical, about endangered animal lives in Deke Weaver's *The Unreliable Bestiary* (2009–...). These strategies are here discussed alongside emerging dramatic genres such as the 'dramatised lecture', centred on experts (Emmott, Rapley and Latour), or Earth Ensemble's 'guerrilla theatre'; two different yet perhaps complementary responses to the insistent reality of ecological upheaval.

In our collection's final chapter, Eliane Beaufils focuses on the immersive installations and scenography through which four contemporary productions (Kris Verdonck's *Exote I*, Pierre

Huyghe's *After ALife Ahead*, Tobias Rausch's *Die Welt Ohne Uns*, and EdgarundAllan's *Beaming Sahara*) convert theatre into a milieu in which spectators are invited to experience their entanglement with other-than-human biological and geological scales. 'Staging Larger Scales and Deep Entanglements: The Choice of Immersion in Four Ecological Performances' discusses how these shows may act as 'diplomats' mediating between humans and ecological domains, by mobilizing a multiplicity of human and nonhuman 'actors' linked by scalar interdependencies. Such mediation takes a variety of shapes: close encounters with the unexpected disruptions created by tourism and commercial exchanges in local and global ecosystems; an installation where the main life forms are aquatic and microscopic, suggesting a future when humans will no longer be *anthropoï*; a 'longterm planttheater', performed over a year and a half in a botanical garden; an encounter with forests, minerals, and ice as hyperactors (as opposed to hyperobjects). These performances create ambiances of fragility and mourning, but also produce vital curiosity and the desire to 'read' other-than-human life through the proximity they allow with biological or abiotic entities. Attentive to the differences between different forms of immersion, Beaufils analyses the epistemological and affective implications of their uses of separation, close-up, darkness, and sensory stimulation, asking how they might enable a 'becoming with'.[66] In response to the crisis of sensibility that enables 'physiocide' in the Anthropocene,[67] Beaufils argues that immersive practices have specific affordances that enable spectators to think 'geobiologically'.

Through its twelve chapters, *Life, Re-Scaled* provides an impression of contemporary American and European literature and performance's engagements with the biological, revealing a common emphasis on interactions between scales and concerns with environmental justice; a desire to encounter animal, vegetal and microbial lives; and a critical engagement with the life sciences and with associated Western ontologies. Through the diversity of genres it studies, this collective

---

66 Donna J. Haraway, *When Species Meet* (Minneapolis and London: University of Minnesota Press, 2008), p. 244.
67 On 'physiocide' as the murder of the vegetal world, see Iain Hamilton Grant, 'Everything Is Primal Germ or Nothing Is: The Deep Field Logic of Nature', *Symposium: Canadian Journal of Continental Philosophy*, 19.1 (2015), 106–24, https://doi.org/10.5840/symposium20151919.

investigation hopes to contribute to a decisive effort, in the arts and humanities of the twenty-first century, to renew relations with the living matter that composes and surrounds us.

Works Cited

Aït-Touati, Frédérique and Bruno Latour, *Inside* (2017), http://www.bruno-latour.fr/node/755.html

Albrecht, Glenn, 'Solastalgia: The Distress Caused by Environmental Change', *Australasian Psychiatry*, 15 (2007), 95–98, https://doi.org/10.1080/10398560701701288

Aldiss, Brian, *Billion Year Spree: The History of Science Fiction* (London: Weidenfeld and Nicolson, 1973).

Arènes, Alexandra, Bruno Latour, and Jérôme Gaillardet, 'Giving Depth to the Surface: An Exercise in the Gaia-Graphy of Critical Zones', *The Anthropocene Review*, 5.2 (2018), 120–35, https://doi.org/10.1177/2053019618782257

Bapteste, Eric et al., 'The Epistemic Revolution Induced by Microbiome Studies: An Interdisciplinary View', *Biology*, 10.7 (2021), 651, https://doi.org/10.3390/biology10070651

Beauregard-Racine, Julie et al., 'Of Woods and Webs: Possible Alternatives to the Tree of Life for Studying Genomic Fluidity in *E. coli*', *Biology Direct*, 6.39 (2011), https://doi.org/10.1186/1745-6150-6-39

Beer, Gillian, 'Preface to the Second Edition', in *Darwin's Plots: Evolutionary Narrative in Darwin, George Eliot and Nineteenth-Century Fiction* (Cambridge: Cambridge University Press, 1983, 2000), pp. xvii–xxxii.

Bendell, Jem, 'Deep Adaptation: A Map for Navigating Climate Tragedy' (2018, 2020), http://lifeworth.com/deepadaptation.pdf

Bird Rose, Deborah, 'Anthropocene Noir', *Arena Journal*, 41/42 (2014), 206–19.

Braidotti, Rosi, *The Posthuman* (Cambridge: Polity, 2013).

Caracciolo, Marco, 'Posthuman Narration as a Test Bed for Experientiality: The Case of Kurt Vonnegut's *Galápagos*', *Partial Answers*, 16.2 (2018), 303–14, https://doi.org/10.1353/pan.2018.0021

Carson, Rachel, *Silent Spring* (Boston: Houghton Mifflin, 1962).

Chakrabarty, Dipesh, 'Brute Force', *Eurozine* (2010), https://www.eurozine.com/brute-force/

Chute, Hillary and Marianne DeKoven, 'Introduction: Graphic Narrative', *Modern Fiction Studies*, 52.4 (2006), 767–82, https://doi.org/10.1353/mfs.2007.0002

Citton, Yves, *The Ecology of Attention*, trans. by Barnaby Norman (Cambridge: Polity, 2017).

Clark, Timothy, *The Value of Ecocriticism* (Cambridge: Cambridge University Press, 2019), https://doi.org/10.1017/9781316155073

Coccia, Emanuele, *The Life of Plants: A Metaphysics of Mixture* (Hoboken: Wiley, 2018).

Crosby, Sarah, 'Beyond Ecophilia: Edgar Allan Poe and the American Tradition of Ecohorror', *Interdisciplinary Studies in Literature and Environment*, 21.3 (Summer 2014), 513–25, https://doi:10.1093/isle/isu080

DeLoughrey, Elizabeth, 'The Myth of Isolates: Ecosystem Ecologies in the Nuclear Pacific', *Cultural Geographies*, 20.2 (2013), 167–84, https://doi.org/10.1177/1474474012463664

Despret, Vinciane and Michel Meuret, 'Cosmoecological Sheep and the Arts of Living on a Damaged Planet', *Environmental Humanities*, 8.1 (2016), 24–36, https://doi.org/10.1215/22011919-3527704

Dickinson, Adam, *Anatomic* (Toronto: Coach House Books, 2018).

Dürbeck, Gabriele and Philip Hüpkes, 'The Anthropocene as an Age of Scalar Complexity', in *Narratives of Scale in the Anthropocene: Imagining Human Responsibility*, ed. by Gabriele Dürbeck and Philip Hüpkes (New York: Routledge, 2022), 1–20, https://doi.org/10.4324/9781003136989

Falletti, Clelia, Gabriele Sofia, and Victor Jacono, eds, *Theatre and Cognitive Neuroscience* (London: Bloomsbury, 2016).

Fludernik, Monica, *Towards a "Natural" Narratology* (London: Routledge, 1996), https://doi.org/10.4324/9780203432501

Foucault, Michel, *Histoire de la sexualité I : La volonté de savoir* (Paris: Gallimard, 1976).

Foucault, Michel, *The History of Sexuality, Volume 1: An Introduction*, trans. by Robert Hurley (New York: Vintage, 1990).

Garland Thomson, Rosemarie, *Extraordinary Bodies: Figuring Physical Disability in American Culture and Literature* (New York: Columbia University Press, 2017).

Gill, Josie, *Biofictions: Race, Genetics and the Contemporary Novel* (London: Bloomsbury, 2020), https://doi.org/10.5040/9781350099869

Ghosh, Amitav, *The Great Derangement: Climate Change and the Unthinkable* (Chicago: Chicago University Press, 2016).

Hamilton Grant, Iain, 'Everything is Primal Germ or Nothing Is: The Deep Field Logic of Nature', *Symposium: Canadian Journal of Continental Philosophy*, 19.1 (2015), 106–24, https://doi.org/10.5840/symposium20151919

Hanson, Claire, *Genetics and the Literary Imagination* (Oxford: Oxford University Press, 2020), https://doi.org/10.1093/oso/9780198813286.001.0001

Haraway, Donna J., *When Species Meet* (Minneapolis and London: University of Minnesota Press, 2008).

Haraway, Donna J., *Staying with the Trouble: Making Kin in the Chthulucene* (Durham: Duke University Press, 2016), https://doi.org/10.1215/9780822373780

Hawkins, Anne H., 'Pathography: Patient Narratives of Illness', *Western Journal of Medicine*, 171.2 (1999), 127–29.

Hayles, N. Katherine, *How We Became Posthuman: Virtual Bodies in Cybernetics, Literature, and Informatics* (Chicago: University of Chicago Press, 1999).

Hayles, N. Katherine, 'Brain Imaging and the Epistemology of Vision: Daniel Suarez's *Daemon* and *Freedom*', *Modern Fiction Studies*, 61.2 (2015), 320–34.

Heise, Ursula K., *Sense of Place and Sense of Planet: The Environmental Imagination of the Global* (Oxford: Oxford University Press, 2008), https://doi.org/10.1093/acprof:oso/9780195335637.001.0001

Heise, Ursula K., *Imagining Extinction: The Cultural Meanings of Endangered Species* (Chicago: University of Chicago Press, 2016), https://doi.org/10.7208/chicago/9780226358338.001.0001

Hendrickx, Sébastien and Kristof van Baarle, 'Décor as Protagonist: Bruno Latour and Frédérique Aït-Touati on Theatre and the New Climate Regime', *The Theatre Times* (18 February 2019), https://thetheatretimes.com/decor-is-not-decor-anymore-bruno-latour-and-frederique-ait-touati-on-theatre-and-the-new-climate-regime/

Herman, David, 'Coda: Toward a Bionarratology; or, Storytelling at Species Scale', in *Narratology beyond the Human: Storytelling and Animal Life* (Oxford: Oxford University Press, 2018), 249–94, https://doi.org/10.1093/oso/9780190850401.001.0001

Holton, Gerald, 'On the Art of Scientific Imagination', *Daedalus* 125.2 (Spring 1996), 183–208.

Horton, Zachary, *The Cosmic Zoom: Scale, Knowledge, and Mediation* (Chicago: University of Chicago Press, 2021), https://doi.org/10.7208/chicago/9780226742588.001.0001

Houser, Heather, *Infowhelm: Environmental Art and Literature in an Age of Data* (New York: Columbia University Press, 2020).

Huxley, Julian, *Evolution: The Modern Synthesis* (London: Allan & Unwin, 1942).

Idema, Tom, *Stages of Transmutation: Science Fiction, Biology, and Environmental Posthumanism* (London: Routledge, 2019).

Kress, W. John and Gary W. Barrett, eds, *A New Century of Biology* (London: Penguin Random House, 2016).

Kucukalic, Lejla, *Biofictions: Literary and Visual Imagination in the Age of Biotechnology* (New York: Routledge, 2021), https://doi.org/10.4324/9781003132325

Levine, Caroline, 'Model Thinking: Generalization, Political Form, and the Common Good', *New Literary History*, 48.4 (2017), 633–53, https://doi:10.1353/nlh.2017.0033

Latour, Bruno, *Facing Gaia*, trans. by Catherine Porter (Cambridge: Polity Press, 2017).

LeDoux, Joseph, *Synaptic Self: How Our Brains Become Who We Are* (New York: Penguin, 2002).

Lonergan, Patrick, 'A Twisted, Looping Form: Staging Dark Ecologies in Ella Hickson's *Oil*', *Performance Research*, 25.2 (2020), 38–44, https://doi.org/10.1080/13528165.2020.1752575

Ngai, Sianne, 'Stuplimity: Shock and Boredom in Twentieth-Century Aesthetics', *Postmodern Culture*, 10.2 (2000), https://doi.org/10.1353/pmc.2000.0013

Meeker, Natania and Antónia Szabari, *Radical Botany: Plants and Speculative Fiction* (New York: Fordham University Press, 2019), https://doi.org/10.5422/fordham/9780823286638.001.0001

Morton, Timothy, 'The Dark Ecology of Elegy', *The Oxford Handbook of the Elegy*, ed. by Karen Weisman (Oxford: Oxford University Press, 2010), 251–71, https://doi.org/10.1093/oxfordhb/9780199228133.013.0015

Morton, Timothy, *Hyperobjects: Philosophy and Ecology After the End of the World* (Minneapolis: University of Minnesota Press, 2013).

Povinelli, Elizabeth, *Geontologies: A Requiem to Late Liberalism* (Durham: Duke University Press, 2016), https://doi.org/10.1215/9780822373810

Puchner, Martin, *Literature for a Changing Planet* (Princeton: Princeton University Press, 2022).

Purdy, Jedediah, 'Thinking Like a Mountain', *N+1*, 29 (2017), https://nplusonemag.com/issue-29/reviews/thinking-like-a-mountain/

Rose, Nikolas, 'Neurochemical Selves', *Society*, 41 (2003), 46–59, https://doi.org/10.1007/BF02688204

Rose, Nikolas, *The Politics of Life Itself* (Princeton: Princeton University Press, 2007).

Shepherd-Barr, Kirsten E., *Theatre and Evolution from Ibsen to Beckett* (New York: Columbia University Press, 2015).

Solnick, Sam, *Poetry and the Anthropocene: Ecology, Biology and Technology in Contemporary British and Irish Poetry* (New York: Routledge, 2016), https://doi.org/10.4324/9781315673578

Spillers, Hortense, 'Mama's Baby, Papa's Maybe: An American Grammar Book', *Diacritics*, 17.2 (1987), 64–81, https://doi.org/10.2307/464747

Thompson, Evan, *Mind in Life: Biology, Phenomenology and the Sciences of Mind* (Cambridge, MA: Harvard University Press, 2007).

Tougaw, Jason, *The Elusive Brain: Literary Experiments in the Age of Neuroscience* (New Haven: Yale University Press, 2018).

Venter, Craig and Daniel Cohen, 'The Century of Biology', *New Perspectives Quarterly*, 21 (2004), 73–77, https://doi.org/10.1111/npqu.11423

Versaci, Rocco, *This Book Contains Graphic Language: Comics as Literature* (New York: Bloomsbury Academic, 2007).

Vesna, Victoria, ed., *Database Aesthetics: Art in the Age of Information Overflow* (Minneapolis: University of Minnesota Press, 2007).

Vint, Sherryl, *Biopolitical Futures in Twenty-First-Century Speculative Fiction* (Cambridge: Cambridge University Press, 2021), https://doi.org/10.1017/9781108979382

Woods, Derek, 'Scale Critique for the Anthropocene', *Minnesota Review*, 83 (2014), 133–42, https://doi.org/10.1215/00265667-2782327

Woods, Derek, 'Scale in Ecological Science Writing', *Routledge Handbook of Ecocriticism and Environmental Communication*, ed. by Scott Slovic, Swarnalatha Rangarajan, and Vidya Sarveswaran (Oxon: Routledge, 2020), 118–28, https://doi.org/10.4324/9781315167343-11

Wynter, Sylvia, 'Unsettling the Coloniality of Being/Power/Truth/Freedom: Towards the Human, After Man, Its Overrepresentation—An Argument', *The New Centennial Review*, 3.3 (Fall 2003), 257–337, https://doi.org/10.1353/ncr.2004.0015

I. INVISIBLE SCALES:
CELLS, MICROBES AND MYCELIUM

2. Human Environmental Aesthetics
The Molecular Sublime and the Molecular Grotesque

Paul Hamann-Rose

Never before has more knowledge existed about the invisible workings of life within and around us. In the twentieth century, biology became molecular. With the molecular make-up of DNA deciphered in 1953, a new mode of thinking biological organisms was afoot, and previously abstract processes in living organisms acquired a new biochemical substance and depth. This complex shift to the molecular level, especially in genetics and microbiology, 'entails a change of scale', as Nikolas Rose underlines: 'It is now at the molecular level that human life is understood, at the molecular level that its processes can be anatomized, and at the molecular level that life can now be engineered'.[1] The completion of the Human Genome Project in 2003 appeared to mark the pinnacle of what Evelyn Fox Keller has aptly described as 'the century of the gene'.[2] The problematic dimension of this shift to the molecular level is that it frequently gives rise to forms of genetic reductionisms that exaggerate the explanatory power of genome sequencing and reduce highly complex and interdependent organismic processes to the protein-coding

---

1   Nikolas Rose, *The Politics of Life Itself: Biomedicine, Power, and Subjectivity in the Twenty-First Century* (Princeton: Princeton University Press, 2006), p. 4.
2   Evelyn Fox Keller, *The Century of the Gene* (Cambridge: Harvard University Press, 2000), p. 1.

nucleic acids of DNA. However, new endeavours like the Earth Genome Project and the Human Microbiome Project make it clear that molecular research, along with its ever-growing impact on social and cultural life, has only accelerated after the turn of the millennium, especially following the development of rapid sequencing technologies in the late 1990s. This mounting store of information about the molecular realms of existence presents the twenty-first century with a historic challenge to the imagination, for molecular biology deals with a scale of being that is inevitably invisible to the naked eye.

Making sense of the molecular realm involves imaginatively crossing the chasm between the domains of the visible and the invisible; this is true for both science and culture. Within the discipline of molecular biology, access to the newly molecular level of life requires specific forms of mediation, from experiments to large-scale computation. Such forms of mediation are imaginative in multiple ways, from the figurative and often metaphorical language of science—think 'genetic book of life', 'genetic editing' or 'genetic blueprint'[3]—to the speculative element of experimental hypotheses. Once biological representations enter cultural discourses, they are subjected to further imaginative mediation as the cultural imaginary attempts to gauge the meaning, relevance and potential of molecular knowledge and discourse. This imaginary takes its cue from the scientific representation of the molecular, but transforms it according to its own conventions, desires, and anxieties. In recent decades, such cultural imaginings of the molecular and microbial world have produced an increasingly rich vision of life at the level of the invisibly small. They also provide a key indicator of the pressing social and political issues around molecular science. My central argument in this chapter is that two aesthetic forms are particularly characteristic of this twenty-first-century vision of the invisible domains of life: what I call the molecular sublime and the molecular grotesque.[4]

The aesthetics of the sublime and the grotesque as they manifest in the cultural imagination of the molecular and the microbial indicate a

---

[3] For an in-depth discussion of the metaphorical status of key genetic concepts see Lily Kay, *Who Wrote the Book of Life? A History of the Genetic Code* (Stanford: Stanford University Press, 2000).

[4] I would like to thank Greg Lynall for his brilliant suggestion of a molecular grotesque when I presented an earlier version of this research at the 2019 annual conference of the British Society for Literature and Science.

particular cluster of concerns and anxieties about the relation between the human and the molecular scales. These concerns are epistemological, representational and, in light of the joint crises of climate change and COVID-19, ontological and existential. The mechanics of contagion have made front-page news during the global pandemic that is poised to define the early twenty-first century, while personalised medicine seems likely to assume an ever-greater role in genetics and immunology research, as well as in people's everyday lives.[5] This context alone signals an untoward need to imaginatively engage with the molecular dimension of human existence. Moreover, imagining the life around us and questioning the often unsettlingly porous boundaries between individual and environment has acquired special urgency as humanity is attempting to re-define its ecological practices and self-understanding. In order to facilitate a new and less myopic perspective on human-environment relations, ecocritic Timothy Clark has suggested large-scale readings of literary texts: inviting the reader to zoom out and consider fictional life at timescales of hundreds of years to reveal characters' imbrication in deep time and the fate of the planet.[6] Contrary to Clark's zooming out, my focus in this chapter will be on zooming in on the hidden *microscopic* rather than *macroscopic* connections between planetary and human life.

Realising that a biologically material continuity exists between the human organism and its environment, both through the imagination and the practices it inspires, has the potential to challenge some of the deep-seated assumptions of human exceptionalism and anthropocentrism that are frequently cited as foundational to anthropogenic climate change.[7] As I will show in the following, the epistemological and affective connotations of the sublime and the grotesque can illuminate such human-environment continuities but also obstruct the self-awareness that would make such connections meaningful to readers and

---

5 Ian Tucker, 'Daniel M Davis: "Unbelievable things will come from biological advances"', *The Guardian*, 3 July 2021.

6 Timothy Clark, 'Derangements of Scale', in *Telemorphosis: Theory in the Era of Climate Change, Vol. 1*, ed. by Tom Cohen (Michigan: Open Humanities Press, 2012), [n.p.], https://dx.doi.org/10.3998/ohp.10539563.0001.001.

7 See, for instance, Timothy Morton, *Ecology Without Nature* (Cambridge: Harvard University Press, 2009), or, for an overview, Andrew Brennan, 'Environmental Ethics', *The Stanford Encyclopedia of Philosophy*, 2021, https://plato.stanford.edu/archives/sum2022/entries/ethics-environmental/.

spectators. In Edmund Burke's canonical theorisation of the sublime, 'the imagination is lost as well as the sense';[8] so can it really represent and afford an understanding of the microscopic life forms that live in, around and on us, not to mention an understanding of their meaning for our macroscopic lives? I will suggest that, while the sublime affords spectacular vistas on the very small scales of the processes of life, it may be the grotesque molecular and microbial distortions of our familiar selves that prove the more productive aesthetic for crossing the distance between the visible and the imagined that often precludes an understanding of how the molecular relates to us.

The forms of the molecular sublime and molecular grotesque which I identify can be found across multiple media and genres. For the purposes of this chapter, I have limited myself to textual instantiations of these forms. I will, however, discuss the sublime and grotesque imaginations of the molecular in a representative, and non-exhaustive, variety of genres. I begin by outlining my conception of the molecular sublime, drawing on examples from recent novels informed by genetic science. As will become apparent, I use the term "molecular" as a shorthand to describe not just such molecular forms of biological organisation as genes, but also slightly larger organisms like microbes. This is partly for the sake of simplicity, and to avoid adding too many new monikers to an already long and possibly endless list of sublimes, and partly because, like all other beings, microbes not only consist of molecules but inhabit a similar scale of being, equally invisible to the human eye. I then move to the genre of popular science writing, especially Ed Yong's *I Contain Multitudes* (2016), to show the particular role of the imagination in these texts' representations of the very small and how the sublime in such descriptions of the molecular and microbial often morphs into the grotesque. In the concluding section I address selected poems from Adam Dickinson's collection *Anatomic* (2018). The last two sections in particular zoom in on the exciting perspective afforded by the sublime and grotesque molecular and microbial landscapes these texts envision, on and off the human body, and the human-environment relations they construct and uncover.

---

8   Edmund Burke, *A Philosophical Enquiry into the Sublime and Beautiful* (Oxford: Oxford University Press, 2015), p. 59.

The Molecular Sublime

When the early pioneers of microscopy, such as Robert Hooke (1635–1703) and Antonie van Leeuwenhoek (1632–1723), confronted their audiences with very small levels of life at previously unknown powers of magnification, they did not just offer the world a better resolution of life at those scales. In many instances, their work revealed that those scales housed living organisms at all. Hooke's account and drawings of the flea, for instance, or the fly, in his 1665 *Micrographia*, depicted them in never-before-seen detail, but Leeuwenhoek's even more powerful microscope showed many micro-organisms for the very first time.[9] The history of microscopy is thus also the history of the discovery of new worlds within the cosmos; at least that is how it was often framed. From Hooke's *Micrographia* to contemporary narratives of the molecular, encountering the miniscule scales of being has repeatedly been aligned with interplanetary travel; Hooke, for example, discusses his microscopic observations of minute organisms alongside observations of the surface of the moon, and the very title of Theodor Rosebury's *Life on Man* frames the human body as a mysterious planet. This rhetoric exemplifies not only that the miniscule is frequently imaginatively associated with or even conflated with the gigantic—microscopes and telescopes rely on similar optical technologies after all—but that the small levels of life on earth also evoke a sense of the unfamiliar more commonly attributed to the uncertainty of outer space.

The notion of the sublime is routinely associated with vast, awe-inspiring magnitudes such as those characterising mountain ridges, the night sky or travel to distant moons. It has often been overlooked that Edmund Burke, when outlining the sublime in his canonical 1757 treatise on the subject, *Philosophical Enquiry into the Origin of Our Ideas of the Sublime and Beautiful*, also includes the very small in his catalogue

---

9 Such scientific representations of the very small levels of life sparked the cultural imagination already in the early modern period, for instance giving rise to such fantastic creatures as the hybrid bear-men in Margaret Cavendish's *The Blazing World* (1666), a text which at the same time sharply criticises the epistemological veracity and utility of the microscope. See Ian Lawson, 'Hybrid Philosophers: Cavendish's Reading of Hooke's *Micrographia*', in *The Palgrave Handbook of Early Modern Literature and Science*, ed. by H. Marchitello and E. Tribble (London: Palgrave, 2017), 467–88, https://doi.org/10.1057/978-1-137-46361-6_22.

of potential sources of sublimity. At first, he generally declares that '[w]hatever is fitted in any sort to excite the ideas of pain, and danger, [...] is a source of the *sublime*'.[10] When experienced not 'too nearly' but 'at certain distances', the ideas of pain and danger can produce a sublime delight which, in its highest form, is expressed as reason-defying 'Astonishment'.[11] Burke's list of sources that evoke such a passionate response includes the widely known examples of infinity, power, magnificence and vastness. Yet, in the midst of these grand iterations of sublimity, and in his section on vastness no less, he also points to the very small as evocative of the sublime:

> However, it may not be amiss to add to these remarks upon magnitude; that, as the great extreme of dimension is sublime, so the last extreme of littleness is in some measure sublime likewise; when we attend to the infinite divisibility of matter, when we pursue animal life into these excessively small, and yet organized beings, that escape the nicest inquisition of the sense, when we push our discoveries yet downward, and consider those creatures so many degrees yet smaller, and the still diminishing scale of existence, in tracing which the imagination is lost as well as the sense, we become amazed and confounded at the wonders of minuteness; nor can we distinguish in its effect this extreme of littleness from the vast itself.[12]

Here, Burke already foreshadows the molecular sublime's central questions and concerns. First of all, the miniscule realms of life, the 'excessively small, and yet organized beings', are granted the same sublime effect as the gigantic and establish vastness as sublime in both visible and invisible dimensions. Further, whether the sublime can be a source of rational self-knowledge about one's relation to the invisible molecular realm is addressed *avant la lettre* in the assertion that at the ever-smaller scales of life 'the imagination is lost as well as the sense'. This denies those who experience the sublime any rational grasp ('sense') of the sublime object, championing an affective reaction instead. Finally, the classification of the excessively small as a source of sublimity implies that these obscure, unimaginably minute levels of life are terrible precisely in their confounding conjuncture of being obscure,

---

10 Burke, *Philosophical Enquiry*, p. 33 (emphasis in the original).
11 Ibid., pp. 34, 47.
12 Ibid., p. 59.

as well as possibly infinite, and at the same time connected to human life. This latter aspect of the sublime as challenging human hubris and confronting the boundaries of the human subject with subversive interconnections to its perceived others, both animate and inanimate, is a central characteristic of the aesthetic of the sublime. This is particularly significant in the case of the molecular and microbial sublime which derives most of its terrible power from imagining and revealing human bodies, to an untoward degree, as part of an environmental ecosystem in which the boundaries between human and nonhuman appear increasingly porous.

As I will shortly demonstrate with two examples taken from recent genetic fictions, the molecular sublime predominantly manifests itself as variations of a Burkean sublime. The most notable theorisation of the sublime besides Burke's, that of Immanuel Kant, fails to capture the molecular sublime's radical challenge to human-environmental boundaries. Kant's conception of the sublime is highly differentiated and evolves throughout his career. What is characteristic of his larger sense of the sublime, however, and particularly of the most widely known version of his theory, as laid out in his *Critique of the Power of Judgement* (1790), is that the sublime marks a transcendence of sensual impressions achieved through the rational powers of the mind.[13] Whatever initially confounds the senses, roughly that which Burke would consider terrible, is eventually recuperated into a sublime rationalisation. The true source of sublimity thus lies not in the natural or imagined object but in 'the superiority of the rational vocation of our cognitive powers'.[14] This emphasis on human superiority effectively eliminates the challenge to human subjectivity and agency that characterises Burke's theory.[15] Moreover, from the beginning, Kant views only the large and powerful as productive of a sublime reaction,[16] neglecting the sublimity of the very small.

---

13 See also Robert Doran, *The Theory of the Sublime from Longinus to Kant* (Cambridge: Cambridge University Press, 2015), p. 201, https://doi.org/10.1017/CBO9781316182017.

14 Immanuel Kant, *Critique of Judgement*, translated by W. Pluhar (Indianapolis: Hackett, 1987), p. 114.

15 See also Vanessa Ryan, 'The Psychological Sublime: Burke's Critique of Reason', *Journal of the History of Ideas*, 62.2 (2001), 265–79, https://doi.org/10.2307/3654358.

16 Doran, *The Theory of the Sublime*, p. 181.

With Burke's emphasis on the excessively small a rare exception, the confrontation of the human with super-humanly grand scales has persistently defined the sublime experience, from Romantic to more recent iterations of sublimity. Peter B. Hales, for instance, suggests an atomic sublime and focuses specifically on the iconography of the mushroom clouds at Hiroshima and Nagasaki. For him, the clouds mark 'the presence of absolute scale', their off-the-charts magnitude leaving the spectator awe-struck.[17] Frances Ferguson, in turn, speaks of a nuclear sublime as the experience of 'the thing that is bigger than any individual, and specifically bigger in terms of being more powerful and, usually, more threatening'.[18] Both Hales and Ferguson prepare the ground for a molecular sublime by finding sublimity on atomic and nuclear scales, yet it is the magnitude of what the atomic can do rather than the atomic scale itself that is seen as sublime. Jos de Mul's sense of a biotechnological sublime similarly identifies the magnitude of biological, especially genetic databases and the infinite possibility of genetic recombination as the source of a new sublime.[19] As I conceive of it here, the molecular sublime can equally be connected to such notions of molecular vastness and endlessness, while twenty-first-century representations of molecular realms showcase a sublime aesthetic that is more specifically driven by the very smallness of the molecular levels of life.

For example, in Simon Mawer's novel *Mendel's Dwarf*, a description of the pioneering biological work of Gregor Mendel suggests the sublimity of a perspective that brings life at the molecular scale into focus:

> He was one of those men whose vision goes beyond what we can perceive with our eyes and touch with our hands, and no one shared his insight. The word *insight* is exact. Mendel had the same perception of nature as Pasteur, who could conceive of a virus without ever being able to see it, or Mendeleyev, who could conceive of elements that had not yet been discovered, or Thomson who could imagine particles yet smaller than the atom. Like them, Mendel looked through the surface of things deep

---

17 Peter B. Hales, 'The Atomic Sublime', *American Studies*, 32.1 (1991), 5–31 (p. 9).
18 Frances Ferguson, 'The Nuclear Sublime', *Diacritics*, 14.2 (1984), 4–10 (p. 6), https://doi.org/10.2307/464754.
19 Jos de Mul, 'The (Bio)Technological Sublime', *Diogenes*, 59.1–2 (2013), 32–40, https://doi.org/10.1177/0392192112469162.

into the fabric of nature, and he saw the atoms of inheritance as clearly as any Dalton or Rutherford saw the atoms of matter.[20]

The passage introduces the sublime vista Mendel uncovered on the processes of what we now know as Mendelian genetic inheritance. The sublimity of this vista derives as much from the image of Mendel gazing 'deep into the fabric of nature' as from the smallness of the level of life he has revealed. Fittingly, because Mendel himself had no idea of the material make-up of the units of inheritance he described, these units are imagined here through the analogy of atoms. Yet, in the overall passage, the physicists John Dalton and Ernest Rutherford are only two in an extended roll-call of important figures from the history of science. This list of scientists and their discoveries has two functions: firstly, by including Mendel in this hagiographic list, he is retrospectively given the recognition he was denied during his lifetime; secondly, and more importantly, the list emphasises the crucial role of the imagination in each of the scientists' sublime revelations about the hidden realms of atoms (Thomson, Dalton, Rutherford), microbes (Pasteur), chemical elements (Mendeleyev), and molecules (Mendel).

Another example from Mawer's novel illustrates how quickly and seamlessly the molecular sublime can morph into the molecular grotesque. Later in the narrative, a discussion about genetics is initially framed through a sublime aesthetic that zooms in on the invisible recesses of the cell nuclei in the human body. In the passage, the geneticist protagonist explains to his friend and soon-to-be lover Jean— yes, pun intended—some of the basics of genetic science:

> The molecule in question—the celebrated double helix, the acronymic DNA—is by now known to all in one way or another. Even high-court judges need to have some idea of it, even readers of the popular press recognise it, if only as a way of catching out a rapist by analysing his sperm. When I speak of this, Miss Piercey makes a face which signifies disgust and disapproval.
>
> "But it's there," I assure her, "whether you like it or not, there in the nuclei of all your cells."
>
> "The sperm?"

---

20 Simon Mawer, *Mendel's Dwarf* (London: Abacus, 2011), p. 4 (emphasis in the original).

> "The DNA. The molecules are there in every cell, carefully folded away like linen in a bottom drawer. Every function of every cell depends on it."
>
> "You mean"—a frown puckers her forehead—"it's there at this moment, wriggling round inside me?" She shifts on her seat, as though things are moving beneath her skirt.[21]

Leaving aside the sexually charged aspect of the dialogue, the conjured image of carefully folded DNA strands 'in every cell' may be argued to gesture towards a genetic network spreading through the human body that, in its ubiquity, evokes the sublimity of the vast. However, I would argue that the main focus in the passage lies on Jean's reaction. She displays astonishment and a certain incredulity in the face of this genetic revelation, as well as an alienating sense of invasion of her physical privacy from within the bounds of her own body. Her complex affective reaction as she struggles with the image of DNA in her cells transforms the molecular sublime into an experience of the molecular grotesque, which, as I will outline in a moment, is characterised precisely by the unsettling of personal agency and of personal—and often taboo—bodily boundaries experienced by Jean in this passage.

Both molecular sublime and molecular grotesque challenge perceived notions of human control over the body and its boundaries. However, it is the molecular sublime in particular which is marked by its power to astonish but confound the understanding. In the first passage quoted from *Mendel's Dwarf*, Mendel's knowledge of the molecular may be the sublime object, but the text's aesthetic and affective emphases lie on the awe-inspiring, but ultimately elusive, molecular 'fabric of nature'. Different theories of the sublime, as well as different representations of the molecular sublime, can be described in terms of how they are positioned towards a dialectic between rational understanding and affect. Whereas Kant suggests the rational transcendence of feeling as the very condition of sublimity, Burke champions the complex passion of the sublime, which explicitly confounds the sense and understanding. Ferguson's nuclear sublime keeps close to Burke by using the sublime to denote precisely what eludes reason about the power of the atom. This dialectic has significant ramifications for the cultural negotiation

---

21 Mawer, *Mendel's Dwarf*, p. 117.

of the molecular through the aesthetic of the sublime. As Jean's amazed but puzzled and uneasy reaction illustrates, the crux of the molecular sublime appears to be that it emphasises wonder—or even disgust, when it tips into the grotesque—over comprehension. And it seems indeed a challenge, at least to this author, to reconcile the sense of anxious astonishment felt at the vision of one's own hand as teeming with DNA, with a rational account of how DNA affects cell processes in that same hand. This sense of wonder affords to undermine the unity of the human body, especially with regard to the environment, as we will see. But if it thrives on a confounded understanding, then this will have implications for how this aesthetic shapes the cultural awareness of genes, viruses, and microbes. In addition, Burke's emphasis on a safe distance as a precondition for the sublime further complicates the possibility of the sublime to really reveal the individual's connection to the molecular. There is hence an inherent and paradoxical tension whenever the sublime object is in fact the body of the observing human subject, or the imbrication of that body in surrounding molecular environments. To the extent that comprehension of this connection is required to understand the concrete relevance of the molecular levels of life for the individual and for communities, the molecular sublime's affective emphasis entails a consequential epistemological limitation.

A purely rational account, however, for example of gene action or of microbes on the skin, is limited in its turn because it rarely achieves to convey the immediate and intimate significance and sense of scale of the molecular afforded by the aesthetic of the sublime. At the same time, sublime representations of the unfathomable scale of molecular life frequently become abstract, imagining the molecular in less concrete terms than Mawer's wriggly, neatly folded DNA. For instance, in Richard Powers' *The Overstory*, genetics' subversive potential to unearth hidden connections between human beings and plants acquires great emotional force, both for the characters and potentially the readers, but the genes themselves become an abstraction of their biochemical specificity:

> You and the tree in your backyard come from a common ancestor. A billion and a half years ago, the two of you parted ways. But even now, after an immense journey in separate directions, that tree and you still share a quarter of your genes...[22]

---

22 Richard Powers, *The Overstory* (New York: Norton, 2018), p. 132.

The human-tree kinship described here, taken from the opening of a book on forests written by one of the characters, is an awe-inspiring, hard-to-compute concept that thrives on genetic science but elevates it to a form of connection that transcends time and space, to the point where it becomes almost immaterial. The sublime here represents the invisible sphere of genetic interconnection through an abstraction that resembles Jean-François Lyotard's sense of the sublime 'absolute', which indicates that which lies outside of representation.[23] Following Lyotard, the sublime affords to show the awe-inspiring nature of trans-species genetic kinship, which is often lost when the concrete materiality of genes is foregrounded, especially through scientific methods of description and observation. Such foregrounding—or making visible—of genetic materiality is prone to emphasise the particular genetic object rather than broader scales of connection because it relies on scientific modes of mediation; the zoom lens of the microscope is necessarily particular. The abstraction of genetics in the excerpt from Powers' novel, in contrast, transcends the particular genetic object in favour of a sublime ecological vision of genetic interconnectedness that extends into the overall aesthetics of the novel, in which multiple storylines come to connect multiple characters across several decades and across the United States. The stress on reciprocal ties between characters and their environments supports Christopher Hitt's claim that an ecological sublime can offer a 'new, more responsible perspective on our relationship with the natural environment'.[24] Powers' sublime perspective on molecular relationality ironically entails affectively overshadowing the more specific and particulate genetic science that inspired it in the first place.

This dynamic of the sublime's dazzling obfuscation of the science that informs it is a common feature not only of the molecular sublime. In Powers' example, genetic knowledge enables the sublime representation but gives way to the challenging experience of imagining a seamless imbrication of humans in the life of plants. Similarly, Lynne Heller argues that in the conception of a microscopic or subatomic sublime,

---

23   Jean-François Lyotard, 'Presenting the Unpresentable: The Sublime', *Artforum*, April 1982, 64–69 (p. 68).
24   Christopher Hitt, 'Toward an Ecological Sublime', *New Literary History*, 30.3 (1999), 603–23 (p. 605).

technology and reason enable the emotion of sublimity, even if that emotion finally transcends rationality.²⁵ Burke himself, in his reference to the 'discoveries' of the 'small, and yet organized beings', already concedes that contemporary biological knowledge is the basis for the imagination of the minute life forms which then confounds the sense. And more generally, as Allen Carlson notes, eighteenth-century conceptions of sublimity were heavily informed by emergent scientific descriptions of the natural world, even if those scientific origins were often glossed over in the final sublime experience.²⁶ The molecular sublime's reliance on—and eventual overshadowing of—molecular science is hence quite a traditional feature of its aesthetics.

An initial sublime experience of molecular levels of life, as represented in both genetic examples above, may lead to a later rationalisation or excavation of the scientific explanations of those molecular levels. Often, the experience of sublimity is reduced with repetition, in any context.²⁷ Whether this holds true for the admittedly provocative realisation that our bodies are ecosystems we share with other molecular and microbial beings, reconciling the affective perception of their scale and presence on the individual body with a rational understanding of their organisation and materiality, remains to be seen.

I will now open my discussion of the molecular sublime to include another genre, popular science, and the aesthetic of the molecular grotesque, which frequently accompanies cultural representations of molecular sublimity. The grotesque, I will argue, can be seen to combine affective anxiety about new forms of human imbrication in living environments with more concrete imaginaries of the materiality of such environments. This also enables the grotesque to proffer not just an emotive but also a more rational understanding and imagination of how the molecular levels, in this instance in the form of microbes, directly touch and shape the human individual.

---

25 Lynne Heller, 'The Intrinsic Irony of the Future Sublime', *Canadian Review of American Studies*, 50.3 (2020), 377–98, https://doi.org/10.3138/cras-2020-006.

26 Allen Carlson, *Aesthetics and the Environment: The Appreciation of Nature, Art and Architecture* (London: Routledge, 2000), p. 85, https://doi.org/10.4324/9780203981405.

27 Ibid., p. 79.

Imagining Microbes: From the Molecular Sublime to the Molecular Grotesque

Ed Yong's 2016 bestselling popular science book, *I Contain Multitudes*, is emblematic of the increasing visibility of microbes in cultural discourses, from reports in the media about new discoveries to supposedly microbiome-stimulating yogurts in the supermarkets. The true pioneer of the microbial popular science book, however, was Theodor Rosebury's 1969 *Life on Man*. Yong's text is driven by a markedly heightened sense of urgency to grasp the impact of microbes on human lives and the environment, which is characteristic of the twenty-first-century fascination with the microbiome. Rosebury's book, in contrast, while no less urgent in its desire to illuminate the significance of microbes, is more locally informed by what he perceives as a contemporary mania for disinfectants, trying to reform his contemporaries' attitudes towards germs and "dirt". I include Rosebury's text here because it prefigures many of the aesthetics which later come to characterise the twenty-first-century treatment of microbes which can be identified in Yong's writing. In addition, Rosebury offers a particularly imaginative style which demonstrates that popular science writing on microbes, while non-fictional, is indeed marked by the same aesthetics as the genetic fictions discussed above. It also provides exceptionally rich examples of both the molecular sublime and the molecular grotesque.

In a particularly striking passage, Rosebury uses a scene from the film adaptation of *Tom Jones* as the medium through which he describes the microbial life in the human mouth:

> The biologist—the old-fashioned kind who uses a light microscope—knows that while the picture changes as we move closer, as we bring a smaller and smaller area into focus, it loses none of its fascination [...]. This is one way in which we see the continuity of the living world, its never-ending wonder and magnificence, from the greatest to the smallest. And so, adding an imaginary zoom microscope to [director Tony] Richardson's color camera, I was able in my mind's eye to zero in on the little fleshy crevices around Tom's and Jenny's teeth as they ate their meal, and to see the turmoil of microbic life there, the spirochetes and vibrios in furious movement, the thicker corkscrew-like spirilla gliding back and forth, and the more sluggish or quiet chains and clusters and colonies of bacilli and cocci, massed around or boiling between detached

epithelial scales and the fibers and debris of cells and food particles. Like the great and beautiful animals in whose mouths they live, these too are organisms, living things; and I could imagine them, quite like Tom and Jenny, making the most of the sudden accession of nourishment after a long fast.[28]

The microbial imaginary evoked here, informed by Rosebury's scientific background, reveals a rich density of life at an otherwise invisible scale. Collapsing visible and invisible scales, Rosebury produces a sublime aesthetic of microbial variety and movement. The uneasiness likely to be experienced by the reader at the previously unknown life forms on seemingly smooth gums and membranes is also likely mixed with the 'wonder' and 'fascination' invoked by Rosebury himself. This is not the abstract sublimity of Powers' genetic vision but a concrete, material sublime perspective on the troublingly and astonishingly diverse life inside the human mouth.

The passage, however, may also be said to present a grotesque aesthetic of the human microbiome. Following Bakhtin, Susan Stewart defines the grotesque body as confusing the established boundaries between 'what is the body and what is not'.[29] Stewart describes fragmentation of the body and the blowing out of proportion of body parts, especially of culturally taboo ones like those marking boundaries between inside and outside, like mouth and genitalia, as typical features of the grotesque.[30] While Stewart identifies this aesthetic as primarily belonging to the gigantic exaggeration, I argue that her sense of the grotesque equally applies to the very small levels of microbes. Even when imagined into the visible spectrum, the microbes are still minute within Rosebury's representation. In typical grotesque fashion, the passage quoted above zooms in on 'fleshy crevices' and in between 'epithelial cells', thus fragmenting the body and turning, in Stewart's words, the 'inside out'.[31] The 'turmoil' and 'furious movement' of the microbial communities depicted here further underline the grotesque alienation of the familiar surfaces of the mouth.

---

28 Theodor Rosebury, *Life on Man* (New York: The Viking Press, 1969), p. 8.
29 Susan Stewart, *On Longing: Narratives of the Miniature, the Gigantic, the Souvenir, the Collection* (Durham: Duke University Press, 1993), p. 105.
30 Ibid.
31 Ibid.

Traditionally, the grotesque often involved the merging of human and animal features, relishing the species' disorder while testing the differences between them. In the context of microbes, the grotesque can be observed to entail a similar boundary challenge, though in this case between human body and its microbial environment. This is already evident in the passage above, where microbes and humans are shown as symbiotic organisms within the 'continuity of the living world', but becomes even more pronounced in another passage from *Life on Man*. Here, Rosebury first invokes what I call the Gulliver trope, which denotes a common technique to imaginatively explore and represent what is usually invisible, namely by embedding a shrunken fictional character within the invisible levels of the universe. Michael Crichton and Richard Preston's 2011 science thriller *Micro*, for example, brings the Gulliver trope to Hawaii, where the novel's shrunken scientist-heroes battle with the sublime wonders of the microbial and insect ecosystems of a tropical rainforest. With such a 'perspective like Gulliver's in Brobdingnag', Rosebury in *Life on Man* invites the reader to

> land on a gently curving plain of relatively hairless skin, say a shoulder blade. We could hardly miss the growing things, but our first impression [...] would be one of plants rather than animals. We would probably find no creeping, crawling, or darting things here. The ground, made up of flat stones like slate but whitish, would show layers of more translucent stones beneath [...]. Among the stones would be standing hairs and the open pits of sweat and sebaceous glands. [...] We would see bundles of soft translucent sticks, stalks, and twiglike things. [...] We would see movements like those of plants growing or flowers unfolding in speeded-up or "time-lapse" movies, because the microbes would be growing before our eyes.[32]

Rosebury presents the surface of the human body as a microbial landscape. Stylistically, Rosebury uses frequent analogies, likening microbes to flowers and plants, and skin-cells to stones, to convey the alien environment of the skin to the reader; a technique, as Stewart points out, typical of representations of the miniature.[33] However, while Stewart identifies the miniature as mostly static, for action would push it into the background,[34] there is distinct movement in this and the

---

32   Rosebury, *Life on Man*, pp. 43–44.
33   Stewart, *On Longing*, pp. 44, 46.
34   Ibid., p. 54.

previous passage while the minute scale remains in the foreground. Rosebury's zooming-in on a minute speck of skin fragments the body and reveals it to be host to a strange world of independently growing organisms. This, together with his focus on minuscule orifices in the whitish translucent skin—the gland pits—clearly mark this microbial landscape as a grotesque representation of the human body, which, besides the mixture of wonder and anxiety it may provoke, bears little trace of the sublime.

The vision of a landscape on the surface of the human body effects an inversion typical of the grotesque, here between body and environment. Even more strongly than the sublime representation of life on man, the grotesque challenges us to reimagine the relation between a human individual and his or her living *umwelt* as one of fluid symbiosis rather than across clear fault lines. The particular strength of the grotesque, in comparison with the sublime, appears to be its retention of the materiality of the microbial world. Its scientifically informed representation of microbial life, while imaginatively portrayed, is not overcome by the complex passion of the sublime. This comparison should not suggest that the sublime and the grotesque are mutually exclusive, just as the sublime should not be reduced to affect and the grotesque to reason. Rather, each aesthetic emphasises a particular constellation of affect, imagination, and rational understanding; and in some texts, sublime and grotesque aesthetics are closely intertwined, as my later analysis of Dickinson's poems will show. In Rosebury's text, the particular affordance of the grotesque is to enable a more rational understanding of the microbial scale of the human body. This, though, requires the grotesque not to become a spectacle, which would cause a 'distancing of the object and a corresponding "aestheticization" of it', marking it as an 'aberration' and thus precluding the reader from relating the microbial vision back to his or her own body.[35] I would argue that the proximity to the grotesque object created through Rosebury's imaginary zoom-lens functions as an equivalent to the proximity characterising the carnivalesque grotesque, which overturns hierarchies between object and onlooker, between inside and outside, and highlights the human imbrication in the microbial environments around and within the body.

---

35   Ibid., p. 107.

In Yong's *I Contain Multitudes*, this emphasis on the human body as microbial environment is even more of a central concern. Yong's description of human bodies as hosts to microbial 'ecosystems' betrays his text's twenty-first-century context of climate change and its foregrounding of ecological discourses, but also the changing epistemological frameworks of microbiology, towards a stronger focus on microbial communities, since the publication of Rosebury's *Life on Man*.[36] This is particularly evident in his comparisons, from a microbiological perspective, of human and animal disease to 'a dying coral reef' or 'a lake that's smothered by algae or a meadow that's overrun with weeds—ecosystems gone awry'.[37] These analogies reiterate the notion of humans as microbial landscapes. Such images of humans as environments evoke the grotesque. 'It can be weird', Yong states, 'to consider [...] our body parts as rolling landscapes'.[38] This weirdness, which is largely due to the fact that 'microbes subvert our notions of individuality', also contains the potential of a sublime perspective on the human body as an 'entire world' of minute microbial life, a perspective more impressed by affect than reason, since these concepts, as Yong himself remarks, 'can be hard to grasp' and 'deeply disconcerting'.[39] Yong's emphasis on the beauty of this 'dizzying shift in perspective' further underlines its sublimity.[40]

In comparison to Rosebury's often flourishingly imaginative style, Yong's overall account of microbial life on and around the human is characterised by a more restrained, matter-of-fact tone, which communicates large quantities of information but gives less weight to imaginatively imbuing the microbes' minuscule scale with either abstract grandeur or particular materiality and form. Even though the book's subtitle promises 'a Grander View of Life', thereby already announcing the sublime, the text's general tone indicates its generic stress on science communication. This arguably limits its ability to rouse an affective response that brings home the conceptual and intimate implications

---

36 Ed Yong, *I Contain Multitudes: The Microbes Within Us and a Grander View of Life* (London: Vintage, 2016), p. 4.
37 Ibid., pp. 4, 23.
38 Ibid., p. 4.
39 Ibid., pp. 3, 24.
40 Ibid., p. 262.

of the human as microbial environment. A rare moment of zooming in on 'our skin', revealing microbes as 'spherical beads, sausage-like rods, and comma-shaped beans', contains traces of a grotesque microbial imaginary but pales in comparison with Rosebury's vision of furious microbial movement.[41]

Towards the end of Yong's book, however, his focus shifts again to microbial human-environment relations which are presented through an extended sublime aesthetic of microbial life. Considering the city of Chicago, Yong explains 'how different everything seemed with microbes in mind'.[42] He then imagines 'the city's microbial underbelly— the rich seam of life that coats it, and moves through it on gusts of wind and currents of water and mobile bags of flesh'.[43] This representation combines elements of sublimity with the grotesque. While the 'mobile bags of flesh' appear grotesque, recalling Yong's earlier description of humans as 'vessels full of microbes', the main effect here, though supported by the grotesque imagery, is of a wondrous and strange vision of seemingly inanimate city space suddenly filled with life. This sublime description includes the image of humans as microbial 'clouds of themselves', where again the human form is likened to a feature of the landscape and metaphorically subsumed under the wealth of its microbial symbionts.[44] While this vision defamiliarizes the city of Chicago as well as the bodies of its inhabitants, its sublimity is particular because it retains a high degree of rational understanding. The passage indeed foregrounds the faculty of the imagination as crucial for access to the ubiquitous but otherwise invisible scales of microbial life. This form of understanding enabled by the imagination is also highlighted by Yong when he states that although the microbial world 'is still invisible to my eyes, I can finally see it'.[45]

Both Yong's and Rosebury's texts imaginatively depict the human body as microbial landscape and ecosystem, revealing the vitally porous boundaries between humans and their microbial and molecular environments. In Rosebury's case, identifying his microbial landscapes

---

41 Ibid., p. 10.
42 Ibid., p. 262.
43 Ibid.
44 Ibid., pp. 26, 262.
45 Ibid., pp. 26, 264.

as ecosystems is the result of a retroactive analytical reading informed by the transformations characterising the study and representations of microbes in the twenty-first century. Microbiology today is not only itself shaped by a new focus on microbial communities and meta-organisms but, as Bapteste and Campos et al. outline, the connective ubiquity of microbes has led to a larger, cross-disciplinary trend away from the analysis of ever-smaller units of enquiry and towards the study of ecological systems and interconnected networks within and between living organisms.[46] Against this background, Yong's extensive use of explicit and analogical references to ecosystems is thus an expression both of the renewed and topical concern with the workings and futures of ecologies in the era of climate change, but also of a distinctly twenty-first-century microbial imaginary.[47]

In the final section of this study, I will now turn to discuss an aesthetics that even more explicitly zooms in on the human impact on their environments.

Molecular Landscapes: New Ways of Reading the Anthropocene

In Crichton and Preston's *Micro*, the natural world comes alive for the shrunken scientists at a previously unknown scale in the Hawaiian rainforest. The soil underneath their boots suddenly heaves with life forms; when they take a bath in a puddle of water, they have the sublime experience of swimming with paramecia, unicellular microbes, a strange but astonishing encounter with life at the level of a single cell that conjures up the origin of life both at a planetary and an individual developmental scale.[48] The novel's rich and scientifically informed imaginary of human-environment encounters and ecological

---

46 Eric Bapteste et al., 'The Epistemic Revolution Induced by Microbiome Studies: An Interdisciplinary View', *Biology*, 10.651 (2021), 1–15, https://doi.org/10.3390/biology10070651.

47 See also Liliane Campos's argument that popular representations of microbes, such as Yong's, increasingly draw on environmental imagery, displacing 'agency away from human individuals' and 'towards collective actors', in Bapteste et al., 'The Epistemic Revolution Induced by Microbiome Studies', p. 4.

48 Michael Crichton and Richard Preston, *Micro* (New York: Harper, 2011), pp. 159, 354–56.

interdependencies is set against the backdrop of the capitalist exploitation of natural resources perpetrated by the text's villain. By highlighting the ecological diversity of soils and puddles and showing the human enmeshment in their diversity of life, *Micro* produces a form of what Heather Sullivan calls 'dirty aesthetics', collapsing perceived boundaries between human and natural environments.[49] In his 2018 poetry collection *Anatomic*, Adam Dickinson takes this concern with the impact of capitalism and industry on environments at microscopic scales even further, constructing a molecular and microbial poetics of environmental connection that explores biochemical ways of reading and writing the Anthropocene.

As part of the writing process of *Anatomic*, Dickinson underwent extensive tests for various chemicals and microbes. In the final collection, the epigraphs signal the chemicals or micro-organisms that are the central concern of each poem. Within the overall aesthetic Dickinson lays out for the collection, the presence and impact of these biochemical entities on his body are framed as ways in which the outside writes onto and into his inside: 'I am an event, a site within which the industrial powers and evolutionary pressures of my time have come to write. I am a spectacular and horrifying crowd'.[50] The mixture of astonishment and terror that characterises his vision of himself as intricately connected with an outside environment, again constitutes a form of the sublime. The whole collection partakes in this aesthetic to the extent that it locates his body in almost endless forms of microbial and biochemical connections to and from the outside world.

More specifically, the sublime is evoked when Dickinson describes the plethora of life on his bodily surface, negotiating the porous outer boundaries of his self:

> My gut is a tropical forest of microbes. Their cells, which cover my entire body, are at least as numerous as my own. These microbiota live on and within me as a giant nonhuman organ, controlling the expression of

---

49 Heather I. Sullivan, 'Dirt Theory and Material Ecocriticism', *Interdisciplinary Studies in Literature and Environment*, 19.3 (2012), 515–31 (p. 515), https://doi.org/10.1093/isle/iss067.

50 Adam Dickinson, *Anatomic* (Toronto: Coach House Books, 2018), p. 9.

genes and the imagined sense of self maintained by my immune system's sensitivity to inside and outside.[51]

The sublime is caused as much by the instability of the self's outer limits, only an 'imagined' by-product of the immune system, as by the vastness of the microbial mesh covering the body as a 'giant nonhuman organ'.[52] The evocation of the microbial dimension of his self, which seems to be both part of and exterior to his body, is infused with the discourse of a global politics of Western imperialism which is also directly related to his metabolism by way of bacteria and food: 'I house bacterial colonies that have become empires of the Western diet, fuelled by sugar, salt, and fat'.[53] Besides microbes and bacteria, Dickinson emphasises the chemical plurality of his body, which is equally influenced by outside forces. In the long poem 'Hormone', sections of which run through the whole collection, this sublime chemical multiplicity of his being, recalling Ed Yong's microbial 'multitudes', is repeatedly emphasised: 'A body/ is a crowd/ getting out/ of bed', and 'with its/ chemicals/ it can never/ be lonely'.[54] In its imagery, this molecular sublime remains largely an abstraction. The materiality of the forest-like microbes in the speaker's gut reinforces the notion of a human microbial landscape rather than revealing the microbes themselves. Whereas, if its familiar materiality is emphasised, the image of a gigantic microbial organ seems more to disrupt the sublime by a grotesque inversion of inside and outside.

The sublimity of the body as a crowd is a paradoxical sublimity; it collapses the safe distance characteristic of Burke's sublime between onlooker and sublime object. This paradoxical sublime reveals even more strongly the instability of the human subject whose physical borders are already undermined by their manifold molecular and microbial environments. This instability is further underlined in Dickinson's text by the agency attributed to the external forces shaping the self and body of the poet—they are shown to be literally 'controlling' vital processes such as gene expression. This agency stands out against other

---

51 Ibid., p. 42.
52 Editors' note: This description of the nonhuman is comparable to the weird 'mycoaesthetics' studied by Derek Woods in this collection (chapter 4).
53 Dickinson, *Anatomic*, p. 10.
54 Ibid., p. 107, lines 1–4, 23–26.

contemporary poetic treatments of microscopic beings, as identified by Sarah Bouttier, which tend to downplay their agential power.[55] This agency of the molecular environment—understood in its most inclusive sense—is further highlighted by Dickinson's poetics, which depicts the environmental impact on his body as a form of writing. The microbes, as well as other natural and artificial biochemical substances, are said to 'enact a form of biochemical writing through their integral involvement in the metabolic processes' of his being.[56]

Dickinson's aesthetics of biochemical writing is most pronouncedly concerned with environmental pollution, especially its origins and impact on the body, when he describes his fat cells as an 'archive of the historical moment':

> Military, industrial, and agricultural history bioaccumulate in adipose tissues. I have found one of the most widely distributed environmental contaminants on the planet in my body: polychlorinated biphenyls, PCBS. Principally manufactured by Monsanto for industrial and commercial applications, these lipophilic pollutants collect like comment sections in the fat of creatures everywhere. If we test for them, we will find them. PCBS constitute a form of writing in the Anthropocene, a recursive script where industrial innovations find their way back into the metabolic messaging systems of the biological bodies that have created them.[57]

Dickinson depicts a circulatory process of writing in which man-made chemicals are released into the environment through industry, where they accumulate in the fatty tissues of human and nonhuman organisms and subtly but manifestly transform their hormonal pathways. This form of writing is marked out as specific to the 'Anthropocene'. In the epoch in which humans have wreaked havoc with the planet's ecosystem, such a chemical form of writing is testimony to the human molecular imprint on the environment, as well as to the intimate human-environmental connections through which this imprint comes back to mark and haunt them.

---

55 Sarah Bouttier, 'The Right Amount of Agency: Microscopic Beings vs Other Nonhuman Creatures in Contemporary Poetic Representations', *Épistémocritique*, 17 (9 May 2018), 1–14, https://epistemocritique.org/the-right-amount-of-agency-microscopic-beings-vs-other-nonhuman-creatures-in-contemporary-poetic-representations/.
56 Dickinson, *Anatomic*, p. 42.
57 Ibid., p. 31.

This molecular poesis of environmental chemospheres turns grotesque when the fatty tissue as archive becomes material and constitutes a carnivalesque inversion of outside and inside. In the poem 'Agents Orange, Yellow, and Red', Dickinson describes how the chemical '2,3,7,8-Tetrachlorodibenzodioxin', a highly toxic by-product of producing bleached paper—as well as the defoliant used during the Vietnam War to reveal enemy combatants hiding under the rich foliage of trees—, has seeped into the writing of his youthful fat cells:

> Northern rivers are warmed
> by the paper mill's piss, which,
> like making the world safe for democracy,
> slowly leaked into my childhood, yellowing
> the lipophilic paperbacks of my
> adipose fat[58]

The complex image of the stained and tautological 'lipophilic paperbacks of my adipose fat' produces a grotesque scriptural embodiment of the fat as written archive. The toxic pollution of rivers and human as well as nonhuman ecosystems finds a record in a body-part-turned-object. This biochemical grotesque, in contrast to Rosebury's grotesque microbial landscapes, arguably appears as spectacle rather than carnival, in which Dickinson's fragmented body is aestheticized as an object and potentially distanced from the reader. The self-exploratory stance of Dickinson as speaker in the poems also instils a certain overall distance between the revelations of his particular body as an environmental archive and the reader's own body; even though Dickinson insists that '[w]hat is inscribed in me is in you, too'.[59] For the reader, the more insightful and affectively intense aesthetic may ultimately be the volume's sublime vista of the environment writing back.

Together, though, the sublime and grotesque molecular landscapes in which Dickinson embeds the human form highlight a powerful imaginary of human environmental aesthetics as transformed by capitalist-industrial molecular writing, and thus a new way of reading the Anthropocene.

---

58 Ibid., p. 15, lines 14–19.
59 Ibid., p. 10.

Conclusion: The Big Moment of the Very Small

The twenty-first-century imaginations of the very small levels of life on and off the human body discussed in this study all share the sense that imagining the genetic, biochemical, and microbial scales of this planet has acquired a new urgency. The big moment of the very small seems to be now. This may seem a provocative claim to make at the tail end of the century of the gene but, as I hope to have shown, the accelerating environmental crisis and increasing interest in molecular forms of ecological interdependence can be observed to drive, at least in some degree, nearly all of the above representations of molecular human environments. The molecular sublime and the molecular grotesque emerge as key aesthetics in the cultural imaginary of this new perspective on the human imbrication in environmental networks, in which the human is revealed as both a polluting and polluted life form.

All of the above imaginations of the molecular exhibit a fascination with the smallest scales of life which have gradually come to light over the centuries, with such discoveries coming to a head in the twenty-first century. While scientific revolutions have revealed smaller and smaller units of life and environmental interconnection, it falls to cultural representations of these levels of life to ask how such insights affect and renegotiate human societies and human bodies in their relation to one another and to their ecological surroundings. Both the molecular sublime and the molecular grotesque entail specific affective and epistemological affordances and limitations, sometimes emphasising the vast scale of the very small and its material proximity to the human body, sometimes distancing the molecular object from the individual human frame, either as abstraction or spectacular aberration. On the whole, however, I would argue that the cultural imaginary has never before been engaged this closely and intensely with the smallest levels of life. The different perspectives opened up here on the human as rolling molecular landscape evince that it remains an awe-inspiring challenge to reconcile the human scale with the strange and astonishing levels of life beyond the visible spectrum.

Works Cited

Bapteste, Eric, Liliane Campos et al., 'The Epistemic Revolution Induced by Microbiome Studies: An Interdisciplinary View', *Biology*, 10.651 (2021), 1–15, https://doi.org/10.3390/biology10070651

Bouttier, Sarah, 'The Right Amount of Agency: Microscopic Beings vs Other Nonhuman Creatures in Contemporary Poetic Representations', *Épistémocritique*, 17 (9 May 2018), 1–14, https://epistemocritique.org/the-right-amount-of-agency-microscopic-beings-vs-other-nonhuman-creatures-in-contemporary-poetic-representations/

Brennan, Andrew, 'Environmental Ethics', *The Stanford Encyclopedia of Philosophy* (2021), https://plato.stanford.edu/archives/fall2021/entries/ethics-environmental/

Burke, Edmund, *A Philosophical Enquiry into the Sublime and Beautiful* (Oxford: Oxford University Press, 2015).

Carlson, Allen, *Aesthetics and the Environment: The Appreciation of Nature, Art and Architecture* (London: Routledge, 2000), https://doi.org/10.4324/9780203981405

Clark, Timothy, 'Derangements of Scale', in *Telemorphosis: Theory in the Era of Climate Change, Vol. 1*, ed. by Tom Cohen (Michigan: Open Humanities Press, 2012), [n.p.], https://dx.doi.org/10.3998/ohp.10539563.0001.001

Crichton, Michael, and Richard Preston, *Micro* (New York: Harper, 2011).

de Mul, Jos, 'The (Bio)Technological Sublime', *Diogenes*, 59.1–2 (2013), 32–40, https://doi.org/10.1177/0392192112469162

Dickinson, Adam, *Anatomic* (Toronto: Coach House Books, 2018).

Doran, Robert, *The Theory of the Sublime from Longinus to Kant* (Cambridge: Cambridge University Press, 2015), https://doi.org/10.1017/CBO9781316182017

Ferguson, Frances, 'The Nuclear Sublime', *Diacritics*, 14.2 (1984), 4–10, https://doi.org/10.2307/464754

Hales, Peter B., 'The Atomic Sublime', *American Studies*, 32.1 (1991), 5–31.

Heller, Lynne, 'The Intrinsic Irony of the Future Sublime', *Canadian Review of American Studies*, 50.3 (2020), 377–98, https://doi.org/10.3138/cras-2020-006

Hitt, Christopher, 'Toward an Ecological Sublime', *New Literary History*, 30.3 (1999), 603–23.

Kant, Immanuel, *Critique of Judgement*, translated by W. Pluhar (Indianapolis: Hackett, 1987).

Kay, Lily, *Who Wrote the Book of Life? A History of the Genetic Code* (Stanford: Stanford University Press, 2000).

Keller, Evelyn Fox, *The Century of the Gene* (Cambridge: Harvard University Press, 2000).

Lawson, Ian, 'Hybrid Philosophers: Cavendish's Reading of Hooke's *Micrographia*', in *The Palgrave Handbook of Early Modern Literature and Science*, ed. by H. Marchitello and E. Tribble (London: Palgrave, 2017), pp. 467–88, https://doi.org/10.1057/978-1-137-46361-6_22

Lyotard, Jean-François, 'Presenting the Unpresentable: The Sublime', *Artforum* (April 1982), 64–69.

Mawer, Simon, *Mendel's Dwarf* (London: Abacus, 2011).

Morton, Timothy, *Ecology Without Nature: Rethinking Environmental Aesthetics* (Cambridge: Harvard University Press, 2009).

Powers, Richard, *The Overstory* (New York: Norton, 2018).

Rose, Nikolas, *The Politics of Life Itself: Biomedicine, Power, and Subjectivity in the Twenty-First Century* (Princeton: Princeton University Press, 2006).

Rosebury, Theodor, *Life on Man* (New York: The Viking Press, 1969).

Ryan, Vanessa, 'The Psychological Sublime: Burke's Critique of Reason', *Journal of the History of Ideas*, 62.2 (2001), 265–79, https://doi.org/10.2307/3654358

Stewart, Susan, *On Longing: Narratives of the Miniature, the Gigantic, the Souvenir, the Collection* (Durham: Duke University Press, 1993).

Sullivan, Heather I., 'Dirt Theory and Material Ecocriticism', *Interdisciplinary Studies in Literature and Environment*, 19.3 (2012), 515–31, https://doi.org/10.1093/isle/iss067

Tucker, Ian, 'Daniel M Davis: "Unbelievable things will come from biological advances"', *The Guardian* (3 July 2021), https://www.theguardian.com/science/2021/jul/03/daniel-m-davis-the-secret-body-interview-immunology-data

Yong, Ed, *I Contain Multitudes: The Microbes Within Us and a Grander View of Life* (London: Vintage, 2016).

3. Still Life and Vital Matter in Gillian Clarke's Poetry

Sophie Laniel-Musitelli

Gillian Clarke's most recent collections of poems—*Making the Beds for the Dead* (2004), *A Recipe for Water* (2009), *Ice* (2012), and *Zoology* (2017)— explore the mutual convergences and cross-metamorphoses between living bodies and inorganic matter. In its endeavour to reveal the vitality of inorganic matter, Clarke's poetry draws on new developments in biological and physical sciences. To do so, it registers the moments when matter changes states, such as the formation and dissolution of rock and ice. Conversely, it tracks the presence of inorganic matter at the heart of living bodies. For example, when, in her poetic autobiography *At the Source*, Clarke muses on the 'shadow-taste of stone' in food and wine, she concludes that 'We are made of stone, metals, stardust'.[1] In Clarke's recent works, living and non-living entities evolve and morph into one another through moments of emergence and disappearance, such as embryogenesis and metamorphosis. Hence, for instance, a tropism towards fossilization and the 'grace of bones / Eloquent in stone',[2] when a living body reverts to the inorganic matter it arose from. The eloquence of stone can be heard when the poet lingers on the 'loosed flotilla of [the] vertebrae' of a fossilized Ichthyosaur and 'the dolphin-flip of her spine', conveying a sense of sleek movement, the image of an agile swimmer who did not lose her grace in death.[3] The fossil also records a moment of emergence as the fossilized sea creature 'dies giving birth':[4] Clarke's

---

1 Gillian Clarke, *At the Source: A Writer's Year* (Manchester: Carcanet, 2008), p. 74.
2 Gillian Clarke, *Zoology* (Manchester: Carcanet, 2017), 'Ichthyosaur', p. 27, 19–20.
3 Clarke, 'Ichthyosaur', 3–4, 8.
4 Ibid., 5.

poetry is drawn towards the reversibility of matter, from living to inert and from inert to living.

That new interest in the sciences corresponds to an environmental turn. In *At the Source*, Clarke explains that 'we were suddenly—it seemed overnight—made aware that the planet could become uninhabitable and that it could die. We all needed a new way to write about the natural world'.[5] Attending to the energies of nonhuman bodies and non-living matter is part of Clarke's commitment to environmental poetics, in an attempt to supplant 'the image of dead matter or thoroughly instrumentalized matter feed[ing] human hubris'.[6] Born in Cardiff in 1937 and currently running a small organic farm in Ceredigion, Clarke has always played a central role in the life of contemporary Welsh poetry: she is the founder of the Writers' Centre in North Wales and was the National Poet of Wales from 2008 to 2016. Thus, she is often categorized as a local poet, especially since most of the poems from her earlier collections, such as *The Sundial* (1978) and *The King of Britain's Daughter* (1993), find their main setting and focus in the Welsh landscape and cultural heritage. Yet, her recent poetry, inspired by climate science, rearticulates the local and the global, inscribing Wales within the broader challenges the Anthropocene poses to poetry. It also connects human temporality and geological times: poetic writing draws on the scientific imagination to alter the scales of space and time prevalent in her earlier work.[7] That is part of her poetry's attempt at inhabiting the Anthropocene, whose 'most difficult challenge [...] is represented by scale effects, that is, phenomena that are invisible at the normal levels of perception but only emerge as one changes the spatial or temporal scale at which the issues are framed'.[8]

This chapter aims at exploring the ways in which Clarke uses shifts in scales to envision the vitality of matter, a vitality simultaneously explored by thinkers such as Jane Bennett, whose *Vibrant Matter: A Political*

---

5   Clarke, *Source*, p. 13.
6   Jane Bennett, *Vibrant Matter: A Political Ecology of Things* (Durham and London: Duke University Press, 2010), p. ix.
7   Editors' note: A similar juxtaposition of geological and human temporalities can be found in *Beaming Sahara* (2019), a performative installation discussed by Eliane Beaufils in chapter 13.
8   Timothy Clark, *Ecocriticism on the Edge: The Anthropocene as a Threshold Concept* (London and New York: Bloomsbury Academic, 2015), p. 22.

Ecology of Things from 2010, provides a key theoretical framework for this analysis. To explore the constant metamorphoses and interactions of organic and inorganic matter, her poetry plays with scientific representations of time, through the heterogenous yet interacting scales of biological time and geological time. Clarke's poetry experiments with modes of sensation inspired by scientific imaging, staging the vitality of matter as it develops over timescales that are imperceptible to the human eye, from geological æons to the microscopic development of new organisms in biology.

The Poetry of Stone

In *At the Source*, Clarke muses on the language and temporality of geology: 'Igneous, metamorphic, sedimentary rock. How I loved my *Guide to Minerals, Rocks and Fossils*. I loved its language, the names of rock. Earth took its time with rock. It took ages. Then life began, fidgeting and wriggling for an unimaginably long, slow time, for ages, æons, chrons'.[9] Clarke surreptitiously moves from the æon, the 'largest division of geological time' to the neologism 'chrons'.[10] From the Greek χρόνος, it leaves the realm of the geological to get closer to lived time: to the temporality of consciousness and to the measured time of everyday life, chronometers, and chronologies. In the seamless passage from æons to chrons, the geological and the biological develop along intersecting timelines. This is also the case in *Making the Beds for the Dead*, in the section entitled 'The Stone Poems'. The title of each poem in the section associates the name of a living being or of a mineral with the name of a geological era: 'Woman washing her hair, Devonian', 'The Stone Hare, Lower Carboniferous', or 'Coal, Upper Carboniferous'.

'The Stone Hare' offers two timescales—stone formation and the emergence of a living body—in one artistic form, that of the stone hare sitting on the poet's desk as she writes:

> In its limbs lies the story of the earth,
> the living ocean, then the slow birth
> of limestone from the long trajectories

---

9 Clarke, *Source*, p. 49.
10 *Oxford English Dictionary*, 2nd ed., version 4.0.0.2 (Oxford: Oxford University Press, 2009) [on CD-ROM], entry 'æon' 3.

> of starfish, feather-stars, crinoids and crushed shells
> that fill with calcite, harden, wait for the quarryman,
> the timed explosion and the sculptor's hand.[11]

The sculpture emerges from stone that was once alive as primitive sea creatures. Those organisms seem to mediate between temporal planes and between scales, since their names—'starfish' and 'feather-stars'—are reminiscent of their origins in the inorganic matter of stars, but also hint at the mineral matter produced by living bodies—such as 'shells'. Carving, like poetry, is the art of reviving the memory of that former life into the figure of a new living body. Clarke's poetry tries to envision affects outside subjectivity: geological times are a way to go beyond subjective memory, towards a form of memory inherent in matter itself. In 'The Stone Hare', the sculpture is 'a premonition of stone'.[12] There are various temporalities at work within the stone, which bears the memory of its past but also the premonition of its future; art simply releases these temporalities. Each moment in the work of art, from stone formation to carving, survives and embraces the other.

'The Stone Hare' explores the formative agency at work within matter, and the common drive towards form in inert matter, living matter, and artistic objects. The stone waiting for the hand of the carver to reveal its form shows that form is latent within matter itself. Hence, Clarke's 'aesthetic is explicitly driven by a Romantic organicism which sees... sculpting as a process, not of construction, but of discovery'.[13] Poetry, like stone carving, is about letting form emerge out of matter, because matter is vital and artistic, and because art is the natural continuation of the active and formative powers within matter. From primitive sea creatures to the stone and the carver, the agents forming the stone hare are animal, mineral, and human: agency is shared by 'an ontologically diverse assemblage of energies and bodies, of simple and complex bodies, of the physical and the physiological'.[14] Clarke's poetry moves from the representation of cross-metamorphoses to the exploration of minerals, living bodies, and artworks as an assemblage sharing agency. The emergence of form is also a sonic process; the more recent poem 'Ice

---

11   Gillian Clarke, *Making the Beds for the Dead* (Manchester: Carcanet, 2004), p. 27, 6–11.
12   Clarke, *Beds*, p. 27, 3.
13   Ian Gregson, *The New Poetry in Wales* (Cardiff: University of Wales Press, 2007), p. 12.
14   Bennett, *Matter*, p. 117.

Music' is a beautiful example of the sonic creativity of matter when it changes states: 'we both hear the music, the high far hum of ice, / strung sound, feather-fall, a sigh of rime, / fog-blurred syllables of trees, sap stilled to stone'.[15] The homophony between rime—which means 'hoarfrost' or 'frozen mist'—and rhyme shows that the vitality of matter is a model for poetic writing. The rhythmic and phonic creativity of matter allows poetry to register its moments of change.

'A Recipe for Water' continues the investigation of minerality in a representation of poetry as a craft, directly hewn out of the conformation of matter. The poem combines explorations of the chemical composition of water, of the origin of the word 'water', and of the organic experience of water on and in the poet's body:

> Calcium, Magnesium, Potassium, Sodium,
> Chloride, Sulphate, Nitrate, Iron.
>
> Sip this, the poetry of stone,
> a mineral Latin in our blood, our bone.
> [...]
>
> *
>
> That drop on the tongue
> was the first word in the world
> head back, eyes closed, mouth open
> to drink the rain
> *wysg, uisc, dŵr, hudra, aqua, eau, wasser*
>
> *
>
> You imagine me writing in the falling rain,
> rain on the roof, writing in whispers
> on the slates' lectern,
> rain spelling out each syllable
> like a child learning to read.[16]

The poem looks for the composition of water in the minerals that water picks up on its journeys through rock formations. The speaker then traces the imagined origin of language in water as the primordial

---

15   Gillian Clarke, *Ice* (Manchester: Carcanet, 2012), p. 14, 10–12.
16   Gillian Clarke, *A Recipe for Water* (Manchester: Carcanet, 2009), pp. 20–21, 11–14, 37–46.

condition of the apparition of life, circulating freely between the organic and the inorganic. It looks for the fabled time when the word for water was born out of a raindrop on a tongue. Language is staged as the result of the action of water on the body: it was born in the throat along with the sensation of thirst and the pleasure of drinking. The poem searches through the material memory of language, because language proceeds from material and organic processes. Poetry tries to access the geological past of language as the result of a sedimentary process, symbolically and quite literally, hence the fact that geology provides a heuristic model for poetic writing. The 'slates' lectern' is represented as a surface moulded by rain, bringing the same movement to geology and writing: both emerge through a long process based on physical contact: the exposure to water, and in particular to rain on the lectern, but also, earlier on, on the tongue and in the throat, connecting the geological and the poetic. Geology provides not only a heuristic model but also a metaphor for the temporality of writing, since Clarke sees her own writing as a process of sedimentation: her poetic material consists in 'layers of experience, story, snapshot, hearsay and imagination, images laid down one on the other like sedimentary rock'.[17] The way geological temporality and the timescale of a human life collide in the poem is part of Clarke's poetic endeavour to inhabit the Anthropocene as a geological era in which human agency plays a central part. Earlier in the poem, Clarke imagines a 'second word for water / Dŵfr. Dŵr. Dyfroedd. Dover'.[18] For her, language is engrained in the land, and human culture draws on the formative forces at work within the material formation of the landscape. This raises the issue of nativism in her poetry. For Ian Gregson, 'Clarke's poetic mission is to champion naturalness, which becomes especially challenging where identity issues are involved [...]. Her equation of language and land implies a very unsettling racial essentialism: in the context of nationality the emphasis on naturalness leads to the ideology of blood and soil'.[19] Gregson's critical analysis participates in an earlier debate surrounding the question of identity as blood and soil in Clarke's poetry, before it was complicated by its environmental turn. Sam Solnick's wider reflection on ecocriticism is helpful to characterize

---

17 Clarke, *Source*, p. 51.
18 Clarke, *Recipe*, p. 21, 25–26.
19 Gregson, *New Poetry*, p. 12.

that turn: Clarke's 'early focus on phenomenological engagement and specific places has been modified by more refined considerations of the complex relationships between local and non-local'.[20] For Meurig Wynn Thomas, gender has always been a crux and a transformative drive in Clarke's identity: 'the traditional Welsh obsession with male ancestor-worship had metamorphosed into Clarke's very differently motivated and very differently orientated search for her distinctive antecedents as a woman'.[21] In 'A Recipe of Water', the origin of Dover in *Dŵr* is thus displaced by the substitution of a male bardic figure by a woman:

> Imagine the moment a man,
> a woman singing in a dark age,
>
> gazed from those chalk heights
> at the vast and broken seas
>
> and sang this word, song and word
> on the tongue, in the throat,
>
> finding a name for the element.[22]

In a reversal of the previous image, the woman no longer stands below the rain, receiving the drop that creates the word 'water', but above the sea, and gives water its name. The substitution does not neutralize the rootedness of the word in the place, but initiates a shift from the imagery of the fixed source back to the element of water and its constant movement, materialized here in the 'vast and broken seas'. In the poem, water is indeed defined by its ability to circulate, picking up its constitutive elements as it goes, 'seeping page by page / through the strata, / run[ning] black in the aquifers'.[23] As a female poet, Clarke had to reinvent her origins within a bardic culture traditionally inherited from man to man, complicating her sense of belonging to an essentialized and localized heritage. Hence her recent poetry's tropism for the circulation of water as a deterritorializing force, to seashores and river banks as

---

20 Sam Solnick, *Poetry and the Anthropocene: Ecology, Biology and Technology in Contemporary British and Irish Poetry* (London and New York: Routledge, 2017), p. 19.
21 Meurig Wynn Thomas, *Corresponding Cultures: The Two Literatures of Wales* (Cardiff: University of Wales Press, 1999), p. 190.
22 Clarke, *Recipe*, p. 21, 27–33.
23 Ibid., p. 20, 4–6.

'ambivalent non-territorial borderland', but also as the sites of a fluid reterritorializing movement of Welsh identity less in blood and soil than in ever-changing waves and poetic language.[24]

Playing with Scale

We can better understand how Clarke's recent environmental poetry recalibrates the scale of local identity by re-examining the mineral figures that traverse it. In *At the Source*, Clarke reminisces about the way she used to imagine stones germinating like seeds, when she was a child: 'Could I suck a pebble until it dissolved like a sweet? Would there be, at the very last moment, a seed, as in an aniseed ball? What would you grow from the seed in a stone? I knew, as a child, that crystals grew, that they accrued, multiplied and made themselves in the dark'.[25] A vital process thus seems at work in the formation of crystals. Crystallization testifies to the ability of inorganic matter to organize into regular structures. The choice of the verb 'to grow' turns the formation of crystals into a process akin to organic development. That vitality is the main drive behind the poem 'Pebble', published four years later, in *Ice*:

> Weigh two hundred million years
> in your hand, the mystery of eras,
> a single syllable
> pulsing in a pebble.
> [...]
> Take in your right hand from the evening sky
> that other sad old stone, the moon.
> You, Earth, pebble, moon-stone,
> held together in the noose of gravity.[26]

A pebble bears within itself the memory of geological æons, 'the mystery of eras'. It is animated by the whole movement of the earth as tides are governed by gravity through the influence of the Moon. Inorganic matter is pulsating with the rhythms of the tide and of geological times, creating the seed of poetic language. From this seed

---

24 Alice Entwistle, *Poetry, Geography, Gender: Women Rewriting Contemporary Wales* (Cardiff: University of Wales Press, 2013), p. 120.
25 Clarke, *Source*, p. 50.
26 Clarke, *Ice*, p. 59, 1–4, 9–12.

emerge the alliterations we can hear in 'single syllable' and 'pulsing in a pebble', reinforcing that assemblage of language, body, and stone. *At the Source* stages the simplicity of the writing process when Clarke asserts that 'language hands [her] a stone'.[27] The verb 'to hand' confers the simulacre of a bodily form to language. In the image of the pebble melting in the mouth like a sweet, the energies of matter congeal into form, for a time, before dissolving. Language, through the pleasure of rhythm and orality, participates in that fluid process. Those energies congeal into syllables when the poet's tongue enters the shared creative process.

Clarke's poetry plays with scales in its endeavour to apprehend the Anthropocene, which 'challenges us to think counter-intuitive relations of scale, effect, perception, knowledge, representation and calculability'.[28] In *At the Source*, Clarke also recalls a game of scales played with pebbles:

> As a child I used to play a game which I called 'big and little', which now seems to me a primitive version of a poet's game, physical and imaginative in nature, yet a child's way into a questioning habit of thought. Half-close your eyes and stare, or blur your ears. A stone becomes a planet. Your breath is the wind, a quarrel is a storm, a storm becomes a war. It works the other way too. Your cupped hand can balance the pebble of the setting sun before it is dropped into the sea. With a finger you can blot out a Neolithic stone, or a planet. Take a magnifying glass to your thumb-print. Place a hair under a microscope. These are geographies. It is a game played with scale and perspective that has always fascinated me.[29]

Clarke's conception of 'geographies' broadens the vision of the narrowly local generally associated with her earlier poetry from a static cartography of Wales to a game of shifting scales moving freely from a pebble on a beach to the night sky, or to a hair under a microscope. Through the analogical interconnection between microcosm and macrocosm in 'your breath is the wind', the poet's body participates in the creative process at work in the elements. It generates a reciprocal relation of creative gift between what is called the subject and what is called the object in classical philosophy. Modes of perception introduced

---

27   Clarke, *Source*, p. 46.
28   Clark, *Ecocriticism*, p. 13.
29   Clarke, *Source*, p. 31.

by scientific representations are essential to the process: to envision the common vitality of the organic and the inorganic, the poet needs to envision incommensurable yet intersecting scales, be they temporal—geological and biological timescales—or spatial, hence the planet in the hand and the hair under the microscope.

Clarke's poetry also makes use of scientific representations when it turns toward carbon, a mineral that mediates between the vegetable realm, animal bodies, and inorganic matter through decomposition. From that perspective, it can be seen as the material manifestation of what Jane Bennett terms 'an ontological field without any unequivocal demarcations between human, animal, vegetable or mineral'.[30] The poem 'Coal—Upper Carboniferous' unfolds along two timescales, from the prehistorical 'tropical swamp that laid down the coal' to the individual story of a miner and his son, exploring the heritage of the coal mining industry in Wales from generation to generation:

> From Abercarn, Gwent,
> from the tropical swamp that laid down the coal
> he cut when he was a boy,
> fourteen years old and a real man now,
> working the stint at his father's side,[31]

The slow process of coal formation from organic remains belongs to geological timescales and participates in historical times through the industrialization of Wales. Local history is reintegrated into planetary time through coal, an entity active both in ecological and economic terms. As one of the main agents of climate change, coal also rearticulates the local and the global as it de-localizes Wales within the wider scale of carbon emissions and their consequences all over the globe. Coal embodies the passage from the living to the dead, and then back not only to carbon-based life forms, but also to a whole civilization based on fossil fuels. The poem looks into the entanglement of human societies and material formations, ever since coal entered an assemblage with human agents during the industrial revolution, in a 'logic [that] encompasses politics as much as physics, economics as much as biology, psychology as much as meteorology [and] recurs at all scales and locations'.[32]

---

30 Bennett, *Matter*, pp. 116–17.
31 Clarke, *Beds*, p. 27, 1–4.
32 Bennett, *Matter*, p. 118.

Clarke also invites us to attend to the interactions and resonances between the mineral and the organic in her poem 'Horsetail', which focuses on a plant that feeds directly on stone. In its essential in-betweenness, horsetail mediates between two vegetable realms—the grass and the tree—and presides over metamorphoses in matter, from stone to metal, and from rock to living body:

> Not a grass. Not a tree. Primitive,
> leafless leftover
>
> from forest giants that fossilised to coal,
> its jointed stems rising in whorls
>
> from coastal salts, stones, ashes, sand,
> colonising ground where the trains once ran.
>
> It feeds on rock, sucks
> metals out of stone,
>
> prospecting for wealth in the ground.[33]

The poem looks into the mineral at the core of living structures. Like rock formations, living bodies bear the memory of the earth, as horsetail appears as a persistence 'from forest giants that fossilized to coal'. The word 'prospecting' connects the agentivity of horsetail with that of miners, recasting human activity as similar to that of plants, thus joining the historical and the ecological. That vision of the plant as living stone offers another model for the relation to minerals in the Anthropocene. By giving the plant the role of the prospector, it places extraction in a cycle of life rather than of exhaustion. Clarke's botanically inspired verse also reflects the assimilation of light in vegetable forms of life, when, for instance, 'Ferns sip sunlight at a rock fissure'.[34] Her poems explore modes of existence akin to plant life, trapping the light, as in 'Ice Harvest', with its 'blocks of luminous blue, sky turned to glass, / each one clear to the needle of light at its core', and in 'Ode to winter', in which humans 'hoard light'.[35] There exist plant-like modes of being within humans and minerals. Her poetry looks into entities that are able to trap light. Like coal, its dark counterpart, light mediates between the

---

33   Clarke, *Recipe*, p. 56, 1–9.
34   Clarke, 'Mine', *Zoology*, p. 36, 4.
35   Clarke, *Beds*, p. 29, 5–6; and *Ice*, p. 75, 1.

organic and the inorganic through the reference to photosynthesis. Like stone, water, and coal, it presides over transformations from one state of matter to another.

Images of Metamorphosis and Development

Clarke's poetry is drawn towards moments when matter changes states, from living to inert, and from inert to living, hence a tropism towards biological metamorphosis, which is understood by contemporary science as a moment of massive birth and massive death at the cells' level. Metamorphosis is a central motif in 'Death's Head Hawkmoth Caterpillar':

> It will spin itself a chrysalis of spittle and clay,
> dissolve, metamorphose, pupate and wait
> for a rearrangement of its molecular being,
> a stirring of self in the sun, a freeing.
> [...]
> the mask of death on its head from the moment of birth.[36]

The 'rearrangement of its molecular being' can be read as a reference to developmental biology, and to the massive death of cells involved when the organism is reborn through metamorphosis. Hence 'the mask of death on its head from the moment of birth'. Following this figurative vein, in 'Marsh Fritillaries', language itself becomes a metaphor for biological metamorphosis: 'I love their language, pupae, chrysalides'.[37] Poetic language strives to emulate metamorphosis in its plasticity and active participation in radical changes of states. Clarke's postromantic vision of the Welsh nation's origins in landscape and language has often been read as a form of nationalistic nostalgia. But in her most recent collections, the quest for the origin morphs into a dialectics between origins and endings where heterogenous temporalities coexist. This is visible in the image of 'the mask of death on its head from the moment of birth', in which birth and death become two faces of the same form of life.

---

36 Clarke, *Recipe*, p. 58, 7–10, 14.
37 (Clarke, *Zoology*, p. 30, 7.

3. Still Life and Vital Matter in Gillian Clarke's Poetry

One of those heterogenous yet colliding timelines on the scale of the organism is cell birth and death in metamorphosis. This is the case in 'Burnet Moths', a poem about the death of a dog:

> By the path, bound to grass stems, spindles of spit,
> chrysalids, papery, golden, torn, unfurling
> sails of damp creased silk, spinnakers filling
> with breath, burnet moth wings of scarlet and black
> like opera stars who live and love and die
> in an hour on the flight of an aria.
>
> Now it's her turn to die [...]
> [...]
> and she crumples to sleep at my feet, folded back
> to before she was born.[38]

Two temporalities close in on each other and become involved in one another: the linear life of humans and dogs alike, and metamorphosis within the life cycle of the burnet moth. Two opposite yet parallel processes meet through the motifs of folding and unfolding. The poem offers a vision of the folding in together of life and death as the two sides of a paper-thin membrane, as the dog is 'folded back', and 'crumples' like the papery chrysalis of burnet moths. In that converging movement, death itself appears as a seamless process of metamorphosis back into inorganic matter, as if that reversion, 'dust to dust, ashes to ashes', was only an inversion of the process presiding over the moths' rebirth. The papery and torn quality of the chrysalis offers a vision of the complex temporality of poetry. It stages poetry's own dependence on its ephemeral material support while at the same time celebrating its own ability to pupate and metamorphose into various layers of meaning on various timescales.

After geology, developmental biology is probably the most important scientific paradigm in Clarke's works, providing a possible homology of methods for her poetry. My reading suggests that this interest is anchored in the shared question of coexisting timescales. Developmental biologists have lately tried to address the various levels of processes unfolding on different timescales within living organisms. Their branch of biology has been defined as 'the science that seeks

---

38 Clarke, *Ice*, p. 38, 7–13, 15–16.

to explain how the structure of organisms changes with time'.[39] It studies various processes shaping the individual through time, such as differentiation, pattern formation, morphogenesis, and growth. These processes used to be studied along a standard timeline, with stages such as fertilization, gastrulation, and, in certain species, metamorphosis. These standardized stages have been increasingly questioned within the field of developmental biology over the past decade, as they make it more difficult to account for distinct but intersecting forms of causation, such as the roles of genetics and of interactions with the environment in development: 'These normal stages are a form of idealization because they intentionally ignore kinds of variation in development, including variation associated with environmental variables'.[40] The tropism in Clarke's poetry towards developmental biology seems less a form of influence and more the recognition of a common problem: how do you attend to the heterogenous yet interacting timescales within bodies and with their environments?

'Oestrus', a poem on embryogenesis, is located at the meeting point of various temporalities: the time of day, the season, the hormonal cycle of the ewe, the encounter with the ram, and the gestational period.

> In shortening days, reducing light,
> her chemistry stirs, sleeping hormones wake
> in her brain's dark chamber, and she's ready,
> restless again for the scent of the ram.
>
> On heat she greets him, sniffs him to be chosen.
> The ewe takes the ram, and something quickens
> in the secret dark, a sensed flowering,
> a difference in the pulse of things,
>
> multiplying and dividing cells,
> ova, zygote, embryo, foetus, lamb,
> an unstoppable force strong
> as the river in the mountain's heart.[41]

---

[39] Jonathan Slack, *Essential Developmental Biology*, 2nd ed. (Malden, MA: Blackwell Publishing, 2006), p. 6.

[40] Alan Love, 'Developmental Biology', *The Stanford Encyclopedia of Philosophy*, Spring (2020), https://plato.stanford.edu/archives/spr2020/entries/biology-developmental/.

[41] Clarke, *Zoology*, p. 46, 5–16.

The poetic voice does not simply enumerate the various stages in the development of an embryo, but also develops a reflection on temporality in poetry and biology. The poem accompanies the emergence of a living form at the conjunction of widely differing yet deeply interacting timescales. For instance, the astronomically determined time of the seasons and of the hour of the day meets the biological clock of an ewe, who also interacts with the rest of the herd to choose a ram. The last lines quoted above, 'an unstoppable force strong / as the river in the mountain's heart', are reminiscent of Conrad Waddington's epigenetic landscapes. To figure the role played by modifications in gene expression during embryonic development, British biologist Conrad Waddington commissioned the drawing of a mountainous landscape for his book *Organisers and Genes* (1940). He then invited his readers to envision the complex interaction of genetics and environment by 'Looking down the main valley towards the sea. As the river flows away into the mountains it passes a hanging valley, and then two branch valleys, on its left bank. In the distance the sides of the valleys are steeper and more canyon-like'.[42] This way, the reader would be able to envision the trajectory of the developing embryo along branching paths, offering a sensorial experience of the three main principles of Waddington's epigenetic theory: 'canalization, homeorhesis, and scaling'.[43] In *Epigenetic Landscapes: Drawing as Metaphor*, Susan M. Squier argues that such creative models 'function kinetically, affectively, and methodologically, as well as epistemologically'.[44] Squier is interested in the way epigenetic landscapes invite us to shift our view of organic development 'from reductionist linearity to situated, kinetic complexity, with ecological and global sociopolitical significance'.[45] One could argue that the poem 'Oestrus' acts as a form of epigenetic landscape; it offers a vision uncannily close to Waddington's visualization of the intersecting causations between genetic code and gene expression as the trajectory of a mountain river, which is always open to bifurcations, though generally canalized. The poem offers a powerfully sensorial model in which the

---

42 Conrad Waddington, *Organisers and Genes* (Cambridge: Cambridge University Press, 1940), p. 93, quoted in Susan M. Squier, *Epigenetic Landscapes: Drawing as Metaphor* (Durham and London: Duke University Press, 2017), p. 11.
43 Squier, *Epigenetic Landscapes*, p. 11.
44 Ibid., p. 16.
45 Ibid., p. 18.

plasticity and creativity of language allow the reader to experience these branching pathways in organic development. Various types of causation, within and outside the body of the ewe, collide to produce a new body, the way geological processes meet human creativity, like the forming of limestone out of sedimented sea organisms to engender a stone hare in Clarke's *Making the Beds for the Dead*. It all begins with the experience of the invisible, 'in [the] brain's dark chamber', and ends with a new visualization of the organic development, a new mode of visibility based on Clarke's scientific imagination, envisioning the inside of the womb, 'like a match struck in the dark'.[46] To register that 'difference in the pulse of things', one needs to combine two modes of vision: poetry as an art of pulse and rhythm, and the scientific ability to play with scales.

Clarke's poetry endeavours to register the 'difference in the pulse of things', that interchange between the organic and the inorganic, in the following poem, entitled 'Virus', which stages the appearance of a virus as a form constantly mediating between life and death in its modes of being and in its potentially destructive interaction with its hosts:

> wanting nothing but a living host
> to practice symmetry
> and cell division.
>
> Brought from space
> on the heel of a star,
> a primitive chemical
> seething in soupy pools,
> its arithmetic heart
> bent on sub-division, multiplication.[47]

The virus is one of the recurring forms of mediation between the organic and the inorganic in Clarke's poetry. Here again, the poem uses the lexicon of developmental biology—cell division and multiplication—to describe an entity at the crossroads of the living and the non-living. The poem moves freely between biological and geological imagery to try and envision the shifting nature of the virus: '[c]ell division' becomes 'its arithmetical heart / bent on sub-division, multiplication', in a process akin to the crystallization in inorganic matter investigated in *At the*

---

46 Clarke, *Zoology*, p. 46, 27.
47 Clarke, 'Virus', *Beds*, p. 56, 4–12.

Source. It looks at the formation of living bodies using mathematical laws shared by inorganic and organic matter. The poem then places its virus under an imaginary microscope: 'On screen, an image / of rotational symmetry / in a box of glass'.[48] Scientific visualization generates poetic imagery: the virus appears 'on screen', as an 'image', as an entity reconstructed by a computer, as a digital construct based on the laws of mathematics. Like developmental biology and crystallization, scientific imaging is based on mathematical laws. Those laws are thus represented as the generative principle at work both in the object and in its modes of representation. Art and scientific visualization then come together in the image of 'still life, / computer generated':

> Or still life,
> computer generated,
> a dandelion head, each seed a field,
> folding, unfolding flower
> smaller than a bacterium,
> butting blind towards the living cell.[49]

Within each microscopic 'dandelion head', each seed becomes a field. The sensory modalities explored in the poem create a game of scales close to the game Clarke used to call 'little and big' when she was a child, with its imaginary geographies based on sensory experiments with astronomical and microscopic scales. 'Virus' thus reveals the potential for new images and new visualizations contained in Clarke's references to science.

Sounding the Flesh

Clarke's poetry is about listening to matter, about attending to its sonic quality: it is about visualizing through sound. In *At the Source*, Clarke reminds us that hearing is about sensing vibrations through direct contact: 'I passed no childhood day without the company of stones, […]. When I put my ear to it I could hear the stone purr like the sea in a shell; I could feel the Neolithic in the stone, like touching the arches when a train crosses a viaduct'.[50] Two poems in *Zoology*, 'Damage' and

---

48 Clarke, *Beds*, p. 56, 13–15.
49 Ibid., p. 56, 19–24.
50 Clarke, *Source*, p. 51.

'Audiology', are about scar tissue formed in the inner ear of the poet when she was an infant, probably from the sound of exploding bombs during World War II.[51] The two poems explore the tactile dimension of sound, which can strike, hurt, and enter the very fabric of the body. In these reflective poems, sound paradoxically becomes what resists the timeline of development; it embraces the temporality of trauma, as it becomes part of the material memory of the body:

> On the screen, a scar, crow on a wire,
> scored word on the cochlea of a baby
> born to war, or a gun fired too close
> that summer of skies crying out loud
>
> when bombers roared the cradled corridors
> spiralling down the ear's conch, scorching
> newborn skin, till a lifetime later the scar
> surfaces, a blemish, blurring sound.[52]

A scar from a very long time ago resurfaces. The scar made by the violent contact of a sound is translated into a visual representation. These images hint that there is something tactile and intimate about poetic imagery and medical imaging alike. In 'Glâs', poetry is also a mode of sounding, trying to form the image of some hidden material construct through the exploratory use of sound:

> and I'm dreaming that secret web of water
> underfoot, down through the storeyed strata
> in Earth's unmappable corridors of stone.
> [...]
> invisible silvers silent as ultrasound.[53]

In places from which sight is excluded, imaging is possible through the use of ultrasound, through the experimental use of sound. Science and poetry are imagined as reconstructing blind landscapes such as the depths of the Earth, in which everything is packed with matter, and there is no distance for the eye to build a sense of perspective. The poem carries out another sensory experiment through touch: the sense of tactility and the feel of gravity built by 'underfoot' and 'down through'

---

51 Clarke, *Zoology*, p. 80 and p. 81.
52 Clarke, 'Audiology', 1–8.
53 Clarke, 'Glâs', *Ice*, p. 43, 6–8, 14.

allow the poem to follow the 'seeping' of water as it 'run[s] black in the aquifers',[54] revisiting the poetics of direct contact that connects the geological and poetic scales in 'A Recipe for Water'.

In 'Scan', dedicated to the use of ultrasounds on pregnant ewes, the homology of method between poetry and scientific imaging is explored further, in the ability to translate sound waves into images:

> The scanner eyes the womb.
> Cells have multiplied,
> the buds of limbs,
> the casket of a skull.[55]

Through innovative imaging, and in particular through the exploratory use of sound, science and poetry alike offer a reconstruction of inner spaces that are inaccessible to the eye, as the young body of the lamb takes shape inside the womb. They also offer a reconstruction of the inner times of ontogeny, as in 'Oestrus'. The forming brain of the lamb is twice concealed: inside the skull, itself inside the womb. The skull and the womb are not only vessels but also formative layers of tegument, as if the young body of the lamb and 'the buds' of its limbs were a germinating bulb. There is a tactile quality to sound in the poem, as it allows for the digital reconstruction of the outlines of the womb. The body of the 'lamb unfolding in her womb'[56] forms in contact with the walls of the uterus, staging the interactions of genetics and epigenetics: in the poem 'Oestrus', fleshly contact appears to be just as formative as the unfolding of the genetic code.

Science in the Landscape

Clarke's interest in science stems from a sense of emergency as several temporalities meet head-on. The colliding timescales of the poet's lifetime and of climate change are present in all her recent collections, from 'Aftermath' to 'Glacier', 'Polar', or 'New Moon', to quote only a few poems.[57] For instance, in 'Glacier' the polar landscape suddenly loses its solidity, in a shift emphasized by the homophony of 'floe' and 'flux':

---

54 Clarke, *Recipe*, p. 20, 4–6.
55 Clarke, *Zoology*, p. 51, 7–10.
56 Ibid., p. 51, 15.
57 Clarke, *Beds*, p. 76, *Recipe*, p. 34, *Ice*, p. 9, and *Zoology*, p. 112.

> Oh, science, with your tricks and alchemies,
> chain the glacier with sun and wind and tide,
> rebuild the gates of ice, halt melt and slide,
> freeze the seas, stay the floe and the flux
> for footfall of polar bear and Arctic fox.[58]

The landscape suddenly acquires the fluidity, provisionalness, and vulnerability of human culture. Conversely, the poem calls for a science powerful enough to act on a planetary scale, hence the poetic voice's appeal to science. Clarke thus wonders in *At the Source*:

> Would Keats, in the light of our knowledge today, have complained about the unweaving of the rainbow? Would he not have found a new nature poetry that praises the way a rainbow is constructed from the seven colours of light split and refracted by a water drop? To combine a curiosity for science with love of the natural world is how humankind must live on earth now, and poetry should speak of it. It is no longer just the concern of those described as 'nature poets' to protest at the spoliation of the earth, or of scientists to show curiosity and concern for the earth.[59]

The need to invent a new poetics is the direct consequence of the altered state of nature. In Clarke's poetry this ambition often takes the form of a dialogue with some of the Romantic poets, who offered renewed visions of nature and of its interactions with humans and their language, in the age of modern science. For instance, in 'A T-Mail to Keats', the poetic voice writes to John Keats to start an imaginary discussion through time (hence the T-mail) about his claim in *Lamia* that science is 'unweav[ing] the rainbow':[60] 'I want to talk with you of the new nature, / of your grief at science for *unweaving the rainbow*'.[61] For Clarke, 'the climate is unweaving the poetry': that unravelling happens both in cultural representations and to the actual rainbow, through the alteration of climate phenomena.[62] In *At the Source*, Clarke meditates on the gradual disappearance of a local species of bluebells: 'Bluebells. *Endymion nonscriptus.* [...] Will Endymion be lost to climate change?'[63] Romantic

---

58 Clarke, *Recipe*, p. 34, 10–14.
59 Clarke, *Source*, pp. 13–14.
60 John Keats, *The Major Works*, ed. by Elizabeth Cook, 2nd ed. (Oxford and New York: Oxford University Press, 2001), p. 321, II, 237.
61 Clarke, *Recipe*, p. 17, 6–7.
62 Clarke, *Source*, p. 92.
63 Ibid., pp. 96–97.

poetry might soon become unreadable since it has built its poetic idiom on living beings doomed to disappear. For Clarke, it becomes urgent to work out a renewed covenant between natural phenomena and poetic language. Her recent poetics is directly inspired by scientific research because it comes from an awareness of the urgency of fighting climate change. As a result, her work is both a poetry of place, deeply rooted in Wales, and also—especially since her shift towards environmental questions—a poetry of migrations and melting glaciers, exploring the complex interconnections between the local and the global. In terms of literary periodization, it seems to unfold along different timelines: it is both deeply postromantic, and uncannily close to new materialist thinking.

The autobiographical poem 'Waves' captures some of the most salient elements that this study has explored in *Making the Beds for the Dead*, *A Recipe for Water*, *Ice*, and *Zoology*. The vibrant poetic language of 'Waves' captures the constant metamorphoses between the organic and the inorganic in ways which are not entirely accessible to the sciences it first drew inspiration from. In *At the Source*, Clarke remembers her conversations with her father during their frequent walks; he used to tell her stories but also scientific facts. Clarke remembers the way science animated her sense of wonder as a child:

> I left education largely in ignorance of science, but I know now that the seeds of excitement about the facts of physics, biology, mathematics, were sown on those westward journeys with my father when, between the stories, he taught me about electricity, gravity, how radio worked, how he sent messages in Morse code during his years at sea as a wireless engineer, where the weather comes from, what the stars are.[64]

The conversations with her father are interwoven with old Welsh stories, so that both take on a cosmological dimension. They deal with the sky and the stars but also with the physics of the transmission of messages. That sense of scientific wonder is revived in 'Waves':

> When long ago my father cast his spell
> with wires and microphones, he told me
> he could send sound on waves the speed of light
> to touch the ionosphere and fall

---

64 Ibid., p. 14.

> home to the wireless on our windowsill.
>
> Sometimes, radio on, half listening, struck still
> by a line of verse, a voice, a chord, a cadence,
> I think of living light in a breaking wave,
> not breath, not fire, not water, but alive,
> the sudden silver of a turning shoal,[65]

The father's signal transmission and the daughter's writing process are interwoven through the motif of the wave. The sound waves bearing the message sent by the radio engineer reflect poetic language: in both cases, the words are made up of sound waves, they are such stuff as the energies of matter are made on, mere disturbances of the air. The waves then materialize on the imagined landscape of a seashore: sound waves become actual waves of water and turn into 'living light in a breaking wave', a vision of light waves and sound waves coming together in the figure of a school of silvery fish, 'the sudden silver of a turning shoal'. This image generates an assemblage of physical forces, living bodies, and poetic language.

In 'Waves', different interacting agencies coexist in one formation: the landscape, the living body and its physical memories, and language embodied in sound. It generates its own complex temporality, weaving together geological, biological, and artistic timescales. In its ability to enter assemblages, poetic language thus performs the vitality of organic and inorganic matter alike.

My thanks to Liliane Campos and Pierre-Louis Patoine for organizing the symposium that led to this chapter, and for their extremely helpful feedback throughout. They greatly improved this chapter with their creative and enlightening suggestions, from the figure of the epigenetic landscape and the subversion of extractive capitalism in Clarke to the possibility of ontogeny as a form of inner temporality. I am also grateful to the Institut Universitaire de France for supporting my research for this chapter.

---

65 Clarke, *Zoology*, p. 19, 1–10.

Works Cited

Bennett, Jane, *Vibrant Matter: A Political Ecology of Things* (Durham and London: Duke University Press, 2010), https://doi.org/10.1215/9780822391623

Timothy Clark, *Ecocriticism on the Edge: The Anthropocene as a Threshold Concept* (London and New York: Bloomsbury Academic, 2015).

Clarke, Gillian, *A Recipe for Water* (Manchester: Carcanet, 2009).

Clarke, Gillian, *At the Source: A Writer's Year* (Manchester: Carcanet, 2008).

Clarke, Gillian, *Ice* (Manchester: Carcanet, 2012).

Clarke, Gillian, *Making the Beds for the Dead* (Manchester: Carcanet, 2004).

Clarke, Gillian, *Zoology* (Manchester: Carcanet, 2017).

Cook, Elizabeth, ed., *John Keats, The Major Works*, 2nd ed. (Oxford and New York: Oxford University Press, 2001).

Entwistle, Alice, *Poetry, Geography, Gender: Women Rewriting Contemporary Wales* (Cardiff: University of Wales Press, 2013).

Gregson, Ian, *The New Poetry in Wales* (Cardiff: University of Wales Press, 2007).

Jarvis, Matthew, *Welsh Environments in Contemporary Poetry* (Cardiff: University of Wales Press, 2008).

Love, Alan, 'Developmental Biology', *The Stanford Encyclopedia of Philosophy*, Spring (2020), https://plato.stanford.edu/archives/spr2020/entries/biology-developmental/

Oxford English Dictionary, 2nd ed., version 4.0.0.2 (Oxford: Oxford University Press 2009) [on CD-ROM].

Slack, Jonathan, *Essential Developmental Biology*, 2nd ed. (Malden, MA: Blackwell Publishing, 2006).

Solnick, Sam. *Poetry and the Anthropocene: Ecology, Biology and Technology in Contemporary British and Irish Poetry* (London and New York: Routledge, 2017).

Squier, Susan M., *Epigenetic Landscapes: Drawing as Metaphor* (Durham and London: Duke University Press, 2017).

Waddington, Conrad, *Organisers and Genes* (Cambridge: Cambridge University Press, 1940).

Wynn Thomas, Meurig, *Corresponding Cultures: The Two Literatures of Wales* (Cardiff: University of Wales Press, 1999).

4. Mycoaesthetics
Weird Fungi and Jeff VanderMeer's *Annihilation*

Derek Woods

> 'Words? Made of fungi?'
> —Jeff VanderMeer, *Annihilation*

The twenty-first century has seen a new wave of interest in the Kingdom Fungi across biology, literature, and visual art. One reason for this has been a shift in both fungi's ecological scale and cultural image driven by the arrival of the 'wood wide web'. This phrase is a punning technomorph coined in the context of Suzanne Simard's research in forest ecology; prior to the phrase, the concept has analogues in indigenous traditions.[1] When Simard published her first paper on the topic, 'Net Transfer of Carbon between Ectomycorrhizal Tree Species in the Field' (1997), Sir David Read, who had shown in 1984 that 'carbon could pass between normal plants through fungal connections', published a commentary at the request of *Nature*'s editors. On the cover of the issue, they placed a

---

[1] As Allison Weir argues in reference to the wood wide webs of Peter Wohlleben and Suzanne Simard, 'it appears that Western science is just discovering what Indigenous scientists have known for many thousands of years'. 'Decolonizing Feminist Freedom: Indigenous Relationalities', in *Decolonizing Feminism: Transnational Feminism and Globalization*, ed. by Margaret A. McLaren (New York: Rowman and Littlefield, 2017), pp. 257–89 (p. 265). Suzanne Simard also suggests that the wood wide web is in accord with indigenous knowledge in *Finding the Mother Tree: Discovering the Wisdom of the Forest* (New York: Alfred A. Knopf, 2021), p. 293.

phrase coined by Read, 'the wood wide web'.² Simard has since become the public face of the idea that trees communicate with one another, nutritionally and semiotically, through networks of fungi in the soil.

The wood-wide web is a new biological scale: it shifts attention from single, familiar mushrooms to the subterranean bodies of fungi known as mycelium—bodies of which mushrooms are only the ephemeral fruit or reproductive structure. Here individuals are hard to define, but bodies might stretch across many square kilometers. As a moving target in twenty-first-century cultures of science, the wood-wide web is also a new biological image: it invites us to see fungi not as individual organisms but as 'technological' networks that grow in the dark, dense, and invisible space of the soil.

One major influence on the new wave of enthusiasm about 'mycology', the study of fungi, was Paul Stamets' book *Mycelium Running: How Mushrooms Can Help Save the World* (2005), which opens with a chapter on 'Mycelium as Nature's Internet'. Ten years later, Anna Lowenhaupt Tsing published a much-cited ethnography of mushroom pickers that cites Stamets, *The Mushroom at the End of the World* (2015). Published in the same year was Peter Wohlleben's *The Hidden Life of Trees: What They Feel, How They Communicate* (2015) which became a bestselling popularization of Simard's work. Dozens of imitative articles followed in digital media. For example, nature writer Robert McFarlane discussed 'The Secrets of the Wood-Wide Web' (2016) in *The New Yorker*. Ed Yong told us that 'Trees Have Their Own Internet' (2016) in *The Atlantic*. Wohlleben and Simard starred in Julia Dordel's documentary *Intelligent Trees* (2016), 'a scientific journey into the "wood wide web."'³ In 2017, Simard gave a talk for TEDx Seattle entitled 'Nature's Internet: How Trees Talk to Each Other in a Healthy Forest' (2017). Soon after, Richard Grant asked whether or not 'Trees Talk to Each Other?' (2018) in *The Smithsonian Magazine*. Claire Marshall reported on how 'Trees Social Networks are Mapped' (2019) for the *BBC*. In the documentary

---

2   Suzanne Simard, 'Net Transfer of Carbon between Ectomycorrhizal Tree Species in the Field', *Nature*, 388 (1997), 579–82, https://doi.org/10.1038/41557. David Read, 'The Ties that bind', *Nature*, 388 (1997), 517–18, https://doi.org/10.1038/41426. Merlin Sheldrake recounts this story of the phrase in *Entangled Life: How Fungi Make Our Worlds, Change Our Minds, and Shape Our Futures* (New York: Random House, 2020), p. 214.

3   *Intelligent Trees*, dir. by Julia Dordel (Dorcon, 2016).

Fantastic Fungi (2019), Louie Schwartzberg's interviewed Simard for a segment on mycelial networks. In an article entitled 'The Wood-Wide Web Can Really Help Trees Talk to One Another' (2020), Josh Gabbatis rehashed these ideas for *Science Focus*. In the same year, Richard Fortey reviewed Merlin Sheldrake's striking book *Entangled Life: How Fungi Make Our Worlds, Change Our Minds & Shape Our Futures* (2020) with the clickbait title 'Wood Wide Web: The Magic of Mycelial Communication' (2021). And the cinematic popularization continues with the German documentary *The Hidden Life of Trees* (2021). The list could go on, embracing a wave of popular science writing and visual culture. This enthusiasm calls for an explanation.

Digital hype about fungi imagined as digital media seems a drastic shift from this taxonomic kingdom's centuries of invisibility. Almost every author who writes about mycology complains that fungi have been ignored by humans, who prefer to notice flowers and charismatic megafauna. Scholars have only begun to study the reasons for this, which include the many ways that fungi have been understood as negative, pathological, vegetable—anything but themselves.[4] And yet, only one of the titles listed above mention fungi despite the fact that fungal mycelia form the very web in question. Even in the context of enthusiasm for fungi, their specificity as a form of life is quickly absorbed by attention to plants and to more familiar concepts of ecological connectedness. As we will see, the wood wide web is often treated as a kind of prosthetic for plant communication rather than a wonderful biological phenomenon in its own right. The goal of this chapter is to explain why this happens through the study of a central structure of twenty-first-century *mycoaesthetics*, or the cultural representation of fungi.

My case study is Jeff VanderMeer's 'weird fiction' novel *Annihilation* (2014)—especially his image of fungal writing in the novel's setting, Area X, an alien ecosystem inexplicably 'terraforming' Earth's biosphere.[5]

---

4   I discuss this history in 'The Fungal Kingdom', *Alienocene: Journal of the First Outernational*, Stratum 8 (2020).

5   Jeff VanderMeer's *Southern Reach Trilogy* includes *Annihilation* (New York: Farrar, Strauss, and Giroux, 2014), *Authority* (New York: Farrar, Strauss, and Giroux, 2014), and *Acceptance* (New York: Farrar, Strauss, and Giroux, 2014). For more on terraforming in relation to literature, philosophy, and ecotheory, see the special issue of *diacritics* edited by myself and Karen Pinkus entitled 'Terraforming', *diacritics*, 47.3 (2019), https://doi.org/10.1353/dia.2019.0023.

This weird ecology is Earth-like yet unearthly/uncanny, possessed by a force defined less by malevolence than mimicry and mutagenesis. When an expedition discovers writing in words made of fungi beneath the ground of Area X, they bring VanderMeer's readers to the core of the novel's critical significance. These subterranean fungal words follow a spiral staircase into the earth; they compose a single endless sentence reminiscent of the final chapter of James Joyce's *Ulysses* (1920), becoming a kind of alien poem nested within the novel.[6] But this is not the only role played by fungal writing in *Annihilation*. The novel's metafiction nests this subterranean fungal script inside a journal written by its first-person narrator, a biologist who enters Area X as part of a doomed expedition.[7] As the novel draws to a close, the biologist leaves her journal on a pile of decomposing journals from previous expeditions, so that the novel ultimately imagines fungi to be infecting, decomposing, and perhaps reading its own narrative. Several layers of form within the novel, the fungal sentence, and the decomposing journal, leave us with a formal complexity that invites careful interpretation.

VanderMeer's image of subterranean fungal writing evokes the wood wide web, but its spiral form also suggests the double helix of DNA. One could read this fungal writing as an ecological genome: Area X is a superorganism and the fungal writing is its DNA: a kind of memory, source code, or nervous system that controls the becoming of the setting. Like the wood wide web, such a reading would take the subterranean writing as a cybernetic information system, where fungi play the roles of media for plants and figures of ecological connectedness.

An alternative way to read the novel's fungal writing is to see it as an expression of the relation between fungi's aesthetic effects and their ontological status as neither plant nor animal. The comparatively recent emergence of fungi as a historically contingent ontology is a major factor

---

6   This similarity between Joyce's Molly Bloom chapter and the fungal sentence raises questions, beyond my scope, about the relationship between science fiction/fantasy and modernist literary form. On modernism and science fiction, see, for example, Ursula Heise, *Sense of Place and Sense of Planet: The Environmental Imagination of the Global* (New York: Oxford University Press, 2008), p. 77 and p. 174; Alison Nikki Sperling, 'Weird Modernisms' (2017), *Theses and Dissertations*, 1542, https://dc.uwm.edu/etd/1542; P. March Russell, *Modernism and Science Fiction* (London: Palgrave Macmillan, 2015).

7   For spelunking investigation of the subterranean in literature and climate change, see Karen Pinkus, *Subsurface* (forthcoming).

in creating the aesthetic 'weirdness' of fungi and making this image of the kingdom prominent in the twenty-first century. Both readings are important, but the second is a needed criticism of the wood wide web in a moment of zeal for anthropomorphism and networks. Fungi are a kingdom no longer conflated with plants or negated as merely parasitic, 'improper life'.[8] They are fungi, not a species of something else; they attract cultural attention for what they alone are and can do.

From the perspective of mycoaesthetics, what *Annihilation* shows so well is that fungi, as a new biological image/scale in the twenty-first century, have both ontological autonomy and a tendency to be captured by more familiar ecological concepts of connectedness. I argue that twenty-first-century mycoaesthetics is constituted by a 'hinge' central to its new prominence. This hinge is an ambivalent movement between the wood wide web and the fungal kingdom as weird life, neither plant nor animal. If the latter answers the question of why fungi are weird, the former tends to dilute this ontological and aesthetic characteristic by shifting plants to center stage or affirming a holism that has a long, troubled history in ecological thought.[9]

My essay begins making this argument by tying VanderMeer's work to a wider cultural field and concludes by asking how literary and aesthetic theory should write with the life forms we find in texts. This is also a question about how ecocriticism and posthumanism should address current debates about formalism, but with an eye to the specific problem of *Annihilation*, where fungi are thematized but also (de)compose the narrative itself. That is, a distinction emerges whereby weak mycoaesthetics indicates the very real ways that human agency can depict fungi one way or another, and strong mycoaesthetics envisions the fungal kingdom's own contribution to its aesthetic imprint: the idea that there is something about life forms, particularly at certain levels of taxonomic abstraction, such as the Kingdom Fungi, the Phylum Mollusca, or the Class Arachnid, that correlates with patterns of

---

8   I borrow this term from Timothy Campbell's *Improper Life: Technology and Biopolitics from Heidegger to Agamben* (Minneapolis: University of Minnesota Press, 2011, https://doi.org/10.5749/minnesota/9780816674640.001.0001), where it refers to life that falls outside the sphere of biopolitical management and nurturing.

9   See Thomas Patrick Pringle, 'The Tech Ecosystem and the Colony', *Heliotrope*, 12 May 2021.

literary and artistic form in ways that should not be reduced to arbitrary construction.

Weird Ecology, Weird Fiction

One way to think ecology is through what I call the transvaluation of weird life. Organisms that once seemed evil, disgusting, useless, small, inferior, or merely strange are said to have some functional role to play.[10] This is already true for one of the earliest ecological concepts, the economy of nature, so named by the parson and naturalist Gilbert White in *The Natural History of Selbourne* (1789). In a passage about worms, White writes that 'earth-worms, though in appearance a small and despicable link in the chain of nature, yet, if lost, would make a lamentable chasm'.[11] A logic of function replaces weirdness and minority. Gross things that live in the dirt are necessary for the whole chain. As Janelle A. Schwartz argues in a book about worms and British Romanticism, in the eighteenth century the meaning of the word worm referred to more than the squiggly annelid of today. For early moderns, the word had a broader sense of lowly life, death, and decay: 'the vermiform as everything from an earthworm to a larva to a maggot, a flying insect, and the unknown'. Indeed, Schwartz's work suggests the worm was a stand-in for weird life, 'a figure through which to consider the origin and progress of life during a period when each new discovery dislodged previously set categories and frustrated attempts to comprehend a totalized life through its unbounded parts'.[12] By valuing the worm, White asks readers to shift their thinking about life from the great chain of being to a proto-ecological view.

Like White's worms, fungi are now evoked as a biological image for ecological functions. Yet this functionalization does not dispel the weird aesthetics of fungi, and not only because of their unstable ontological status as neither plant nor animal. In recent years, weirdness has become

---

10 More on this argument in Derek Woods, 'Scale in Ecological Science Writing', in *The Routledge Handbook of Ecocriticism and Environmental Communication* (New York: Routledge, 2019), pp. 118–29.
11 Gilbert White, *The Natural History of Selborne* (New York: Penguin, 1977), p. 196.
12 Janelle A. Schwartz, *Worm Work: Recasting Romanticism* (Minneapolis: University of Minnesota Press, 2012), p. 11, https://doi.org/10.5749/minnesota/9780816673209.001.0001.

a descriptor of global warming and attendant ecological mutation. Environmentalist Hunter Lovins calls it 'global weirding'.[13] For Jonathan Turnbull, 'recent scientific discoveries [...] are often accompanied by a simultaneous sense of estrangement and fascination, which are often associated with the *weird*. On our terraformed planet, the weird is unearthly, gesturing towards and veering away from Earth. [...] The weird involves (un)earthly belonging'.[14]

In the twenty-first century, weird life has become a new bioaesthetic category, with fungi as one of its central representatives.[15] In his essay about ecological and climatic estrangement, Turnbull cites the 'radiotrophic' fungus *Cladosporium sphaerospermum*. This radiation-eating fungus has been found throughout the ruins of the Chernobyl nuclear reactor, 'the most radioactive place on Earth'.[16] Evidently this fungus is able to use gamma radiation to grow while protecting itself from mutagenic effects with the pigment melanin.[17] Stranger still, there are more than 200 species of fungi huddling around the reactor.[18] In the same brief piece, Turnbull also cites VanderMeer's exemplary trilogy. Area X clearly resembles both the rewilded, ominous, and mutant ecology of Chernobyl and Russian director Andrei Tarkovsky's film *Stalker* (1979), where an expedition enters 'the zone', a mysterious place of idyllic fields, forests, and ruins haunted by a psychoactive force.[19] Tarkovsky's famous film was, in turn, and adaptation of Arkady and Boris Strugatsky's novel *Roadside Picnic* (1972). As this citation path from novel to film to ethnographic essay on the climatic weird clearly suggests, the weird morphs readily between genres and contexts, so that it is difficult to establish distinctions between the contexts of fiction

---

13 Thomas L. Friedman popularized 'global weirding' in 'The People We Have Been Waiting For', *The New York Times*, 2 December 2007.
14 Jonathon Turnbull, 'Weird', *Environmental Humanities*, 13.1 (2021), 275–80 (p. 275), https://doi.org/10.1215/22011919-8867329.
15 See, for example, David Toomey, *Weird Life: The Search for Life that Is Very, Very Different from Our Own* (New York: W.W. Norton & Co., 2013).
16 Johnathon Turnbull, 'Weird', p. 277.
17 Ekaterina Dadachova and Arturo Casadevall, 'Ionizing Radiation: how fungi cope, adapt, and exploit with the help of melanin', *Current Opinion in Microbiology*, 11.6 (2008), 525–31, https://doi.org/10.1016/j.mib.2008.09.013.
18 N.N. Zhdanova et. al., 'Ionizing radiation attracts soil fungi', *Mycological Research*, 108.9 (2004), 1089–96, https://doi.org/10.1017/S0953756204000966.
19 *Stalker*, dir. by Andrei Tarkovsky (Mosfilm, 1979).

and non-fiction. When it comes to weird ecology, both grapple with the speculative defamiliarization of life.

Nevertheless, critics have done valuable work on the weird as a specifically literary and aesthetic category. From Jeff and Ann VanderMeer to Mark Fisher, Graham Harman, S.T. Joshi, Kate Marshall, Timothy Morton, Benjamin Noys, Alison Sperling, and Eugene Thacker, among others, the history of weird fiction is defined by relations between the weird and the uncanny, the weird and the queer, the old weird and the new, the weird and horror, the weird and the body, the weird and ecological thought.[20] These critics share the conclusion that the weird is an aesthetic at play among science fiction, fantasy and horror, one with close ties to both literary and philosophical realism. They share with Turnbull and others the idea that weird aesthetics deserves attention in new ways because of global warming and ecological violence. As Sperling writes in an article on the *Southern Reach Trilogy*, 'a particularly 'weird' ecology is one explicitly linked to modes of embodiment specific to the environmental conditions of the twenty-first century'.[21] With this line of argument, Sperling deepens the relation between this literary mode and the mutation of ecosystems.

This recent and environmental weird also determines what Noys and Murphy distinguish as the last of three stages in the history of weird fiction. For them, the third stage is characterized by 'a new sensibility of welcoming the alien and the monstrous as sites of affirmation and becoming', a transvaluation that invites comparison with White's important worm. Noys and Murphy find a contrast, in this affirmative repurposing, to 'Lovecraft's horror at the alien, influenced by his racism'. Disgust at human otherness, biopolitical hierarchies, and what Calvin L.

---

20 See Ann VanderMeer and Jeff VanderMeer, *The New Weird* (Ashland, OH: Tachyon Publications, 2008); Graham Harman, *Weird Realism: Lovecraft and Philosophy* (Washington: Zero Books, 2012); S.T. Joshi, 'Establishing the Canon of Weird Fiction', *The Journal of the Fantastic in the Arts*, 14.3 (2003), 333–41; Timothy Morton, *Dark Ecology: For a Logic of Future Coexistence* (New York: Columbia University Press, 2016, https://doi.org/10.7312/mort17752); Kate Marshall, 'The Old Weird', *Modernism/modernity*, 23.3 (2016), 117–34, https://doi.org/10.1353/mod.2016.0055; Timothy S. Murphy and Benjamin Noys, 'Introduction: Old and New Weird', *Genre*, 49.2 (2016), 117–34, https://doi.org/10.1215/00166928-3512285; Alison Sperling, 'Second Skins: A Body-Ecology of Jeff VanderMeer's *The Southern Reach Trilogy*', *Paradoxa*, 28 (2016), 230–55; Eugene Thacker, *In the Dust of this Planet: Horror of Philosophy, Vol. 1* (Washington: Zero Books, 2011).

21 Alison Sperling, 'Second Skins', p. 230.

Warren would call anti-black metaphysics, saturated the early stages of the weird.[22] By contrast, 'the new weird adopts a more radical politics, with 'the alien, the hybrid, and the chaotic as subversions of the various normalizations of power and subjectivity'.[23] If horror and the weird were initially indistinguishable, new weird fiction lowers the volume of horror enough for it to become positive while retaining an experience of otherness. If 'monstrous' bodies are essential to the generic conventions of horror, which also evoke conventional biological objects of disgust such as insects and decomposition (and thus fungi), then the new weird posits that monstrosity is better than the proper life of furry pets, charismatic megafauna, and privileged human bodies.

Mark Fisher may be right to say that 'any discussion of weird fiction must begin with Lovecraft'.[24] But fungi were already established at the core of weird fiction by the work of Edgar Allan Poe and other American writers in the nineteenth century. Kate Marshall indicates these writers in her argument about the earlier, Gothic origins of weird fiction. For her, 'an expanded sense of what might constitute weird writing beyond the *Weird Tales* writers or the boundaries of the New Weird offers in turn an expanded set of literary resources through which to think the nonhuman'.[25] If the structure of mycoaesthetics that I introduced in the first section is new to the twenty-first century, the relation between fungi and weird fiction is not.

One example that supports this argument on the terrain of the fungal weird is the use of fungal imagery in Poe's *The Fall of the House of Usher* (1839). It appears in a key moment for establishing the ominous setting, when the narrator's host first welcomes him to the gloomy house, where 'minute fungi overspread the whole exterior, hanging in a fine tangled web-work from the eaves'.[26] Further examples exist both in and out of the category of weird fiction, forming a minor tradition in Anglophone literature: the magic mushroom in Lewis Carroll's *Alice in Wonderland* (1865) that gives Alice the power to shrink or grow; the

---

22 Calvin L. Warren, *Ontological Terror: Blackness, Nihilism, and Emancipation* (Durham: Duke University Press, 2018).
23 Benjamin Noys and Timothy S. Murphy, 'Introduction', p. 125.
24 Mark Fisher, *The Weird and the Eerie*, p. 16.
25 Kate Marshall, 'The Old Weird', p. 634.
26 Edgar Allan Poe, *The Fall of the House of Usher, and Other Tales*, 1839 (New York: Signet Classics, 2006), p. 120.

fungal underworld of John Urri Lloyd's old weird novel *Etidorpha; or, the End of the Earth* (1895); the 'fungoid'[27] skin of H.G. Wells' aliens in *The War of the Worlds* (1898); Philip K. Dick's telepathic slime mold in *Clans of the Alphane Moon* (1964); the gentle fungal plague in Ling Ma's *Severance* (2018); the serial killer, named after mycologist Paul Stamets, who uses fungi to digest people in Bryan Fuller's TV series *Hannibal* (2013–2015).[28] Fungi have a long standing association with the weird as a literary mode, not accidentally but as important elements of its aesthetic effect.

During the first two decades of the twenty-first century, the appearance of the ecological weird due to heightened awareness of the Anthropocene has dovetailed with the reinvigoration of weird fiction, as VanderMeer's work shows, and the role of fungi has only amplified.[29] This new context for weird fiction has, in turn, led critics to look back and read the roots of the genre in a new light. For example, Fisher reads Lovecraft as a case of the naturalistic rather than the supernatural weird. In this frame, briefly put, 'a natural phenomenon such as a black hole is more weird than a vampire'.[30] One naturalistic story that seems to have influenced VanderMeer's trilogy is Lovecraft's story 'The Colour Out of Space' (1927), where an alien substance arrives with a meteorite in a placid New England town. This substance soon takes the form of an unknown color—a new band in the electromagnetic spectrum. As Lovecraft's narrator puts it, 'the colour, which resembled some of the bands in the meteor's strange spectrum, was almost impossible to describe; and it was only by analogy that they called it colour at all'.[31] As

---

27 H.G. Wells, *The War of the Worlds*, 1998 (New York: Penguin, 2005), p. 22.
28 John Urri Lloyd, *Etidorpha; or, the End of Earth, the Strange History of a Mysterious Being and the Account of a Remarkable Journey* (Cincinnati: John Urri Lloyd, 1895); H.G. Wells, *The War of the Worlds* (Peterborough, ON: Broadview, 2003), p. 55; Philip K. Dick, *Clans of the Alphane Moon* (New York: Mariner Books, 2013), Ling Ma, *Severance* (New York: Farrar, Strauss, and Giroux, 2018); 'Amuse-Bouche', *Hannibal*, NBC, 11 April 2013.
29 The weird would have been an apt fourth chapter in Sianne Ngai's elaboration of Kantian aesthetics in *Our Aesthetic Categories: Zany, Cute, Interesting* (Cambridge, MA: Harvard University Press, 2012), where she adds to his familiar notions of the beautiful and the sublime.
30 Mark Fisher, *The Weird and the Eerie*, p. 15. See also Eugene Thacker, 'Naturhorror and the Weird', in *Spaces and Fictions of the Weird and the Fantastic*, ed. by Julius Greve and Florian Zappe (London: Palgrave Macmillan, 2019), 13–24, https://doi.org/10.1007/978-3-030-28116-8_2.
31 H.P. Lovecraft, *The Call of Cthulhu and Other Weird Stories* (New York: Penguin, 1999), pp. 175–76.

the 'baffling bands' of this 'queer colour' with an 'unknown spectrum' permeates the environment of the town, flowers and leaves take on its hue.[32] In a beautiful biological image, the colors of autumn include it too. Over time, however, this color becomes a malevolent force that devours bodies like a cosmic parasite, turning farms and their inhabitants into grey dust. But Lovecraft's story is weirdest in its initial premise, where the new color estranges the pastoral landscape without destroying it. When the color out of space becomes a devastating and 'shapeless horror',[33] this glimmer of the weird as opposed to the horrible gets reabsorbed, as it were, by horror, with its reliance on graphic violence, death, and the supernatural. Here, the weird is an aesthetic phase that approaches horror without reaching its intensity, much like weird life is a category that revalues organisms considered repulsive and disgusting by viewing them as exotic, and, if still disturbing, as more desirable than repulsive.

Wood Wide Web as Ecological Genome

In an allusion to 'The Colour Out of Space', Area X begins with a fragment of light. In the third novel of *Southern Reach*, *Acceptance* (2014), the 'sliver' of light falls in the lawn of a lighthouse before an invisible barrier separates it from the rest of the Earth. There is no explanation of the mysterious terraforming that creates Area X, with all its beauty, psychotropy, and mutagenic power. Like rural New England in Lovecraft's story or the Earth as terraformed by the Oankali in Octavia Butler's *Xenogenesis Trilogy*, Area X is ambiguously (extra)terrestrial, both Earth and another planet. But if the light that falls outside the lighthouse in the third volume looks like 'glass', 'a key', 'a gleam', and a 'shifting spiral of light',[34] the first volume's narrator also characterizes Area X as thorn and parasite:

> Think of it as a thorn, perhaps, a long, thick thorn so large it is buried deep in the side of the world. Emanating from the side of this thorn is an endless, perhaps automatic, need to assimilate and mimic. Assimilator and assimilated interact *through the catalyst of a script of words, which*

---

32   Ibid., p. 176.
33   Ibid., p. 197.
34   Jeff VanderMeer, *Acceptance* (New York: Farrar, Strauss, and Giroux, 2014), pp. 24–25.

> *powers the engine of transformation.* Perhaps it is a creature living in a perfect symbiosis with a host of other creatures. Perhaps it is 'merely' a machine. But in either case, if it has intelligence, that intelligence is far different from our own. *It creates out of our ecosystem a new world, whose processes and aims are utterly alien*—one that works through supreme acts of mirroring, and by remaining hidden in so many other ways, all without surrendering the foundations of its otherness as it becomes what it encounters.[35]

For my claim that twenty-first-century mycoaesthetics works as a 'hinge' between fungal weirdness and the wood wide web, the importance of this passage lies in the fact that the biologist comes to see the fungal writing, 'the catalyst of a script of words', as the agency that 'powers the engine of transformation'. Such a causal script evokes philosophical work on performativity in the sense of linguistic action, notions of virality between biology and digital culture, and the visions of language as an alien parasite that we find in earlier experimental science fiction writers such as William S. Burroughs.[36] From the chemical sound of this 'catalyzing' script of words, it is easy to make the connection to the genome, which leads me to read both the helical fungal script of *Annihilation* and the wood wide web as a kind of ecological genome.

The biologist begins to think that the tower may be a 'living creature of some sort', and thus that the expedition is 'descending into the living organism'. If the tower is an organism, the fungal writing becomes its DNA.[37] The narrator of *Annihilation* is a female biologist whose partner died on a previous expedition into Area X. Focalization corresponds with specialization; the unnamed characters are referred to by their occupations. This focalization is also metafiction in that the narrative of annihilation takes the form of a journal or scientific report about the expedition. Before offering any other background about this character, however, the plot takes us quickly to Area X's subterranean 'tower', a spiral staircase made of stone that leads down into the Earth. The zone's second major architecture is a lighthouse. But the lighthouse is a human artefact left from the time before the boundary separated Earth from Eaarth[38] and the tower is a product of Area X itself.

---

35 Jeff VanderMeer, *Annihilation*, pp. 190–91.
36 See for example William S. Burroughs, *The Soft Machine* (New York: Olympia Press, 1961).
37 Jeff VanderMeer, *Annihilation*, p. 41.
38 Bill McKibben's term for Earth under climate change is *Eaarth: Making a Life on a Tough New Planet* (Toronto: Vintage, 2011).

As the characters enter the tower, the expedition begins to break down. The biologist discovers words made of fungi on the wall—the first sign that the place they have entered will not be easy to explain or comprehend. On the first descent, she inhales spores released by the words, which begin her transformation into what becomes a 'leviathan', 'a monumental storm', 'a mountain' in the third novel.[39] She will ultimately become Area X at some more distributed level than that of a single bounded organism.

Inhabiting the stairway is a being called the 'Crawler'. Readers eventually learn that this being is the source of the subterranean tower's fungal words. For the biologist, seeing the Crawler for the first time is 'a similar experience at a thousand times the magnitude' of seeing for the first time a rare starfish named the destroyer of worlds. At the core of Area X, she encounters the crawler as a life form completely beyond analogy, 'a figure within a series of refracted panes of glass', 'a series of layers in the shape of an archway', 'a great sluglike monster ringed by satellites of even odder creatures, 'a wall of flesh that resembled light [...] things lazily floating in the air around it like soft tadpoles'.[40] This nearly unimaginable image of the crawler moves the narrative toward a limit case of weirdness. Despite the comically extreme description, however, both Lovecraft's color and the Crawler are depicted by radicalizing natural phenomena—they answer to Fisher's argument about the naturalist weird and Marshall's interest in the relation between weird fiction and speculative realism. Both VanderMeer and Lovecraft use analogies with other bands of electromagnetic spectrum and with terrestrial organisms like slugs and tadpoles. Considered spatially, then, the center of Area X is a limit case because it is the weirdest life that can still be understood in a naturalist frame. The fungal writing is contiguous with the crawler, but less radically alien or external. Like ripples in disturbed water, the setting's weirdness diminishes as the narration moves away from this central unimaginable entity.

If the fungal words are the first indication of Area X's fundamental weirdness, they are also something more, considering the biologist's closing theory of the place. As a genome, the spiral staircase inscribed with writing evokes the spiral helix of DNA, the information molecule

---

39 Jeff VanderMeer, *Acceptance*, pp. 194–95.
40 Ibid., pp. 176–77.

of life. From this biological perspective, the Crawler is at least as much a reader as a writer, moving along an unbroken spiral like the ribosomes often described as 'crawling' along strands of mRNA as they decode genes for protein synthesis.[41] The passage from the surrounding natural landscape into the paranatural zone thus recalls twenty-first-century digital animations discussed by Adam Nocek in *Molecular Capture: The Animation of Biology* (2021), such as *The Inner Life of the Cell* (2006), which visualize biochemical reactions invisible to both microscopes and the naked eye.[42]

The weird ecology of Area X is an alienation that brings out what seems unnatural about nature itself—as for 'speculative realist' readers such as Harman and Thacker, for whom weird fiction narrates realities that are unreal because so different from what human senses can perceive and from our scales of time and space. For her part, Marshall seeks 'an expanded set of literary resources through which to think the nonhuman and to think beyond some of the paradoxes that thought presents'.[43] The biologist's descent into the tower then becomes an allegory of scaling 'down' into the world of molecules, while Area X becomes a stand in for the otherness of the microscopic scale and its putative ability to control what happens at the scales of human senses and social systems.

That the Crawler's words are fungal words offers a tempting connection between this ecological genome and the wood wide web. This analogy across scale raises the possibility that the mycelial internet is not only about trophic relation among plants and fungi, but also a means of control, memory, and reproduction like DNA is for organisms. But the analogy only goes so far. If the wood wide web is a kind of memory system, then it would also be radically different from the function of a genome. Just as the superorganism analogy broke down in the history of ecology, so the wood wide web can only be loosely

---

41 Dieter Beyer et. al., 'How the Ribosome Moves Along the mRNA during Protein Synthesis', *The Journal of Biological Chemistry*, 269.48 (1993), 30713–17 (p. 30714), https://doi.org/10.1016/S0021-9258(18)43872-0.

42 Adam Nocek, *Molecular Capture: The Animation of Biology* (Minneapolis: University of Minnesota Press, 2021, https://doi.org/10.5749/j.ctv1cdxg6p); XVIVO, *The Inner Life of the Cell* (Harvard University Department of Molecular and Cellular Biology, 2006).

43 Kate Marshall, 'The Old Weird', p. 634.

compared to a genome in the biological sense. Like the connected tree roots of Ursula K. Le Guin's story *Vaster than Empires and More Slow*,[44] the wood wide web is something else, a kind of horizontal vehicle of communication among species, the nervous system or communication system of a biome. In this reading, fungi are not only fungi but the informative fiber of ecological connectedness. They express a shift, in twenty-first-century cultures of science, from 'bio' to 'eco', from concern with genomes and DNA to Anthropocene ecosystems, climate change, and weird ecologies. And this raises the question of whether the holist connectedness of the wood wide web has more to do with the reductionist DNA than most scholars seem to expect.

The success of the wood wide web could almost be explained by how the twenty-first century media environment selects life forms that most resemble its own structure: through our seemingly autonomous posts and retweets, platforms seek their mirror image in nature. For Jedediah Purdy, nature answers well to 'the imaginative imperatives and limitations of its observers'. It follows that we should not be surprised that 'after centuries of viewing forests as kingdoms, then as factories (and, along the way, as cathedrals for Romantic sentiment), the 21st century would discover a networked information system under the leaves and humus'.[45] Purdy is right to be sceptical of this latest conceptual metaphor for nature. The question of whether the better analogy is DNA or the nervous system for the internet is less important than the fact that both converge on a predictably cybernetic logic of information and transmission. Well before digital modernity, mutual influence between cybernetics and ecology during the second half of the twentieth century made this convergence possible.[46]

Notions of web, mesh, network, entanglement, symbiosis, and assemblage have been essential for countering overly individualist, liberal, and Neo-Darwinian ideas of competition among bounded organisms in a

---

44 Ursula K. Le Guin, 'Vaster than Empires and More Slow', in *The Wind's Twelve Quarters* (New York: Harper Perennial, 2004 [1970]), pp. 181–217.

45 Jedediah Purdy, 'Thinking Like a Mountain', *N+1*, 29 (2017). Sheldrake also worries about the repurposing of 'starry-eyed fantasies of the internet' and 'digital utopia' in the form of the wood wide web's horizontality (*Entangled Life*, p. 162).

46 For example, see Fred Turner, *From Counterculture to Cyberculture: Stewart Brand, The Whole Earth Network, and the Rise of Digital Utopianism* (Chicago: University of Chicago Press, 2006), pp. 43–44 and p. 203, https://doi.org/10.7208/chicago/9780226817439.001.0001.

struggle for life with the non-negotiable interdependence of life on Earth. Given the success of network concepts and their mainstreaming in images like the wood wide web, however, critics can now learn more by seeing where they break down than by celebrating them. If we have already seen that fungi are quickly taken up as figures of ecological relation through both cybernetic rhetoric and the transvaluation of weird life (in both ecology and fiction), in the concluding section of this chapter, I argue that twenty-first-century mycoaesthetics is also about the ontology of fungi alone as an autonomous kingdom that is neither plant, animal, nor exemplar of the ecological thought. In *Annihilation*, the image of fungal writing lends itself to both interpretations. The hinge between ecological relation and fungal autonomy that VanderMeer foregrounds in his trilogy is the central structure of twenty-first-century mycoaesthetics.

The Fungal Kingdom

One downside of the wood wide web is that it risks reducing mycelia to a tool used by plants. For Sheldrake, there is thus an insidious 'plant-centrism' at work in many discussions of the wood wide web, as shown by the titles of the books and articles I mentioned above (*The Hidden Life of Trees*; 'Trees Have Their Own Internet'; 'Do Trees Talk to Each Other?').[47] Sheldrake argues in a chapter on the wood wide web that 'plants have been the protagonists' in stories about shared fungal networks. Within prevailing instrumentalist logics of technology, technomorphism can have the effect of reducing organisms to tools: 'fungi have featured inasmuch as they connect plants and serve as a conduit between them', so that they become 'little more than a system of

---

47 As in the interconnected plant-planet of Le Guin's *Vaster than Empires and More Slow* and Jeffrey Nealon's tendency to collapse the Deleuzo-Guattarean rhizome into plant life in *Plant Theory: Biopower and Vegetal Life* (Stanford: Stanford University Press, 2015, https://doi.org/10.1515/9780804796781). I am not the first to point out that mycelium seems better suited than the roots of plants to the kind of 'distributed territory of rhizomatic plant life' that Nealon sees as an alternative biopolitical model to organic wholeness and plant/animal binaries (p. 118), but fungi are never mobilized to help in this deconstruction. Instead, Nealon follows the Aristotelian and Linnean tradition of collapsing fungi into the plant kingdom or into an expansive category of the 'vegetal' of which plants are the only exemplary life form.

plumbing that plants can use to pump material between one another'.⁴⁸ Without a detour through post-instrumental theories of technology, the wood wide web, as an image and figure of speech, comes with the risk of falling back on metaphysical hierarchies of life. For Sheldrake, 'plant-centric perspectives can distort. Paying more attention to animals than plants contributes to humans' plant-blindness. Paying more attention to plants than to fungi makes us fungus-blind'. The wood wide web implies 'that plants are equivalent to the web pages, or nodes, in the network, and fungi are the hyperlinks joining the nodes to one another'.⁴⁹

While these comments come in the context of a discussion of Simard's work and its robust public reception, they can be generalized as a lesson about the invisibility of fungi—an irony of mycoaesthetics given that the motive behind the wood wide web is to make the invisible subterranean scale of fungi visible. The wood wide web becomes an example of how fungi can be ignored or reduced to plant prosthetics despite that fact that 'every link in the wood wide web is a fungus with a life of its own', so that fungi are 'active participants'⁵⁰ rather than instruments or altruists. In this way, Sheldrake's critique of the wood wide web from the perspective of someone fundamentally invested in mycelium is an interpretation of mycoaesthetics that shows how writing about fungi can easily blur away into the most general concept of ecological thought: for Timothy Morton, the idea that 'everything is interconnected'.⁵¹

Sheldrake does not seem to doubt that plants might communicate with one another through fungi or negotiate symbioses through subsurface mycelial media. He is interested in the metaphors we use and the baggage they bring along with them. Despite the title of his book about fungi, *Entangled Life*, he emphasizes the question of what makes fungi different from other life. He raises a question useful for understanding the hinge between ecological genome and fungal kingdom: faced with the soil's internet, 'are we able to stand back, look at the system, and let the polyphonic swarm of plants and fungi and

---

48 Merlin Sheldrake, *Entangled Life*, p. 160.
49 Ibid., p. 160.
50 Merlin Sheldrake, *Entangled Life*, p. 161.
51 Timothy Morton, *The Ecological Thought* (Cambridge, MA: Harvard University Press, 2010), p. 1.

bacteria that make up our homes and our worlds be themselves, and quite *unlike* anything else? What would that do to our minds?'[52]

At least since the turn of the century, VanderMeer seems to have been interested in such questions in relation to literary form. Before writing the *Southern Reach* trilogy, he published two books more closely aligned with the hollow earth novels of writers like Jules Verne and, more obscurely, John Urri Lloyd. Like Lloyd's *Etidorpha*, VanderMeer's *The City of Saints and Madmen* (2001), *Shriek* (2006), and *Finch* (2009) are about a world that contains its own negative image, an underworld kingdom of sentient fungi.[53] As the twentieth century gave way to the twenty-first, VanderMeer centered his weird fiction on the image and underworld scale of the fungal kingdom. His interest in these weird life forms helps me sustain my argument about mycoaesthetics as a hinge between the wood wide web and fungal autonomy. While the fungal writing in *Annihilation* can be read as an ecological genome, VanderMeer's prior interest in fungal underworlds suggests his fungal words are more than just accidental figures of connectedness.

The alternative to reading the 'tower' as the genome of Area X becomes clear in what the biologist sees on her second descent, when she examines the words more closely under the influence of their spores:

> Things only I could see: That the walls minutely rose and fell with the tower's breathing. That the colors of the words shifted with a rippling effect, like the strobing of a squid. That, with a variation of about three inches above the current words and three inches below, there existed a ghosting of *prior words*, written in the same cursive script. Effectively, these layers of words formed a watermark, for they were just an impression against the wall, a pale hint of green or sometimes purple the only sign that once they might have been raised letters.[54]

So much could be said about how VanderMeer uses a kind of life-form rhetoric in passages like this one, where the light media of squid amplify the aesthetic effect of the fungal script, as though the subsurface tower were not only underground but underwater. The squid simile evokes the chiasmic history of naming terrestrial life after aquatic (oyster

---

52 Merlin Sheldrake, *Entangled Life*, p. 174.
53 Jeff VanderMeer, *The City of Saints and Madmen: The Book of Ambergris* (Rockville, MD: Cosmos, 2001); *Shriek: An Afterword* (New York: Tor, 2006); *Finch* (Portland, OR: Underland, 2009).
54 Jeff VanderMeer, *Annihilation*, p. 48.

mushrooms) and vice versa (catfish). The rippling colors allude to the rich tradition of imagery, in science fiction, that draws on sea life to visualize extra-terrestrials.

While the fungal words compose an endless modernist sentence about death, decay, darkness, and worms (thus recalling White and Schwartz on weird life and associating worms with the fungal weird), the fact that these words are themselves composed of words leads the narrator to wonder whether the meaning matters or whether these words are building material or a process of fertilization rather than any kind of purposeful communication.[55] The biologist's theories give the fungal writing causal roles in the production of Area X, but the formal qualities of the Crawler's poem suggest a different interpretation.

From the perspective of strong mycoaesthetics, the essence of this passage is the 'ghosting' of words that makes the tower's fungal writing a palimpsest. Like a medieval manuscript that has been erased and overwritten, the words on the wall of the tower are inscribed in a medium made of similar words that are now fading or decaying into unreadability. The relationship between medium and form is relative; what was once the form, words with meaning, is now the medium for another form.[56] This entails recursivity because the pattern is self-similar. When we shift from one level to another, from language to the inscription surface, we find language again. The pattern repeats, self-similar at the level of form and its material substrate. But the fungal writing is also recursive in the sense of fractal repetition through scaling. It recalls Leibniz's pond filled with fish, where 'each portion of matter can be conceived as like a garden full of plants, or like a pond full of fish', but a pond or a garden in which 'each branch of a plant, each organ of an animal, each drop of its bodily fluids is also a similar garden or a similar pond'.[57]

The turn to modernist form with the endless sentence amplifies the recursivity of the fungal words through allusion to the aesthetics of art

---

55 Ibid., pp. 91–93.
56 Niklas Luhmann explores this relationship between medium and form in *Theory of Society, Vol. 1*, 1997 (Stanford: Stanford University Press, 2012), pp. 113–20.
57 Gottfried Wilhelm von Leibniz, *Monadology and Other Philosophical Essays*, trans. Paul Schrecker and Anne Martin Schrecker (Indianapolis: Bobbs-Merrill Educational Publishing, 1965), p. 159. See also Gilles Deleuze, *The Fold: Leibniz and the Baroque*, 1998 (Minneapolis: University of Minnesota Press, 1992).

for art's sake. Like Marcel DuChamp's famous urinal, *Fountain* (1917), modernist art constantly refers to itself by eschewing representation and questioning what counts as art. While this idea of modernism is as familiar as it is foundational, its role here is to add another layer of recursion to the scenes of fungal writing in *Annihilation*. If the reading of the fungal writing as a kind of ecological genome imagines the tower's connectedness with the rest of Area X, then the recursivity of the fungal writing makes it seem separate and self-generating.

The image of words within words or inscribed in a medium of words also has an intensifying effect commensurate with the narrator's own heightened perception, which is itself produced by the psychedelic effect of the words. The narrative doubles down on the weirdness of fungi, which leads us to think less about what kind of place Area X might be or what it might have to say about the estranging nature of ecological relation and more about the meaning of fungi in the narrative. What are fungi and why does the fungal kingdom appear in the historically contingent ways it does? Answering questions such as this might be better served by what Frédéric Neyrat calls an 'ecology of separation' than by an ecology of connectedness. Where he fears that the latter tends to make everything available for dynamic transformation according to the logics of neoliberal resilience, 'the net of a flat world, rendering all beings equivalent and annulling all exteriority', I share the position that ecological thought needs to reincorporate the 'ontological separation necessary for any relation'.[58] When we turn from one side to the other of the ambivalent mycoaesthetics at work in VanderMeer's prose and, more broadly, in the fungal image as a twenty-first-century culture of biology, the question becomes how to think fungi as a distinct kingdom.

For many, no life form should be considered in isolation. The relations between fungi and other organisms, including humans, are what should really interest us. Yet the story of what fungi are and how they took on their current ontological status as a taxonomic kingdom is a surprising one. For example, fungi are more closely related to animals than to plants: DNA sequencing has shown that the distance between plants and fungi

---

58 Frédéric Neyrat, *The Unconstructable Earth: An Ecology of Separation* (New York: Fordham University Press, 2019), https://doi.org/10.5422/fordham/9780823282586.001.0001.

is greater than that between fungi and animals.⁵⁹ As Lynn Margulis notes in *Five Kingdoms*, fungi did not become a kingdom in their own right until 1969, when R. H. Whittaker argued to make them one.⁶⁰ Five is no longer the 'right' number of kingdoms, and if you search for biology's latest accepted number you will not find clear and easy answers. Carolus Linnaeus wrote in keeping with ancient tradition when he proposed his new system of classification in *Systema Naturae* (1735), dividing nature into animal, vegetable, and mineral. In Linnaeus's still-current schema of Kingdom, Phylum, Class, Order, Family, Genus, and Species, fungi figured as members of the plant kingdom. As G. C. Ainsworth argues in his *Introduction to the History of Mycology*, 'from the time of the herbalists, fungi, even if confused with corals and other organisms, have been associated with plants'. It was in this form that modern natural history carried forward the premodern idea that fungi are lesser versions of something else. Ainsworth enumerates the many ways in which natural history characterized fungi as negative or epiphenomenal.⁶¹ Parasite, secretion, accumulation of moisture, negative ontology, incomplete plant, or the primitive ancestor of plants—as plants, fungi could never compete with the world of flourishing leaves and colorful flowers. For those who observed through the lens of plant/animal metaphysics, they seemed pale and sickly by comparison. The systematic classification of life existed for almost two and a half centuries before fungi were given clear and separate status.

The newness of the fungal kingdom's separate status compared to the continuity of plant and animal life from ancient categories to modern kingdoms goes some way toward explaining why fungi are weird—why they so often emanate an aesthetics of the horrible, queer, psychedelic, or eerie, as in VanderMeer's new weird fiction. The emergence of fungi between the plant and animal kingdoms is a different explanation of fungal weirdness from the classical one offered by Gordon Wasson's 1957

---

59 Cavalier-Smith, Thomas, and E.E. Chao, 'The Opalozoan Apusomonas Is Related to the Common Ancestor of Animals, Fungi, and Choanoflagellates', *Proceedings of the Royal Society B*, 261.1360 (1995), 1–6, p. 1, https://doi.org/10.1098/rspb.1995.0108.

60 R.H. Whittaker, 'New Concepts of Kingdoms of Organisms', *Science*, 163.3863 (1969), 150–60, https://doi.org/10.1126/science.163.3863.150.

61 See G. C. Ainsworth, *Introduction to the History of Mycology* (Cambridge: Cambridge University Press, 1976), p. 13. Histories of zoology and botany abound, but this appears to be the only scholarly history of mycology available in English.

concept of *mycophobia* that divides cultures into the mycophilic and the mycophobic.[62] In his famous field guide to mushrooms, *All That the Rain Promises and More...* (1991), David Arora finds that in a 'fungophobic (mushroom-loathing) society such as ours', 'it takes a certain boldness and curiosity to seek mushrooms'.[63] He goes on to say that eccentric, bold, and curious fungophiles have shaped the form of his book. The book's photographs document the weird antics of foragers, starting with the cover image of a trombonist in a tuxedo, leering at the camera from beneath an oak tree as he cradles his instrument in one hand and a pile of chanterelles in the other.

The weirdness or eccentricity at large in amateur mycology might well be a symptom of widespread fear of fungi in the Anglo-imperial world. Wasson and Arora are right that other cultures are more inclined to love the fungi, and there may be a way to explain fungal weirdness in terms of cultural relativism and colonial history. It would then be a mistake to think my ontological and scientific account of kingdoms applies universally.

In future work, much more should be said about the relation between kingdom and ontology about the historical contingencies through which certain humans came to know about fungi, but also about the ontological status of taxonomic kingdoms. To do fungal ontology is to think through the significance, for literature and science, of this process of abstraction whereby a third category emerges between plants and animals. If viruses famously deconstruct the opposition between life and nonlife, then fungi do the same for the thin bright line between plants and animals. But the point of this deconstruction is not to dissolve all categories into indistinction, flux, or plasticity. The point is to show that the plant/animal binary, as a persistent ontology of life, cracks open to yield a multiplicity—which is not to say an open-ended or unlimited diversity of categories. In this reading of twenty-first-century fungal scales and images, turning from the wood wide web to fungal ontology,

---

62 Gordon Wasson first popularized psychotropic mushrooms of the genus *Psilocybe* in 'Seeking the Magic Mushroom', *Life*, 13 May 1957. See also Erik Davis, "Mushroom Magick: A Visionary Field Guide." 2 April 2009. https://techgnosis.com/mushroom-magic/.

63 David Arora, *All that the Rain Promises and More* (Berkeley: Ten Speed Press, 1991), p. 3.

the fungal weird is an effect of this fitful emergence in histories of literature and science.

Where does this leave us with the idea that fungi shape their own representation in non-arbitrary ways? Certainly, the desire for strong mycoaesthetics is palpable and urgent in the form of *Annihilation*. The diegetic narrator breathes in spores from the fungal words, which affect her perception and thus make her a uniquely psychotropic unreliable narrator; the spores that help her see the words more clearly come from the words themselves, and the biologist reports that the fungal words 'infected our sentences', the dialogue of the novel, 'when we spoke'.[64] When she goes to the lighthouse, she discovers a large pile of decomposing journals left by previous expeditions, 'rife with striations of mold', so that 'the history of exploring Area X could be said to be turning into Area X';[65] thus also turning into the biologist's fate. But the narrative of *Annihilation* also presents itself as the biologist's journal, which she leaves on the pile of journals to molder and decay with the others, becoming fungal words of a different kind; if the novel begins with fungi that turn into words, it ends with words that turn into fungi. No doubt this chiasmus, along with the other examples of embeddedness and recursiveness given here show a desire for literature to incorporate weird life into the infrastructure of meaning, or for weird life to express itself through literature, proliferating from the biosphere into the 'semiosphere'.[66]

Yet all of this happens in the pages of a novel, leaving us at the border between weak mycoaesthetics and strong. Even if I agree with scholars who embrace nonhuman agency, semiosis that precedes or breaches species boundaries, and, as Tobias Menely puts it, the need 'to identify textual symptoms that express not historical but socioecological and even geohistorical contradiction',[67] it remains difficult to move past the objection that we can only know nonhuman life through mediating constructions that have little to do with the object itself. It is easy to

---

64 Jeff VanderMeer, *Annihilation*, p. 47.
65 Ibid., p. 112.
66 Yuri M. Lotman, *Universe of the Mind: A Semiotic Theory of Culture*. 1990. Translated by Ann Shukman (Bloomington: Indiana University Press, 2000), p. 125.
67 Tobias Menely, *Climate and the Making of Worlds: Towards a Geohistorical Poetics* (Chicago: University of Chicago Press, 2021), p. 20, https://doi.org/10.7208/chicago/9780226776316.001.0001.

agree with Sheldrake that metaphors are about more than the inevitable literariness of scientific knowledge, because they show how hard it is 'to make sense of something without a little part of that something rubbing off on you'.[68] But it is more difficult to offer a watertight argument for this stain. For the process of understanding both contemporary enthusiasm about fungi (especially the wood wide web) and the role of fungi in weird fiction such as VanderMeer's novels, strong mycoaesthetics would theorize the difference between simply studying representations of x, y, or z life forms in literature and something more significant. For future work, it will be crucial to continue to think about how conversations regarding form, mode, genre, and reading practice can and should shape how we read texts in ecocriticism and other fields that address the nonhuman in the humanities—put differently, what might an eco-formalism look like that would have the same influence as materialisms in our fields? At stake here is a concern with how nonhuman agency plays itself out through texts, but also the more specific question of how bio-ontologies such as kingdom or phylum, levels of abstraction different from concrete organisms, have already structured the texts we read at the level of form as much as content.

Works Cited

Ainsworth, G.C., *Introduction to the History of Mycology* (Cambridge: Cambridge University Press, 1976).

'Amuse-Bouche', *Hannibal*, NBC, 11 April 2013.

Arora, David, *All that the Rain Promises and More* (Berkeley: Ten Speed Press, 1991).

Beyer, Dieter, and others, 'How the Ribosome Moves along the mRNA during Protein Synthesis', *Journal of Biological Chemistry*, 269.48 (1994), 30713–17, https://doi.org/10.1016/S0021-9258(18)43872-0

Burroughs, William S., *The Soft Machine* (New York: Olympia Press, 1961).

Butler, Octavia E., *Lilith's Brood* (New York: Warner Books, 1989).

Campbell, Timothy, *Improper Life: Technology and Biopolitics from Heidegger to Agamben* (Minneapolis: University of Minnesota Press, 2011), https://doi.org/10.5749/minnesota/9780816674640.001.0001

---

68 Merlin Sheldrake, *Entangled Life*, p. 214.

Carroll, Lewis, *Alice's Adventures in Wonderland and Through the Looking Glass* (New York: Oxford University Press, 2009).

Cavalier-Smith, Thomas, and E. E. Chao, 'The Opalozoan Apusomonas Is Related to the Common Ancestor of Animals, Fungi, and Choanoflagellates', *Proceedings of the Royal Society B*, 261.1360 (1995), 1–6, https://doi.org/10.1098/rspb.1995.0108

Dadachova, Ekaterina, and Arturo Casadevall, 'Ionizing Radiation: How Fungi Cope, Adapt, and Exploit with the Help of Melanin', *Current Opinion in Microbiology*, 11.6 (2008), 525–31, https://doi.org/10.1016/j.mib.2008.09.013

Davis, Erik. "Mushroom Magick: A Visionary Field Guide" (2 April 2009), https://techgnosis.com/mushroom-magic/

Deleuze, Gilles, *The Fold: Leibniz and the Baroque*, trans. Tom Conley (Minneapolis: University of Minnesota Press, 1993).

Dick, Philip K., *Clans of the Alphane Moon* (New York: Mariner Books, 2013).

Fantastic Fungi, dir. by Louie Schwartzberg (Moving Art, 2019).

Fisher, Mark, *The Weird and the Eerie* (Los Angeles: LA Review of Books, 2016).

Fortey, Richard, 'Wood Wide Web: The Magic of Mycelial Communication', *Times Literary Supplement*, 6144 (2021).

Friedman, Thomas L., 'The People We Have Been Waiting For', *New York Times* (2 December 2007).

Gabbatis, Josh, 'Can the Wood-Wide Web Really Help Trees Talk to Each Other?', *Science Focus* (15 May 2020), https://www.sciencefocus.com/nature/mycorrhizal-networks-wood-wide-web/

Grant, Richard, 'Do Trees Talk to Each Other?', *Smithsonian Magazine* (March 2018), htttps://www.smithsonianmag.com/science-nature/the-whispering-trees-180968084/

Harman, Graham, *Weird Realism: Lovecraft and Philosophy* (Washington: Zero Books, 2012).

Heise, Ursula, *Sense of Place and Sense of Planet: The Environmental Imagination of the Global* (New York: Oxford University Press, 2008).

The Hidden Life of Trees, dir. by Jörg Adolph and Jan Haft (MPI, 2020).

Intelligent Trees, dir. by Julia Dordel (Dorcon, 2016).

Joshi, S. T., 'Establishing the Canon of Weird Fiction', *The Journal of the Fantastic in the Arts*, 14.3 (2003), 333–41.

Le Guin, Ursula K., 'Vaster than Empires and More Slow', in *The Wind's Twelve Quarters* (New York: Harper Perennial, 2004 [1970]), pp. 181–217.Leibniz, Gottfried Wilhem von, *Monadology and Other Philosophical Essays*, trans. Paul Schrecker and Anne Martin Schrecker (Indianapolis: Bobbs-Merrill, 1965).

Lloyd, John Urri, *Etidorpha; or, the End of Earth, the Strange History of a Mysterious Being and the Account of a Remarkable Journey* (Cincinnati: John Urri Lloyd, 1895).

Lotman, Yuri M., *Universe of the Mind: A Semiotic Theory of Culture*, trans. by Ann Shukman (Bloomington: Indiana University Press, 2000).

Lovecraft, H. P., *The Call of Cthulhu and Other Weird Stories* (New York: Penguin, 1999).

Luhmann, Niklas, *Theory of Society, Volume 1*, trans. Rhodes Barrett (Stanford: Stanford University Press, 2012).

Ma, Ling, *Severance* (New York: Farrar, Strauss, and Giroux, 2018).

March-Russell, P., *Modernism and Science Fiction* (London: Palgrave Macmillan,, 2015).

Marshall, Claire, 'Wood Wide Web: Trees' Social Networks Are Mapped', *BBC* (15 May 2019), htttps://www.bbc.com/news/science-environment-48257315

Marshall, Kate, 'The Old Weird', *Modernism/modernity*, 23.3 (2016), 117–34, https://doi.org/10.1353/mod.2016.0055

Macfarlane, Robert, 'The Secrets of the Wood Wide Web', *New Yorker* (7 August 2016), htttps://www.newyorker.com/tech/annals-of-technology/the-secrets-of-the-wood-wide-web

McKibben, Bill, *Eaarth: Making a Life on a Tough New Planet* (Toronto: Vintage, 2011).

Menely, Tobias, *Climate and the Making of Worlds: Towards a Geohistorical Poetics* (Chicago: University of Chicago Press, 2021), https://doi.org/10.7208/chicago/9780226776316.001.0001

Morton, Timothy, *Dark Ecology: For a Logic of Future Coexistence* (New York: Columbia University Press, 2016), https://doi.org/10.7312/mort17752

Murphy, Timothy S. and Benjamin Noys, 'Introduction: Old and New Weird', *Genre*, 49.2 (2016), https://doi.org/10.1215/00166928-3512285

Nealon, Jeffrey T., *Plant Theory: Biopower and Vegetal Life* (Stanford: Stanford University Press, 2015), https://doi.org/10.1515/9780804796781

Neyrat, Frédéric, *The Unconstructable Earth: An Ecology of Separation* (New York: Fordham University Press, 2019), https://doi.org/10.5422/fordham/9780823282586.001.0001

Ngai, Sianne, *Our Aesthetic Categories: Zany, Cute, Interesting* (Cambridge, MA: Harvard University Press, 2012).

Nocek, Adam, *Molecular Capture: The Animation of Biology* (Minneapolis: University of Minnesota Press, 2021), https://doi.org/10.5749/j.ctv1cdxg6p

Pinkus, Karen and Derek Woods, 'From the Editors: Terraforming', *Diacritics*, 47.3 (2019), https://doi.org/10.1353/dia.2019.0023

Poe, Edgar Allan, 'The Fall of the House of Usher', in *The Fall of the House of Usher and Other Tales* (New York: Signet Classics, 2006 [1839]), 117–37.

Pringle, Thomas Patrick, 'The Tech Ecosystem and the Colony', *Heliotrope* (12 May 2021).

Purdy, Jedediah, 'Thinking Like a Mountain', *N+1*, 29 (2017).

Read, David 'The Ties that Bind', *Nature*, 388 (1997), 517–18, https://doi.org/10.1038/41426

Sheldrake, Merlin, *Entangled Life: How Fungi Make Our Worlds, Change Our Minds, and Shape Our Futures* (New York: Random House, 2020).

Schwartz, Janelle A., *Worm Work: Recasting Romanticism* (Minneapolis: University of Minnesota Press, 2012), https://doi.org/10.5749/Minnesota/9780816673209.001.0001

Simard, Suzanne, *Finding the Mother Tree: Discovering the Wisdom of the Forest* (New York: Alfred A. Knopf, 2021).

Simard, Suzanne, *Nature's Internet: How Trees Talk to Each Other in a Healthy Forest*, online video recording, YouTube, 2 February 2017, htttps://www.youtube.com/watch?v=breDQqrkikM

Simard, Suzanne, 'Net Transfer of Carbon between Ectomycorrhizal Tree Species in the Field', *Nature*, 388 (1997), 579–82, https://doi.org/10.1038/41557

Sperling, Alison, 'Second Skins: A Body-Ecology of Jeff VanderMeer's *The Southern Reach Trilogy*', *Paradoxa*, 28 (2016), 230–55.

Sperling Alison, 'Weird Modernisms' (unpublished doctoral thesis, University of Wisconsin-Milwaukee, 2017).

Stalker, dir. by Andrei Tarkovsky (Mosfilm, 1979).

Stamets, Paul, *Mycelium Running: How Mushrooms Can Save the World* (Berkley: Ten Speed Press, 2005).

Thacker, Eugene, *In the Dust of this Planet: Horror of Philosophy, Vol. 1* (Washington: Zero Books, 2011).

Thacker, Eugene, 'Naturhorror and the Weird', in *Spaces and Fictions of the Weird and the Fantastic*, ed. by Julius Greve and Florian Zappe (London: Palgrave Macmillan,, 2019), pp. 13–24, https://doi.org/10.1007/978-3-030-28116-8_2

Toomey, David, *Weird Life: The Search for Life that Is Very, Very Different from Our Own* (New York: W.W. Norton & Co., 2013).

Tsing, Anne Lowenhaupt, *The Mushroom at the End of the World: On the Possibility of Life in Capitalist Ruins* (Princeton, NJ: Princeton University Press, 2015).

Turnbull, Johnathon, 'Weird', *Environmental Humanities*, 13.1 (2021), 275–80, https://doi.org/10.1215/22011919-8867329

Turner, Fred, *From Counterculture to Cyberculture: Stewart Brand, the Whole Earth Network, and the Rise of Digital Utopianism* (Chicago: University of Chicago Press, 2006), https://doi.org/10.7208/chicago/9780226817439.001.0001

VanderMeer, Ann and Jeff VanderMeer, *The New Weird* (Ashland, OH: Tachyon Publications, 2008).

VanderMeer, Jeff, *Acceptance* (New York: Farrar, Strauss, and Giroux, 2014).

VanderMeer, Jeff, *Annihilation* (New York: Farrar, Strauss, and Giroux, 2014).

VanderMeer, Jeff, *Authority* (New York: Farrar, Strauss, and Giroux, 2014).

VanderMeer, Jeff, *The City of Saints and Madmen: The Book of Ambergris* (Rockville, MD: Cosmos, 2001).

VanderMeer, Jeff, *Finch* (Portland, OR: Underland, 2009).

VanderMeer, Jeff, *Shriek: An Afterword* (New York: Tor, 2006).

Warren, Calvin L., *Ontological Terror: Blackness, Nihilism, and Emancipation* (Durham: Duke University Press, 2018).

Wasson, Gordon, 'Seeking the Magic Mushroom', *Life* (13 May 1957).

Weir, Alison, 'Decolonizing Feminist Freedom: Indigenous Relationalities', in *Decolonizing Feminism: Transnational Feminism and Globalization*, ed. by Margaret A. McLaren (New York: Rowman and Littlefield, 2017)), pp. 257–89.

Wells, H.G., *The War of the Worlds* (Peterborough, ON: Broadview, 2003).

White, Gilbert, *The Natural History of Selbourne* (New York: Penguin, 1977).

Whittaker, R.H., 'New Concepts of Kingdoms of Organisms', *Science*, 163.3863 (1969), 150–60.

Wohllenben, *The Hidden Life of Trees: What They Feel, How They Communicate*, trans. by Jane Billinghurst (Vancouver: Greystone Books, 2015).

Woods, Derek, 'The Fungal Kingdom', *Alienocene: Journal of the First Outernational*, Stratum 8 (2020), 1–14.

Woods, Derek, 'Scale in Ecological Science Writing', in *The Routledge Handbook of Ecocriticism and Environmental Communication*, ed. by Scott Slovic, Swarnalatha Rangarajan, and Vidya Sarveswaran (New York: Routledge, 2019), pp. 118–29, https://doi.org/10.4324/9781315167343-11/

Yong, Ed, 'The Wood Wide Web', *Atlantic* (14 April 2016), htttps://www.theatlantic.com/science/archive/2016/04/the-wood-wide-web/478224/

XVIVO, *The Inner Life of the Cell* (Harvard University Department of Molecular and Cellular Biology, 2006).

Zhdanova, N.N., and others, 'Ionizing Radiation Attracts Soil Fungi', *Mycological Research*, 108.9 (2004), 1089-1096, https://doi.org/10.1017/s0953756204000966

II. NEURO-MEDICAL IMAGING AND DIAGNOSIS

5. To Be or Not to Be a Patient

Challenging Biomedical Categories in Joshua Ferris's *The Unnamed*

Pascale Antolin

In *Wanderlust: A History of Walking*, historian Rebecca Solnit defines walking as 'ideally [...] a state in which the mind, the body and the world are aligned, as though they were three characters finally in conversation together [...]. Walking allows us to be in our bodies and in the world without being made busy by them. It leaves us free to think without being wholly lost in our thoughts'.[1] However, the pedestrian experience she describes has very little to do with Tim Farnsworth's own experience of walking in Joshua Ferris's novel, *The Unnamed*. A successful Manhattan lawyer, Tim, develops a condition that causes him to walk, against his will, until he is so exhausted that he collapses. No matter how many doctors he consults, the disorder is not identified, let alone treated.

Reflecting on disability in contemporary literature, Stuart Murray writes that '[I]n a world governed by new neurological knowledge, any unusual activity can be seen as a syndrome'.[2] It is true that since the 1990s, declared the 'Decade of the Brain' by President George H. W. Bush, a wide and rapidly expanding spectrum of neuroimaging

---

[1] Rebecca Solnit, *Wanderlust: A History of Walking*, 2nd ed. (New York: Penguin Books, 2001), p. 5.

[2] Stuart Murray, 'The Ambiguities of Inclusion. Disability in Contemporary Literature', in *The Cambridge Companion to Literature and Disability*, ed. by Claire Barker and Stuart Murray (Cambridge: Cambridge University Press, 2018), p. 99, https://doi.org/10.1017/9781316104316.008.

technologies has become available, resulting in a neuro-technological revolution. Popular books using the findings of neuroscience were published, and soon became bestsellers: the most famous included *Consciousness Explained* (1991) by philosopher Daniel Dennett and *How the Mind Works* (1997) by psychologist Steven Pinker. They propounded neural theories of mind and explained mental phenomena in terms of brain processes. Since then, not only has neuroscience developed but its application to fields beyond medicine has been widespread. Literature is one of them, hence the emergence both of the 'brain memoir'[3] and of the 'neuronovel' (Roth), or 'syndrome novel' (Lustig and Peacock), or 'neuronarrative'[4] (Jonhson). 'Both genres engage brain research, translating neurobiological theories into literary experiments'.[5] The neuronovel—and I will use this word for the sake of simplicity— draws on neuroscience and, often, neurological syndromes, to explore the complex relations between body, mind, self, and world. In the introduction to *Diseases and Disorders in Contemporary Fiction*, T. J. Lustig and James Peacock write that most syndrome novels rely on 'the heightened presence of scientifically-defined conditions'[6]—Tourette's in *Motherless Brooklyn*, Capgras in *The Echo Maker*, amnesia in *Man Walks into a Room*, to quote only a few. Like all these books, *The Unnamed* 'vividly illustrates the contemporary fascination with both the workings and the sciences of the mind'.[7] But Ferris' narrative stands out because Tim's condition is never 'scientifically-defined'. It challenges medical classifications and clinical expertise. Hence, the protagonist turns into 'a medical orphan',[8] that is, a patient—in the etymological sense of the

---

3   Jason Tougaw, *The Elusive Brain: Literary Experiments in the Age of Neuroscience* (New Haven: Yale University Press, 2018), p. 3, pp. 74–92, https://doi.org/10.12987/yale/9780300221176.001.0001.
4   In the field of literature, 'Neuro Lit Crit', or cognitive literary studies, should be mentioned as well, with authors like Gabrielle Starr, Kay Young and Mary Thomas Crane, among others.
5   Tougaw, *The Elusive Brain*, p. 3.
6   T. J. Lustig, and James Peacock, 'Introduction', in *Diseases and Disorders in Contemporary Fiction: The Syndrome Syndrome*, ed. by T. J. Lustig, and James Peacock (London: Routledge, 2013), p. 1, https://doi.org/10.4324/9780203067314.
7   Tanja Reiffenrath, 'Mind over Matter? Joshua Ferris's *The Unnamed* as Counternarrative', *Literary Refractions*, 5.1 (December 2014), p. 2, https://doi.org/10.15291/sic/1.5.lc.10.
8   Robert Aronowitz, 'When Do Symptoms Become a Disease?', *Annals of Internal Medicine*, 134.9 (Part 2), May 2001, 803, https://doi.org/10.7326/0003-4819-134-9_part_2-200105011-00002.

word, from the Latin *patior, pati*, to suffer—without any clinically based diagnosis.

According to Annemarie Golstein Jutel, ideally,

> [t]he diagnosis provides structure—sorting out the real from the imagined, the valid from the feigned, the significant from the insignificant, the physical from the psychological. Once the diagnosis is made, the concordant treatment can be planned, the prognosis reflected on, and resources allocated [...]. The diagnosis is both rudder and anchor: its pursuit guides the individual to the doctor's consulting rooms, while its assignment positions identity and behavior.
>
> Being diagnosed gives permission to be ill.[9]

In Ferris' novel, Tim is denied this permission, and thereby he is doomed to the chaos created by his condition. It consumes his life, erodes his personal and social identities until it destroys him altogether. It is true that diagnosis—of a chronic illness, in particular—is also likely to disturb and '*re*arrange individual identity, threatening previous self-definition as the individual now "inhabits" an illness'.[10] In his memoir, *When Breath Becomes Air*, for instance, neurosurgeon Paul Kalanithi describes the moment when he was diagnosed with cancer 'as if a sandstorm had erased all trace of familiarity'.[11] In Ferris' novel, however, a diagnosis could also have provided Tim with a new identity, the collective identity that brings together all the patients suffering from the same disease.

Instead, Tim's sickness is never conferred legitimacy: 'when a doctor deems a patient's condition to be medical, the latter receives previously unauthorized privileges such as permission to be absent from work, priority parking, insurance benefits, reimbursement for treatment, or access to services'.[12] Tim's condition remains an illness, that is, a personal experience of sickness, and it is never recognized as a disease, i.e., 'what Western medicine considers biological or psychophysiological dysfunction'.[13] In other words, Tim never becomes a patient in the medical sense of the word, as he does not experience the 'narrative

---

9   Annemarie Golstein Jutel, *Putting a Name to it: Diagnosis in Contemporary Society* (Baltimore: Johns Hopkins University, 2011), p. 4.
10 Ibid., p. 11.
11 Paul Kalanithi, *When Breath Becomes Air* (New York: Random House, 2016), p. 121.
12 Golstein Jutel, *Putting a Name to it*, p. 67.
13 Ibid., p. 64. Also see Arthur Frank, *The Wounded Storyteller: Body, Illness and Ethics* (Chicago: The University of Chicago Press, 1995), pp. 5–6.

surrender' mentioned by Arthur Frank in *The Wounded Storyteller*, that is, 'surrendering oneself to the care of a physician [...] [who] becomes the spokesperson for the disease'.¹⁴ Tim is left to 'the isolation of his [...] suffering',¹⁵ as he loses his job, his wife, and his family, and eventually finds himself alone walking across America. Goldstein Jutel explains that the doctor's inability to explain the symptom shifts responsibility to the patient, 'positing the cause [of the condition] as the patient's mental health'.¹⁶ But this is a responsibility that Tim declines in an early passage where his wife, Jane, is the focal character:

> The health professionals suggested clinical delusion, hallucinations, even multiple personality disorder. But he said, 'I know myself'. He said, 'I'm not in control, Jane'. His mind was intact. His mind was unimpeachable. If he could not gain dominion over his body, that was not 'his' doing.¹⁷

The passage discloses Tim's denial of a psychological origin to his disorder. It also prefigures his increasingly disrupted sense of self, and the split between his body and his mind, which is one of the major themes of the novel.

While materialist or reductionist conceptions have developed since the 1990s, producing a view of the self as no longer free but determined by brain processes—with Joseph LeDoux's 'synaptic self' and Nikolas Rose's 'neurochemical self' as significant examples—other neuroscientists 'take the position that we don't know enough yet, but we will one day. Literary writers tend not to engage the debate so much as use it as a source for material. The unresolved debate makes space for the literary representation of the relation between matter and mind'.¹⁸ This 'representation of the relation between matter and mind' is at the heart of Ferris' *The Unnamed*. Challenging the materialist view of the self in particular, the novelist shows to what extent the relations between body, brain, and mind are anything but simple or stable, particularly in critical moments such as Tim's undiagnosed condition. According to Tanja Reiffenrath, 'many illness narratives elucidate that particularly in moments of crisis, the congruence body/brain and mind is contested

---

14 A. Frank, *The Wounded Storyteller*, pp. 5–6.
15 Golstein Jutel, *Putting a Name to it*, p. 11.
16 Ibid., p. 84.
17 Joshua Ferris, *The Unnamed* (New York: Little, Brown & Co., 2010), p. 24.
18 Tougaw, *The Elusive Brain*, p. 4.

and often re-conceptualized in a dualist fashion',[19] which leads her to analyze Ferris' novel in the light of Cartesian dualism. As for Nathan D. Frank, he argues that 'even if we decide that neurology does not directly apply to *The Unnamed*'s disabled Tim Farnsworth, it provides enough of a meditation on what neuroscience *could* mean for the protagonist that any other interpretation will be colored by these potential ramifications'.[20]

The degradation of Tim's health is highly suggestive of the embodied chaos described by A. Frank in his exploration of illness narratives: 'The chaos story presupposes *lack* of control, and the ill person's loss of control is complemented by medicine's inability to control the disease'.[21] It is true that A. Frank analyzes illness memoirs rather than fiction, but the dramatic disruption of Tim's sense of self is also represented by an increasingly chaotic, fragmented narrative form. This strategy is a powerful means for Ferris not just to suggest the deterioration of Tim's condition mimetically, but also to question categories. As chaos prevails indeed, traditional references and conventions—whether medical, lexical, or literary—are necessarily disrupted and questioned.

While in his famous 2009 essay, 'The Rise of the Neuronovel', Marco Roth argued that in the neuronovel the brain has replaced the mind, in Ferris' book, by contrast, it is the body that threatens the mind, while the missing name of Tim's sickness creates chaos. This essay will thus focus on the challenging of medical knowledge, of neurological reduction, and of social and literary categories.

Challenging Medical Knowledge and Classifications

At the beginning of the novel, when Tim is working at his desk one night and the lights in his office switch off, he experiences the moment as a 're-entry into the physical world. Self-awareness. Himself as something more than mind thinking'.[22] This episode is not the first instance where the dichotomy between Tim's body and his mind has been mentioned in the novel. However, it is a very special moment: as Tim becomes aware

---

19 Reiffenrath, 'Mind over Matter', p. 4.
20 Nathan D. Frank, 'Of Non-Mice and Non-Men: Against Essentialism in Joshua Ferris's *The Unnamed*', in *Disability Studies Quarterly*, 40. 2 (2020).
21 A. Frank, *The Wounded Storyteller*, p. 100.
22 Ferris, *The Unnamed*, p. 37.

of his body again,[23] he seems to feel whole and happy. The short sentence 'That was happiness'[24] concludes the episode.

A few minutes later, Tim is faced with a brutal crisis: 'He looked back at Peter, who stood in the doorway, but his body kept moving forward'.[25] The dramatic last sentence signals the end of the congruence between body and mind—with the 'body' as the subject of the sentence—and the return of Tim's alienating condition. However, the extract is not followed by a description of Tim's compulsive walking—as often happens in the novel—but by an analepsis, instead, where Tim remembers a doctor's words in direct speech, as if they were so unacceptable that he had been unable to appropriate them: '"There is no laboratory examination to confirm the presence or absence of the condition, [...] so there is no reason to believe the disease has a defined physical cause or, I suppose, even exists at all"'.[26] The juxtaposition on the same page of the disorder and its denial by a doctor[27] exposes the limits of the empiricist frame of reference. Besides, the doctor's words exemplify what Goldstein Jutel describes as the doctor 'shift[ing] responsibility (for the inability to explain the symptom) from the doctor to the patient',[28] who is not just denied credibility but turned into a fraud. While the doctor's name Regis—from the Latin, *rex, regis*, king—corroborates the authority of his words, the passive form, by contrast, describes Tim's reception of the message: 'I was told'.[29] The phrase 'there is' turns the doctor's opinion into a universal statement. It also represents the objective third-person of scientific discourse, even though it is ironically challenged by the subsequent comment clause, 'I suppose', disclosing the doctor's logic and conclusion as unscientific. What the passage especially reveals is Tim's doctors' refusal 'to conceptualize the disease in other than

---

23 The passage also suggests that it was a characteristic—or requirement—of Tim's job that his mind and his intellect, should work intensely. Meanwhile, his body was forgotten. This interpretation establishes a connection between Tim's job and his condition, but it is never clearly confirmed.
24 Ferris, *The Unnamed*, p. 37.
25 Ibid., p. 40.
26 Ibid., p. 42.
27 The same strategy is used on page 46: as Tim experiences his compulsive walking ('some failsafe mechanism moved him around red lights and speeding cars'), two doctors are mentioned: 'One located the disease in his mind, the other in his body'.
28 Goldstein Jutel, *Putting a Name to it*, p. 84.
29 Ferris, *The Unnamed*, p. 41.

physical, neuroscientific terms'.³⁰ Since the first edition of the *Diagnostic and Statistical Manual of Mental Disorders (DSM-I)* in 1952, the mind has been increasingly medicalized. However, it was in 1980, with the publication of DSM-III, that the 'biomedical syndrome' emerged and 'the Age of the Syndrome began'. Since then, 'mental illnesses [have been] diseases of the brain or central nervous system'.³¹

Whenever the different physicians Tim has consulted are mentioned in the narrative, their names and more or less preposterous diagnoses and treatments—re-enacting his birth, for instance, or 'seven days of colonics and grass-and-carrot smoothies'³²—are listed in disorderly fashion, thus highlighting Tim's confusing experience as a medical orphan. This confusion is enhanced syntactically either by long sentences,³³ where medical doctors and healers are lumped together as equally ineffective, or by very short, fragmented paragraphs³⁴ underlining their diverging views. The confusion reaches a climax, paradoxically, when Tim is eventually given a diagnosis by a doctor aptly named Klum,³⁵ whose medical specialty is not specified: 'benign idiopathic perambulation'.³⁶ With their Greek or Latin roots, the three increasingly long, polysyllabic words debunk the medical authority since they convey no meaning at all, and merely amount to an admission of impotence. As Tim repeats them to himself, they are italicized, thus turning into signifiers without any signified, which Tim can even play with: from 'idiopathic', for instance, he coins '*idiopaths*'. No wonder that the list of doctors in that particular passage should end up with a quack, a 'genealogical healer'.³⁷

Another doctor then offers to take a 'clean image of [Tim's] brain [...] *in situ*'³⁸—resorting to Latin words again—using a prototypical headgear equipped with sensors to register Tim's brain activity. But the

---

30 Reiffenrath, 'Mind over Matter', p. 6.
31 Patricia Waugh, 'The Naturalistic Turn, the Syndrome, and the Rise of the Neo-Phenomenological Novel', in *Diseases and Disorders in Contemporary Fiction: The Syndrome Syndrome*, ed. by T. J. Lustig and James Peacock (London: Routledge, 2018), p. 18, https://doi.org/10.4324/9780203067314.
32 Ferris, *The Unnamed*, p. 38.
33 Ibid. p. 38, p. 46, p. 48.
34 Ibid. p. 39, p. 41.
35 The doctor is ironically named Klum, from the old German *klumm* (knapp), meaning short, limited.
36 Ferris, *The Unnamed*, p. 41.
37 Ibid., pp. 41–42.
38 Ibid., p. 64.

device is strangely ambivalent. While it is described in highly emphatic terms, it looks very pedestrian: 'It had been retrofitted to perform an *extraordinary* purpose and manufactured *exclusively* and *at great expense*, but he wondered how such an *everyday object* could serve to advance an understanding of his mystery'.[39] The grandiloquence suggests both the doctor's confident speech—as he talks Tim into trying the latest technological development—and maybe Tim's renewed hope of a diagnosis. Ironically, however, 'the very medicalization that would "exonerate"[40] Tim Farnsworth is also the institution that brands and stigmatizes him to the point that he must evade and deflect social inquiry [...] thus a shaved head and a curious helmet deploy Tim's liberation as his non-liberation [...]'.[41]

All the passages dealing with Tim's doctors, therefore, are highly critical of medicine and its classifications. For instance, they tend to focus on the debates surrounding psychiatry and neurology,[42] and pit them against each other:

> The psychiatrists believed his situation came from a physical malfunction of the body, something organic and diseased, while the neurologists pointed to the scans and the tests that revealed nothing and concluded that he had to be suffering something psychological. Each camp passed the responsibility for his diagnosis to the other, from the mind to the body, back to the mind [...].[43]

It is even Tim's psychiatrist who recommends 'genealogical healing',[44] when she realizes she cannot make sense of his condition. In most cases, exaggeration and irony are used to arouse suspicion of doctors' scientific expertise, and undermine the medical authority. Ferris' physicians also propose diagnoses and treatment methods that are as ludicrous as

---

39   Ibid., p. 86 (emphasis added).
40   This is an allusion to a passage on page 65, where Dr Bagdasarian is talking to Tim: '"I know you've fallen into depression because no empirical evidence has emerged to *exonerate* you—I use your word, which I have remembered many years—to *exonerate* you from the charge of being mentally ill. You hate it when people say it is something all in your head."' (emphasis added).
41   N. Frank, 'Of Non-Mice and Non-Men'.
42   For a detailed summary of these debates, see Nikolas Rose's article, 'Neuroscience and the Future for Mental Health', *Epidemiology and Psychiatric Sciences* 25 (2016), 95–100, https://doi.org/10.1017/S2045796015000621.
43   Ferris, *The Unnamed*, pp. 100–1.
44   Ibid., p. 42.

those suggested by healers and gurus—the headgear that looks like a bicycle helmet, for instance—and, thereby, they all find themselves on an equal footing, equally disparaged. Eventually, to conceal the limits of their knowledge, doctors put the blame on Tim—either the disorder 'is something all in [his] head'[45] or he is a freak of nature, with an unheard-of condition. The failure of the headgear to produce any results also shows the limits of the neuro-technological revolution.

This episode of eccentric neuroimaging is somehow reminiscent of the 'pivotal event' in the Naturalist novel, as it was defined by Frank Norris in his 1901 essay, 'The Mechanics of Fiction': 'It is the peg upon which the shifting drifts and currents must—suddenly—coagulate, the sudden releasing of the brake to permit for one instant to labor full steam ahead. Up to that point the action must lead; from it, it must decline'.[46] This new failure of medicine literally concludes a chapter in Tim's life, and in the novel. From then on, he no longer seeks a diagnosis or treatment, that is, he no longer seeks help from outside—particularly from doctors—hence the beginning of part two with its ominous title, 'The Hour of Lead',[47] borrowed from Emily Dickinson's poem 'After great pain, a formal feeling comes—'.[48] The failure of the headgear represents the beginning of a significant process of degradation in Tim's condition and life that will end with his death. However, the three parts that follow this failure in Ferris' novel do not show the 'increase in speed' mentioned by Norris as a result of the pivotal event.[49] The structure of *The Unnamed* is far more complex, therefore, than that recommended by Norris for a Naturalist novel. Ferris both borrows tools, and departs, from that tradition and its sympathy for medicine—as evidenced by Claude Bernard's influence on Emile Zola, for instance.

In the opening pages of part two, Tim has lost his job as a lawyer and his big office; he has also given up hope of ever receiving a diagnosis,

---

45 Ibid., p. 65.
46 Frank Norris, 'The Mechanics of Fiction' (December 1901) in *The Literary Criticism of Frank Norris*, ed. by Donald Pizer (Austin: University of Texas Press, 1964), p. 59.
47 Ferris, *The Unnamed*, p. 119.
48 The four parts of the novel bear titles borrowed from the same poem by Dickinson: 'The Feet, mechanical', 'The Hour of Lead', 'First—Chill—then Stupor', and 'then the letting go'. Dickinson describes how moments of intense suffering or grief are followed by periods of numbness, as the emotional trauma can hardly be processed. Ferris's narrative follows the evolution of the poem, which thus structures the novel.
49 Norris, 'The Mechanics of Fiction', p. 60.

let alone a cure. Therefore, he turns his back on medical expertise, and takes matters into his own hands, so to speak. As he has failed to find a 'spokesperson for [his] disease',[50] Tim starts to describe his condition in his own words,[51] and his self is increasingly reduced to the binary of the body and the mind.

Challenging Neurological Reduction

Early in the narrative, Tim's condition is described by his wife, in a chapter where she is the focal character, as 'a hijacking of some obscure order of the body, the frightened soul inside the runaway train of mindless matter, peering out from the conductor's car in horror'.[52] The description is suggestive of the traditional metaphor of the body as machine, dating back to Descartes' *Traité de l'homme* (1648). Tim's disorder seems to have taken control of his body from inside, while he has turned into a powerless spectator in a modern tragedy. The same figurative language is used in another passage focalized by Tim: 'He looked down at his legs. It was like watching footage of legs walking from the point of view of the walker. [...] the brakes are gone, the steering wheel has locked, I am at the mercy of this wayward machine'.[53] Here Tim's confusion is underlined by the shifts from the past to the present tense, and from the third to the first-person narrative. However, the metaphoric representation evolves in the following passage: 'His body wouldn't be contained or corralled. It had, it seemed, a will of its own'.[54] Here, 'corralled' suggests a wild animal, instead. Tim also moves from the image of one entity manipulating another from inside to a fight between two independent opponents: 'his body [...] spoke a persuasive language of its own [...] these two opposite wills worked to gain the better of each other in a struggle so primitive that it could not be named'.[55]

The primitive struggle is reminiscent of Robert Louis Stevenson's *Dr Jekyll and Mr. Hyde*[56] (1896) and Frank Norris' *Vandover and The Brute*,

---

50   A. Frank, *The Wounded Storyteller*, p. 6.
51   Ferris, *The Unnamed*, p. 126.
52   Ibid., p. 24.
53   Ibid., p. 33.
54   Ibid., p. 44.
55   Ibid., p. 109.
56   See Marina Ludwigs, 'Walking as a Metaphor for Narrativity', in *Studia Neophilologica*. Special Issue 'The Futures of the Present: New Directions in (American) Literature

written around 1894–1895, but only published posthumously in 1914. These novels stage forms of dualism—suggested by their titles—as they take up the image of the beast within the civilized man. While Stevenson's brute emerges under the influence of a drug, Norris' beast appears as a consequence of mental illness. Vandover, the eponymous hero, suffers from lycanthropy and, as each crisis develops, his perception of himself changes: 'His intellectual parts dropped away one by one, leaving only the instincts, the blind unreasoning impulses of the animal'. Norris does not mention Vandover's mind or any resistance on the part of his character's intellect, and the 'beast'[57] prevails. By contrast, resistance is always present in Ferris' narrative, as suggested by the following passage: 'He continued to think, "I'm winning," or "Today, he won," depending on how well *his mind, his will, his soul* (he did not know the best name for it) fought against the lesser instincts of his body'.[58] While the word 'instincts' features in both extracts, Ferris' protagonist also expresses 'a yearning to achieve some transcendent spiritual meaning presumed to be absent from the postmodern world'[59]—as suggested by the evolution from the mind to the will to the soul. Ferris borrows from nineteenth-century naturalism to question contemporary materialism, as they both deny the mind and free will: '[Tim] revolted against the disproportionate power enjoyed by chemical imbalances and shorting neural circuits. He could say the words "autonomic nervous system," whereas the autonomic nervous system just was; therefore, he was superior to the autonomic nervous system'.[60] From beginning to end, his hero clings to his mind as self and dismisses his body. Tim's position, however, is paradoxical: on the one hand, he adheres to the traditional mind/body dualism against contemporary reductionism; on the other, he denies mental illness as the cause of his condition.[61]

---

and Culture', 87 (2015), 116–28, https://doi.org/10.1080/00393274.2014.981962.
57 Frank Norris, *Vandover and the Brute*, 2nd ed. (Lincoln: University of Nebraska Press, 1978), p. 309.
58 Ferris, *The Unnamed*, p. 252 (emphasis added).
59 Stephen J. Burn, 'Mapping the Syndrome Novel', in *Diseases and Disorders in Contemporary Fiction: The Syndrome Syndrome*, ed. by T. J. Lustig and James Peacock (London: Routledge, 2018), p. 45, https://doi.org/10.4324/9780203067314.
60 Ferris, *The Unnamed*, p. 214.
61 See Mathew H. Gendle, 'The Problem of Dualism in Modern Western Medicine', *Mens Sana Monographs*, 14.1 (Jan-Dec. 2016), 141–51, https://doi.org/10.4103/0973-1229.193074.

Tim's view of his body changes dramatically in the second part of the novel. While his body turned on him when he first experienced the disorder, Tim now turns against his body. He no longer blames his legs for the compulsory walking, but sets himself against his whole body, which becomes a synecdoche for his condition. Since no sick part can be identified, the whole body is rejected, or I should say, 'ab-jected'. This approach indeed brings to mind Julia Kristeva's theory of the abject:

> There looms, within abjection, one of those violent, dark revolts of being, directed against a threat that seems to emanate from an exorbitant outside or inside, ejected beyond the scope of the possible, the tolerable, the thinkable. It lies there, quite close, but it cannot be assimilated. [...] And yet, from its place of banishment, the abject does not cease challenging its master.[62]

Tim cannot get rid of his body, he can only 'abject' it, because it is part and parcel of who he is—it even prevents him from committing suicide.[63] With a diagnosis, the abject would have been located in one body part, and framed. Instead, Tim's abject body testifies to his attempt to distinguish between sick and healthy without the tool of diagnosis. Tim's split self, therefore, represents his desperate effort to define his self merely along the lines of his thinking mind, without the threat that his alien body poses to his identity.

As the novel unfolds and the gap between body and mind widens, Tim experiences a sort of descent into hell with a tragic decline of his mental health. First, there is the 'constant fear of a recurrence'[64] throughout the second part, and then the relapse in part three: 'His body moved him down the sidewalk',[65] which is highlighted again by the reversal of the traditional roles of subject and object. In the same chapter, Tim calls his wife and says: '"Well, I've fed the son of a bitch" [...] "I've fed the son of a bitch and now we are standing outside the mini-mart" [...] "We're feeling better" [...]'.[66] The major shift in narrative persons—from the first-person singular to the first-person plural—signals Tim's loss of

---

62 Julia Kristeva, *Powers of Horror: An Essay on Abjection*, trans. by Leon S. Roudiez (New York: Columbia University Press, 1982), p. 1, p. 2.
63 Ferris, *The Unnamed*, p. 109.
64 Ibid., p. 149.
65 Ibid., p. 195.
66 Ibid., p. 200.

a unified identity, and his dramatic recognition of the presence of 'the other' for the very first time. In the struggle for control that follows, Tim always tries to resist and weaken his body: 'The other stopped saying food, food, and started saying leg, leg—but he continued to eat the doughnuts and ignored him'.[67] The war between Tim's mind and his body also turns into a war between the syntax of a narrative self and the parataxis of instinct, stressed here by the repetitions.

At this stage, the narrative, too, undergoes a dramatic evolution as it starts mimicking Tim's increasingly degraded (mental) health. Not only are the last three parts in the novel significantly shorter than the first—a sign of the protagonist's more and more fragmented self—but, in part four, the chapter division disappears, only longer and shorter paragraphs remain, and down to the end Tim is nearly the only focal character. The novel has turned into 'a chaos narrative'.

Challenging Social and Literary Categories

The phrase 'chaos narrative' is used by A. Frank to refer to personal narratives of illness that are so 'threatening' they can hardly be told.[68] According to Frank, in these narratives,

> the body is so degraded by an overdetermination of disease and social mistreatment that survival depends on the self's *dissociation* from the body, even while the body's suffering determines whatever life the person can lead. [...] A person who has recently started to experience pain speaks of 'it' hurting 'me' and can dissociate from that 'it'. The chaos narrative is lived when 'it' has hammered 'me' out of self-recognition.[69]

While *The Unnamed* is not an illness memoir, the experience of 'chaos embodied'[70] described by A. Frank has a lot in common with Tim's own experience. 'But', A. Frank adds, 'in the lived chaos there is no mediation, only immediacy'.[71] This maybe one of the reasons for the use of an omniscient third-person narrator in the novel: it allows for some critical distance to prevail throughout. While zero focalization[72] is employed

---

67 Ibid., p. 207.
68 A. Frank, *The Wounded Storyteller*, p. 98.
69 Ibid., p. 103.
70 Ibid., p. 102.
71 Ibid., p. 98.
72 According to Gerard Genette, zero focalization 'corresponds to what English-language criticism calls narrative with omniscient narrator [...] (where the narrator

occasionally, Ferris mostly uses the main characters as focalizers—Tim especially, as well as his wife Jane and their daughter Becka—giving the reader access to both the sick protagonist's and his family's viewpoints and emotions. Each of them focalizes specific chapters, in disorderly fashion, down to part three, where only the first chapter is focalized by Jane, and part four, where Tim is the main focalizer. From then on, the representation of his degraded mental health turns increasingly dramatic—as testified, for instance, by the body's repeated intrusions into the narrative: monosyllabic words[73] at first, 'then progressing to simple sentences ("Leg is hurting" [213]) and sarcasm ("Deficiency of copper causes anemia, just so you know" [216]) before arriving at full-blown taunting [...] [Ferris 223–24]'.[74] Other no less dramatic strategies are also employed to elucidate these intrusions: a fill-in-the-blanks passage,[75] dramatic dialogues[76] and recurrent typographic variations.[77] The distorted text turns into a mirror image of Tim's increasing mental alienation. But his body's growing decay is nevertheless documented, particularly by the use of parataxis and long lists of more or less serious ailments, equally intruding upon the narrative: 'He had renal failure, an enlarged spleen, sepsis-induced hypotension, cellular damage to the heart. He had trench foot and a case of dysentery. He required assisted breathing and intravenous antibiotics'.[78] Tim seems to enjoy these dry lists of medical names as they contrast with the absent name of his major condition and represent rare moments when his body can at last be labelled and classified. Ferris' novel, therefore, combines a mostly conventional narrative strategy with experimentation and fragmentation in the concluding chapter. As Tim is denied the 'narrative surrender'[79] experienced by patients who have been given a diagnosis and fails to

---

      knows more than the character, or more exactly, says more than any of the characters knows). In [internal focalization], Narrator = Character (the narrator says only what a given character knows) [...]'. Gerard Genette, *Narrative Discourse: An Essay in Method*, trans. by Jane E. Lewin (Oxford: Blackwell, 1980), pp. 188–89.

73 Ferris, *The Unnamed*, p. 207.
74 Chingshun J. Sheu, 'Forced Excursion: Walking as Disability in Joshua Ferris's *The Unnamed*', M/C Journal, 21.4 (2018), https://doi.org/10.5204/mcj.1403.
75 Ferris, *The Unnamed*, p. 221.
76 Ibid., pp. 222–25.
77 Ibid., p. 233.
78 Ibid., p. 222. Also p. 269, pp. 278–79.
79 A. Frank, *The Wounded Storyteller*, p. 6.

speak for himself at the beginning of the second part,[80] literature—that is, Ferris' *The Unnamed*—seems to take over and offer at least a successful narrative for Tim's condition, if not a diagnosis.

However, the end of the book does not bring about any clear resolution, as suggested by the last sentence and the conflicting interpretations it has provoked: 'the exquisite thought of his eternal rest was how delicious that cup of water was going to taste the instant it touched his lips'.[81] It seems to bring body and mind back together through the co-presence in the sentence of 'thought' and 'lips'. Chingshun J. Sheu argues that Ferris 'manages to grant victory to both mind and body without uniting them: his mind keeps working after physical death, but its last thought is of a "delicious [...] cup of water" [310]. Mind and body are two, but indivisible'.[82] But Reiffenrath considers that 'the protagonist is indeed able to win the protracted war against his body'.[83] These diverging views highlight the ambiguity at the heart of the narrative,[84] and the author's disruption of readerly expectations and literary conventions alike.

For instance, while the title of Ferris' novel seems to refer to his protagonist's unnamed condition, it is also an allusion to Beckett's novel, *The Unnamable* (1953), as Ferris himself confirmed in an interview.[85] Ironically, while Beckett's narrator-protagonist cannot move, he is affected by another compulsion than Tim: he cannot stop talking. As for the intertextual allusions to Dickinson's poetry in the titles of the four parts, they are programmatic since the poem leads the reader from the disorientation provoked by loss to death. They also yoke the contemporary (neuro)novel with nineteenth-century poetry, a poem moreover which neither follows a regular rhyming scheme nor a regular metric pattern, in other words, a poem questioning conventional poetic rules. *The Unnamed* also borrows from detective fiction, a recurrent

---

80 Tim 'sp[eaks] a language only he underst[ands]'. Ferris, *The Unnamed*, p. 126.
81 Ibid., p. 310.
82 Sheu, 'Forced Excursion'.
83 Reiffenrath, 'Mind over Matter', p. 14.
84 For a detailed analysis of the conflicting views of critics on *The Unnamed*, see Nathan Frank, 'Of Non-Mice and Non-Men: Against Essentialism in Joshua Ferris's *The Unnamed*'.
85 Joshua Ferris, 'Involuntary Walking; the Joshua Ferris Interview'. *ReadRollShow*. Created by David Weich. Sheepscot Creative, 2010. *Vimeo*, 9 Mar. 2010, https://vimeo.com/10026925.

feature of the neuronovel,[86] but it is only a subplot, playing a minor role in the narrative.

The dramatic degradation of Tim's health and of his personal and social lives is reminiscent of the Naturalist novel, too—as seen earlier. Tim's return to a primitive life in the wilderness, in particular, recalls the eponymous character's departure from San Francisco after the murder of his wife in Norris' 1899 *McTeague*. However, Tim's wilderness has nothing to do with McTeague's, or even the mythical American wilderness, as evidenced by the opening lines of part four: 'everywhere was a wilderness to him who had known only the interiors of homes and offices and school buildings and restaurants and courthouses and hotels'.[87] What is suggested here, and underlined by the polysyndeton, is that, no matter how successful, Tim's whole life relied on a fundamental loss, both of meaning and traditional references. This is confirmed in a scene at the bank late in the novel, when the protagonist has supposedly returned to a primitive life:

> When the banker took him back and accessed his portfolio with the *various* websites and passwords he'd been given, he saw an *inordinate amount* of money diversified across a *wide* spectrum of investment vehicles. This caused him to turn away from his computer screen and stare at the man across from him. His foot was perched on the edge of the desk and he was picking dried blood from his leg and collecting the flakes in the palm of his hand. The banker fished the garbage pail out from under the desk; 'Do you need this?' he asked.[88]

The emphatic evocation of the data on the computer screen—that is, of Tim's wealth—stands in sharp contrast with the grotesque description of the protagonist as a sort of tramp 'in a soiled T-shirt and ripped chino',[89] sitting across from the banker, with his foot on the desk. Despite the circumstances, Tim's body language suggests that he has remained a hegemonic white male, so that the American myth of the return to a primitive life in the wilderness is completely distorted. The passage also elucidates the reason for the presence in the novel of numerous episodes and anecdotes about Tim's corporate life. They draw a satirical portrait of urban professionals and the American work ethos.

---

86   See Tougaw, *The Elusive Brain*, pp. 132–35.
87   Ferris, *The Unnamed*, p. 247.
88   Ibid., p. 208 (emphasis added).
89   Ibid.

Tim's profession as a successful lawyer suggests the power of civilization—with the law as the bedrock of the social contract. This is confirmed in a passage focalized by Jane, where she imagines Tim at work: 'she could still picture him in a climate-controlled conference room [...] drinking civilized lattes and assessing the other side's evidence. It was what he wanted, this corporate pastoral'.[90] The first hypallage (or transferred epithet) 'civilized lattes' showcases the association, in Jane's mind, of civilization with a safe, indoors environment peopled with rational beings. As for the second hypallage, 'corporate pastoral', it highlights the subversion of the pastoral ideal[91] in contemporary American society. This subversion is also illustrated incidentally by Lev Wittig, a former partner at Tim's law firm, who needs a snake in the room when he has sex, because this is how he gets excited.[92] As for Mike Kronish, another partner, he is such a workaholic that his children hardly ever see him and end up calling him 'Uncle Daddy'.[93] Episodes of this kind are scattered in the novel, and challenge both the 'perfect' appearance of Tim's corporate life, the so-called rational beings surrounding him, and the very notion of civilization they supposedly embody as lawyers. To a certain extent, the novel creates a parallel image of doctors and lawyers as illusions of civilization.

Eventually, Tim's condition questions the concept of disability as his (compulsive) walking can be associated with the figure of the *flâneur* originating in nineteenth-century Paris,[94] and Ferris said in an interview that he had the *flâneur* in mind when he wrote the novel.[95] According to David Serlin, however, disabled bodies are usually excluded from the literature on *flânerie* because

> the embodied experience of disability challenges and even thwarts cultural expectations of the firm division between public and private spheres. [...] The disabled *flâneur* visibly alters perceptions of public space by exposing that which has typically pertained to the 'interieur'—visible

---

90 Ibid., p. 25.
91 See Leo Marx, *The Machine in the Garden. Technology and the Pastoral Ideal*, 2nd ed. (Oxford: Oxford University Press, 2000).
92 Ferris, *The Unnamed*, p. 143.
93 Ibid., p. 148.
94 See Peter Ferry, 'Reading Manhattan, Reading American Masculinity. Reintroducing the Flâneur with E. B. White's *Here Is New York* and Joshua Ferris's *The Unnamed*', *Culture, Society & Masculinities*, 3.1 (2011), 49–61 (p. 50), https://doi.org/10.3149/CSM.0301.49.
95 Ferris, 'Involuntary Walking; the Joshua Ferris Interview'.

bodily differences as well as the invisible effects of institutionalization or, in more contemporary circumstances, networks of care giving and mutual support—to the outside world in ways that are anathema to narratives of modern autonomy.[96]

By setting his disabled character in a modern urban landscape—at least at the beginning of the novel—Ferris also destabilizes 'the apparent position of power of the male figure in the city of modernity'.[97] By becoming a *flâneur*, Tim falls out of the corporate order and into the feminized field of the inactive and dependent.

Disability is traditionally associated with inferiority and feminization,[98] deficit, absence and loss.[99] The challenge of disability is evidenced in particular by Jane's description of Tim's body when they are about to make love, after one of his compulsory walks: 'She felt all along his lean walker's body, the legs that were all muscle now and the torso that had slimmed down to the ribs as if he were a boy again'.[100] Tim's condition here does not involve deficit and loss but, instead, muscle gain and rejuvenation. His boy's body, however, can also suggest a loss of virility, and the feminization mentioned by Rosemarie Garland-Thomson. A white protestant male and a successful professional, Tim initially stands for the 'normate', as it is defined by Garland Thomson: 'The term *normate* usefully designates the social figure through which people can represent themselves as definitive human beings. Normate, then, is the constructed identity of those who, by way of the bodily considerations and cultural capital they assume, can step into a position of authority and wield the power it grants them'.[101] While his compulsive walking disables Tim, both physically and socially, 'it revolves around walking, the paradigmatic act of ability in popular culture, as connoted in the phrase "to stand up and walk."' Tim is 'able-bodied—in fact, we might say he is "over-able."'[102] Both 'over-able' and disabled, normate, and deviant, wealthy and homeless, a lawyer, and a tramp, Tim defies

---

96 David Serlin, 'Disabling the Flâneur', *Journal of Visual Culture*, 5.2 (2006), p. 200, https://doi.org/10.1177/1470412906066905.
97 Ferry, 'Reading Manhattan, Reading American Masculinity', p. 50.
98 Rosemarie Garland Thomson, *Extraordinary Bodies: Figuring Physical Disability in American Culture and Literature* (New York: Columbia University Press, 2017), pp. 8–9.
99 Murray, 'The Ambiguities of Inclusion: Disability in Contemporary Literature', p. 96.
100 Ferris, *The Unnamed*, pp. 104–5.
101 Garland Thomson, *Extraordinary Bodies*, p. 8.
102 Sheu, 'Forced Excursion'.

social categories and turns into a figure of resistance in the context of the American capitalistic society.

At the end of the novel, his ailments are so numerous that, paradoxically, the diagnostic list tends to exceed the disability category. Tim is no longer 'over-able' but overly disabled, so to speak. His compulsive walking seems to have taken a back seat as he is diagnosed with 'conjunctivitis', 'leg cramps', 'myositis', 'kidney failure', 'chafing and blisters', 'shingles', 'back pain', 'bug bites, ticks, fleas and lice', 'sun blisters', 'heatstroke and dehydration', 'rhabdomyolysis', 'excess [blood] potassium', 'burning tongue', 'head cold', 'pneumonia', 'pleurisy', 'acute respiratory distress syndrome', 'excess fluid [in] his peritoneal cavity', and 'brain swelling'.[103] The proliferation of signifiers may suggest an ironic attempt to compensate for the lack of the major signified—the name of Tim's condition—and at last provide him with the 'empirical evidence […] [of] a legitimate physical malfunction'[104] he has been desperately looking for for years.

While disability is questioned,[105] Ferris' *The Unnamed*, however, 'leads to the "lesson" of syndrome novels: that the body/brain interface […] is more enigmatic than lawyers can imagine and even scientists explain. Or, to quote Emily Dickinson, […] "the brain is wider than the sky."'[106] Ferris thus seems to delight in blurring traditional categories: not only is the body said to have 'a mind of its own'[107] but, in the end, 'the soul is the mind is the brain is the body'.[108] The passage certainly illustrates Tim's degraded mental health, but it also suggests that 'the relationship between brain and self'—or brain and soul and body and mind—remains 'a fascinatingly complex conundrum',[109] as Tougaw writes. The repeated allusions to the soul in the narrative could suggest

---

103 Ferris, *The Unnamed*, pp. 278–80.
104 Ferris, *The Unnamed*, p. 65.
105 Disability is also questioned in another neuronovel, *Motherless Brooklyn*. The protagonist's Tourette's turns into a benefit and not a blemish. See Pascale Antolin '"I am a freak of nature": Tourette's and the Grotesque in Jonathan Lethem's *Motherless Brooklyn*', *Transatlantica* 1, 2019, https://doi.org/10.4000/transatlantica.13941.
106 Tom LeClair, '*The Unnamed*', Review of *The Unnamed*, by Joshua Ferris. *Barnes and Noble Review*, 18 January 2010, https://www.barnesandnoble.com/review/the-unnamed. *Wider than the Sky. The Phenomenal Gift of Consciousness* is also the title of an essay by neurologist Gerard Edelman (New Haven: Yale University Press, 2004).
107 Ferris, *The Unnamed*, p. 44.
108 Ibid., p. 233.
109 Tougaw, *The Elusive Brain*, p. 12.

resistance to 'biology and strict materialism'[110] and 'a partial return to religion and spirituality'[111]—as in Jonathan Franzen's essay 'My Father's Brain'. But Tim's views on the soul change throughout the novel, depending on the circumstances. First, his conception of the soul seems uncertain—'[...] his mind, his will, his soul (he did not know the best name for it)'[112]—then he announces that 'there's no soul [...] No God',[113] before changing positions again when he is faced with his wife's cancer: 'The soul was inside her doing the work of angels to repulse the atheistic forces of biology and strict materialism'.[114] Stephen Burn interprets these conflicting views as 'Ferris's exploration of the novel's dialogic capabilities simultaneously to endorse the authoritative languages of "chemical imbalances and shorting neural circuits" and the most mystical "work of the divine" (214, 305)'.[115] Tim's diverging positions also illustrate Ferris' challenge of conventional binaries—religion or science, health or disability, the body or the mind—as his unnamed disorder questions categorization. They show Ferris 'explor[ing]' the power of language and the novel to promote inclusion over exclusion, 'synthesis over rupture, compromise over raw polarities'[116]—a characteristic, Burn argues, of the contemporary syndrome novel. In the end, considering the limits of medicine, literature emerges as an alternative solution to make sense of illness.

A syndrome novel staging a character without any diagnosed syndrome, *The Unnamed* stands apart, and challenges the (sub-)genre from the start. It is 'precisely because Tim Farnsworth's condition can be read as mental and/or as physical, as neurological and/or as non-neurological, [that] *The Unnamed* sets up as a radically individuating novel [...]'.[117] With his hero's undiagnosed disorder, Ferris also questions medical classifications, and his doctors turn out to be as powerless, even pathetic sometimes, as the quacks his hero has also consulted. No matter what medical experts may say, however, Tim denies the materialist conception of the self to such an extent that he rejects or 'abject[s]' his

---

110 Ferris, *The Unnamed*, p. 304.
111 Burn, 'Mapping the Syndrome Novel', p. 46.
112 Ferris, *The Unnamed*, p. 252.
113 Ibid., p. 300.
114 Ibid., p. 304.
115 Burn, 'Mapping the Syndrome Novel', p. 46.
116 Ibid., p. 47.
117 N. Frank, 'Of Non-Mice and Non-Men'.

whole chaotic body. But the blank at the heart of the narrative disturbs traditional references and questions social categories—particularly disability and health.

As for Ferris' introduction of the soul into his fiction, it suggests both a strategy of resistance to the 'totalizing claims of contemporary neuroscience',[118] and a synthetic approach promoting a dialogue between science and spirituality. Ferris introduces another dialogue, between contemporary and conventional literary genres, as he borrows from traditional genres like detective fiction, nineteenth-century poetry, and the Naturalist novel, so that he challenges both generic and historical categorization. This dialogical tendency is strengthened by intertextuality, as Dickinson's poem structures his narrative, breaking textual barriers and generating another 'dialogue among several writings'.[119]

While it is and is not a neuronovel, as Tim is and is not a patient, *The Unnamed* is certainly a 'textual syndrome', to quote LeClair, referring to the etymology of syndrome as 'a place where several roads meet'.[120] It is also an 'open work', in the sense of Umberto Eco, that readers are invited to interpret and 'conclude',[121] and a 'literary laboratory',[122] that is, a site for experiments and investigation, particularly of the relations between brains, minds, bodies, and world.

Works Cited

Antolin, Pascale, '"I am a freak of nature": Tourette's and the Grotesque in Jonathan Lethem's *Motherless Brooklyn*', *Transatlantica*, 1 (2019), https://doi.org/10.4000/transatlantica.13941

Aronowitz, Robert, 'When Do Symptoms Become a Disease?' *Annals of Internal Medicine*, 134.9 (Part 2), May 2001, 803–08, https://doi.org/10.7326/0003-4819-134-9_part_2-200105011-00002

---

118 Burn, 'Mapping the Syndrome Novel', p. 47.
119 Julia Kristeva, 'Word, Dialogue, and Novel', in *Desire in Language: A Semiotic Approach to Literature and Art*, ed. by Leon S. Roudiez, trans. by Thomas Gora et al. (New York: Columbia University Press, 1980), p. 65.
120 *Online Etymology Dictionary*, https://www.etymonline.com/word/syndrome.
121 Umberto Eco, *The Open Work*, trans. by Anna Cancogni (Cambridge: Harvard University Press, 1989), p. 19.
122 Tougaw, *The Elusive Brain*, p. 5. Tougaw borrows the image from Heather Houser's *Ecosickness in Contemporary U.S. Fiction: Environment and Affect* (New York, Columbia University Press, 2014).

Burn, Stephen J., 'Mapping the Syndrome Novel', in *Diseases and Disorders in Contemporary Fiction: The Syndrome Syndrome*, ed. by T. J. Lustig and James Peacock (London: Routledge, 2018), pp. 35–52, https://doi.org/10.4324/9780203067314

Dickinson, Emily, 'After great pain, a formal feeling comes—', in *The Complete Poems* (Global Grey ebooks: 2020), p. 357.

Eco, Umberto, *The Open Work*, trans. by Anna Cancogni (Cambridge: Harvard University Press, 1989).

Ferris, Joshua, *The Unnamed* (New York: Little, Brown & Co., 2010).

Ferris, Joshua, 'Involuntary Walking; the Joshua Ferris Interview', *ReadRollShow*, created by David Weich, Sheepscot Creative, 2010. *Vimeo*, 9 Mar. 2010. https://vimeo.com/10026925

Ferry, Peter, 'Reading Manhattan, Reading American Masculinity: Reintroducing the *Flâneur* with E. B. White's *Here Is New York* and Joshua Ferris's *The Unnamed*', *Culture, Society & Masculinities*, 3.1 (2011), 49–61, https://doi.org/10.3149/CSM.0301.49

Frank, Arthur, *The Wounded Storyteller: Body, Illness and Ethics* (Chicago: The University of Chicago Press, 1995).

Frank, Nathan D., 'Of Non-Mice and Non-Men: Against Essentialism in Joshua Ferris's *The Unnamed*', *Disability Studies Quarterly*, 40.2 (2020).

Garland Thomson, Rosemarie, *Extraordinary Bodies: Figuring Physical Disability in American Culture and Literature* (New York: Columbia University Press, 2017).

Gendle, Mathew H., 'The Problem of Dualism in Modern Western Medicine', *Mens Sana Monographs*, 14.1 (Jan-Dec. 2016), 141–51, https://doi.org/10.4103/0973-1229.193074

Gerard Genette, *Narrative Discourse: An Essay in Method*, trans. by Jane E. Lewin (Oxford: Blackwell, 1980).

Golstein Jutel, Anne-Marie, *Putting a Name to it: Diagnosis in Contemporary Society* (Baltimore: Johns Hopkins University, 2011).

Johnson, Gary, 'Consciousness as Content: Neuronarratives and the Redemption of Fiction', *Mosaic*, 41.1 (March 2008), 169–84.

Kalanithi, Paul, *When Breath Becomes Air* (New York: Random House, 2016).

Kristeva, Julia, *Powers of Horror: An Essay on Abjection*, trans. by Leon S. Roudiez (New York: Columbia University Press, 1982).

Kristeva, Julia, 'Word, Dialogue, and Novel', in *Desire in Language: A Semiotic Approach to Literature and Art*, ed. Leon by S. Roudiez, trans. by Thomas Gora et al. (New York: Columbia University Press, 1980), pp. 64–91.

LeClair, Tom, 'The Unnamed', review of The Unnamed, by Joshua Ferris, Barnes and Noble Review (18 January 2010), https://www.barnesandnoble.com/review/the-unnamed

LeDoux, Joseph, Synaptic Self: How Our Brains Become Who We Are (New York: Penguin, 2002).

Ludwigs, Marina, 'Walking as a Metaphor for Narrativity', Studia Neophilologica. Special Issue 'The Futures of the Present: New Directions in (American) Literature and Culture', 87 (2015), 116–28, https://doi.org/10.1080/00393274.2014.981962

Lustig, T. J. and James Peacock, 'Introduction', in Diseases and Disorders in Contemporary Fiction: The Syndrome Syndrome (London: Routledge, 2018), pp. 1–16, https://doi.org/10.4324/9780203067314

Murray, Stuart, 'The Ambiguities of Inclusion: Disability in Contemporary Literature', in The Cambridge Companion to Literature and Disability, ed. by Claire Barker and Stuart Murray (Cambridge: Cambridge University Press, 2018), pp. 90–103, https://doi.org/10.1017/9781316104316.008

Norris, Frank, 'The Mechanics of Fiction' (December 1901) in The Literary Criticism of Frank Norris, ed. by Donald Pizer (Austin: University of Texas Press, 1964), pp. 58–61.

Norris, Frank, Vandover and the Brute, 2nd ed. (Lincoln: University of Nebraska Press, 1978).

Online Etymology Dictionary, https://www.etymonline.com/word/syndrome

Reiffenrath, Tanja, 'Mind over Matter? Joshua Ferris's The Unnamed as Counternarrative', Literary Refractions, 5.1 (December 2014), 1–20, https://doi.org/10.15291/sic/1.5.lc.10

Rose, Nikolas, 'Neuroscience and the Future for Mental Health?', Epidemiology and Psychiatric Sciences, 25 (2016), 95–100, https://doi.org/10.1017/S2045796015000621

Roth, Marco, 'The Rise of the Neuronovel', n + 1, issue 8: Recessional (Fall 2009), https://nplusonemag.com/issue-8/essays/the-rise-of-the-neuronovel/

Serlin, David, 'Disabling the Flâneur', Journal of Visual Culture, 5.2 (2006), 193–208, https://doi.org/10.1177/1470412906066905

Sheu, Chingshun J., 'Forced Excursion: Walking as Disability in Joshua Ferris's The Unnamed', M/C Journal, 21.4 (2018), https://doi.org/10.5204/mcj.1403

Solnit, Rebecca, Wanderlust: A History of Walking, 2nd ed. (New York: Penguin Books, 2001).

Tougaw, Jason, The Elusive Brain: Literary Experiments in the Age of Neuroscience (New Haven: Yale University Press, 2018), https://doi.org/10.12987/yale/9780300221176.001.0001

Waugh, Patricia, 'The Naturalistic Turn, the Syndrome, and the Rise of the Neo-Phenomenological Novel', in *Diseases and Disorders in Contemporary Fiction: The Syndrome Syndrome*, ed. by T. J. Lustig and James Peacock (London: Routledge, 2018), pp. 17–34, https://doi.org/10.4324/9780203067314

6. Neurocomics and Neuroimaging

David B.'s *Epileptic* and Matteo Farinella and Hana Roš's *Neurocomic*[1]

Jason Tougaw

> Overturning the age-old axiom that a picture is worth a thousand words, perhaps these PET images require millions of words to be understood![2]
>
> Joseph Dumit, *Picturing Personhood:*
> *Brain Scans and Biomedical Identity* (2004)

In his graphic memoir *Epileptic*, David B. portrays an impossible fantasy: that he might find a doctor who could 'transfer' his brother Jean-Christophe's epilepsy into him.[3] He fantasizes that an exchange of brain matter might enable him to feel what it is like to be his brother. It is a fantasy of overcoming *the explanatory gap*, a term coined by philosopher Phillip Levine to describe a persistent obstacle to understanding consciousness from a neurobiological point of view: nobody can explain how immaterial experience—self, consciousness, cognition, memory,

---

1 This chapter is a revised version of the chapter 'Neurocomics and Neuroimaging', in Jason Tougaw, *The Elusive Brain: Literary Experiments in the Age of Neuroscience* (New Haven: Yale University Press, 2018), pp. 186–227, https://doi.org/10.12987/9780300235609-012.
2 Joseph Dumit, *Picturing Personhood: Brain Scans and Biomedical Identity* (Princeton: Princeton University Press, 2004), p. 24.
3 David B., *Epileptic* (New York: Pantheon, 2006); the original French title is *L'Ascension du Haut Mal*.

imagination, affect—emerges from brain physiology, from synaptic networks and brain regions, groups of neurons oscillating in and out of sync, stimulating each other with varying amplitudes of electricity, circulating chemicals that change each other's behaviour.[4]

Fig. 1 David B., *L'Ascension du Haut Mal* (1999) © David B. and L'Association. All rights reserved. English translation: 'Armed with my newfound strength, I fantasize that I could take on my brother's disease if a resourceful scientist were to transfer it into my skull'.

'I fantasize', David B. writes, 'that I could take on my brother's disease if a resourceful scientist were to transfer it into my skull'.[5] Throughout *Epileptic*, he struggles to empathize with Jean-Christophe. In his fantasy, brain science will rewrite his failure and undo the mutual alienation of two siblings. But no scientist is that resourceful. Readers don't need to see David B.'s imaginary scientist to know he is a quack (though versions of him *do* appear in other panels). The trappings of his steampunk lab undercut the fantasy with irony and despair. David B. draws Jean-Christophe's epilepsy as a serpent that slinks from panel to panel and page to page. In this particular panel, the serpent fuses with the characters' brain matter and the wires of retro machinery of a mad scientist's laboratory. In others, it slips out of the brain through scenes of family domesticity, medical clinics, and characters' dreams

---

4   Joseph Levine, 'Materialism and Qualia: The Explanatory Gap', *Pacific Philosophical Quarterly*, 64 (October 1983), 354–61, https://doi.org/10.1111/j.1468-0114.1983.tb00207.x.

5   B., *Epileptic*, p. 168.

and memories, to become a recurring image of epilepsy's reach beyond the brain. Epilepsy is a brain disorder, but it affects the whole of Jean-Christophe's body and life. He suffers physically and emotionally; his family is defined by the helplessness it makes them feel; his community ostracizes him. Neurology cannot eliminate Jean-Christophe's seizures, and the constellation of their effects is surely beyond the power of biomedicine. *Epileptic* reminds readers of a fact that is easy to overlook: the brain reaches through the whole body, through selfhood, touching identity, family, social life, and the physical environment.

Matteo Farinella and Hana Roš's *Neurocomic* (2014) offers a similar set of philosophical ideas about the relation between the brain and the self—in the form of a basic neuroscience lesson wrapped loosely in a fictional visual narrative.[6] Farinella is an illustrator with a Ph.D. in neuroscience and Roš a research associate in neuroscience and pharmacology at the University College, London. Together they have created a hybrid text of literary neuroscience—a graphic neurology fairytale primer. *Neurocomic*'s quest narrative could not be more explicit, linking a search for self directly with the protagonist's journey through the human brain. A generic man finds himself trapped in a book read by a generic woman who attracts him. The characters are allegorical composites, unlike those in most brain narratives, which tend to represent individual, even idiosyncratic, experience. Like many works of contemporary literature, *Neurocomic* and *Epileptic* ask, in philosopher Catherine Malabou's words, *what should we do with our brain?*[7] How might we understand its relation to identity? How should we live with it, study it, or write about it? Like so many twenty-first-century brain narratives, both texts conceive the physical brain as central to the stories they tell, the conflicts they plot, and the characters they portray; both genres engage brain research, translating neurobiological theories into literary experiments. Their creators experiment with narrative forms that may frame new views on the relationship between brain matter and the immaterial experiences that compose a self—what philosophers call phenomenology.

---

6   Matteo Farinella and Hana Roš, *Neurocomic* (London: Nobrow Press, 2014).
7   Catherine Malabou, *What Should We Do with Our Brain?*, trans. by Sebastian Rand (New York: Fordham University Press, 2008), p. 63.

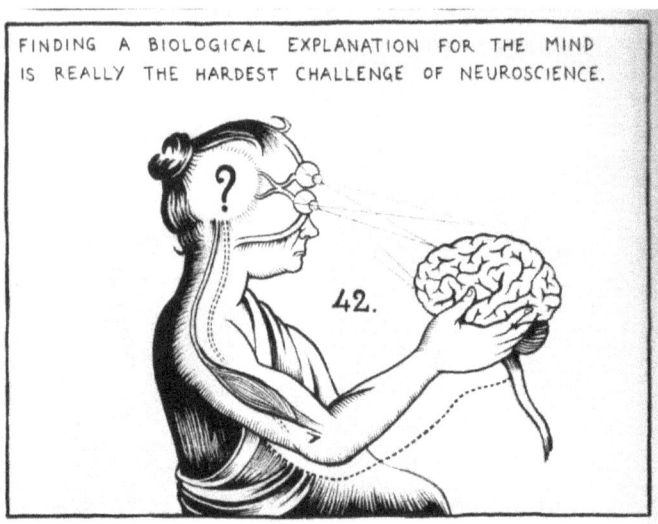

Fig. 2 Matteo Farinella and Hana Roš, *Neurocomic* (2014) © Matteo Farinella and Hana Roš. CC BY-NC-ND 4.0

Whereas *Neurocomic* creates characters whose brains become a vehicle for telling a story about the current state of neuroscientific knowledge, *Epileptic* portrays its characters' brains as part of an ensemble of images that define their search for what it means to suffer as a result of neurological conditions beyond their control. Like so many brain narratives, both books offer alternatives to 'you are your brain'/'you are not your brain' debates. The interplay of image and text in these graphic narratives becomes an analogue to the inexorably unraveling binary between physiology and subjectivity. That interplay offers constant reminders that we can see physiology—from the macro view of a whole brain to the micro views available through neuroimaging technologies—but we cannot see subjectivity. Nonetheless, an artist can represent it, just as developers and practitioners of brain scanning technologies hope they might be able to.

Neurocomic tells a representative story, not a particular one. Like Alice through her rabbit hole, the composite man falls through the book into what appears to be his own brain (or a composite one).

He meets a series of guides—famous figures from the history of neurology, including Santiago Ramón y Cajal, Charles Sherrington, and Eric Kandel—who lead him through the bewildering and often

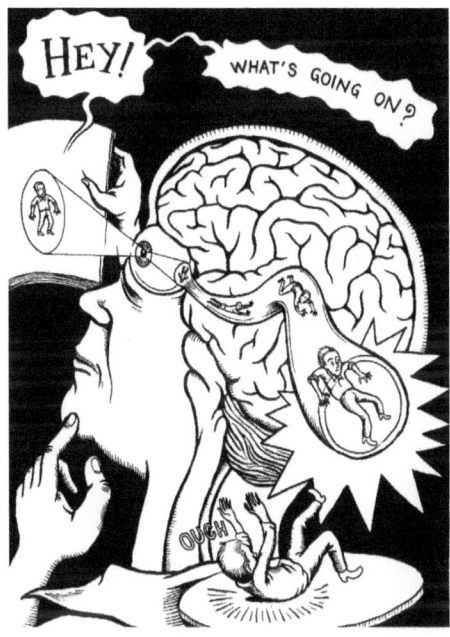

Fig. 3 Matteo Farinella and Hana Roš, *Neurocomic* (2014) © Matteo Farinella and Hana Roš. CC BY-NC-ND 4.0

terrifying 'forest' of his own brain and finally into the 'castle of our consciousness' (where he is reunited with the reading woman who initiated his ambivalent quest).[8] But does the generic man find his *self* through the journey? Not quite. Instead he discovers he is an object of representation, twice over. He is a character in a book, made of pen strokes, panels, shapes, and words; and he is an animated being, made of bones, flesh, cells, electricity, and proteins whose continuous inter-relations would seem to create him—though, as in most literature inspired by neuroscience, it is not clear how his identity emerges from these inter-relations. The making of self through the tools of artistic representation become a substitute for the more elusive making of self through physiology. In many ways, *Neurocomic* would appear to tell a simple—and even simplistic—story about a series of great men who made great discoveries in the history or brain research. But its frame-tale structure complicates that story through its emphasis on representation. The frame embeds a history of neuroscience within the emerging

---

8   Farinella and Roš, *Neurocomic*, p. 113.

tradition of literary experiments that entangle neurological questions with aesthetic experiments.

Neurocomic's opening scenario and central conceit gives fictional form to a fantasy Michael Phelps, one of the key developers of the brain scanning technology positron emission tomography (PET), described in an interview with anthropologist Joseph Dumit:

> Your body looks like it is a physical, anatomical substance, but inside there are all kinds of cells that are metabolizing things, or moving around and doing things, signaling to each other. We'd like to be able to watch this action. That is the objective. You know the activity is there, and you'd like to build a camera that can watch it. Well, one way to do that is first to say, "Well, if I was really little, I could go in there, move around, and watch those things. But since you cannot go in there, you can send a messenger. So you do that. So you take a molecule that will go and participate in that portion. [...] That is really what PET does. It reveals to us something that we know is going on in your body, but that we can't get to.[9]

With *Neurocomic*, Farinella and Roš fulfill Phelps' impossible wish, creating a protagonist who shrinks, to become 'really little', who can 'go in there, move around, and watch' the 'action' of his own brain. A human cannot shrink to enter a brain, but a character in a comic can. In that sense, the representational tools of a comic enable what is not possible in life—as might happen in a dream. When they give visual form to Phelps' fantasy scenario, Farinella and Roš emphasize a host of disparities between comics and brain scanning technologies—differences in aims, techniques, and cultural status. Their literary experiment involves play, irony, fantasy, and the breaking of boundaries, while scientific experiments involve observation, truth, and the boundaries of method. But Farinella and Roš's experiment also suggests some relations between literary and scientific experiment, which share the fundamental aims that motivate Phelps' fantasy: to get inside, to 'know something [...] we can't get to'. The impetus for these very different enterprises is rooted in personal suffering, the mysteries of physiology, and the making of knowledge. Finally—and perhaps most obvious, though little discussed—they both require human practitioners. A PET image and a frame of a graphic narrative are both designed by people whose judgments shape their results and their meanings. When

---

9   Dumit, *Picturing Personhood*, pp. 2–3.

Neurocomic's protagonist falls into the brain, readers see him tumbling into the explanatory gap.

The Tools of Comics

Literary criticism of graphic narratives has exploded with the genre's growing popularity and circulation in the last decade. Its most influential critics emphasize the possibilities for representing fluid identity and experience made possible by the distinctive features of the genre, starting with the interplay of text and image, but also including the creation of visual voice, frames that contain meaning, disrupted frames that loosen it, and the gutters between pages that guide and pace the reading experience. Rocco Versaci argues the interplay of text and image 'reminds us at every turn (or panel) that what we are experiencing is a representation'.[10] In their introduction to an influential 2006 issue of *Modern Fiction Studies*, Hilary Chute and Marianne DeKoven argue that graphic narrative 'calls a reader's attention visually and spatially to the act, process, and duration of interpretation' because 'it refuses a problematic transparency, through an explicit awareness of its own surfaces'.[11] The hybrid form of graphic narrative enables modes of representation that bypass linear narrative. As Chute and DeKoven note, 'the form's fundamental syntactical operation is the representation of time as space on the page'. Following on this idea, they make several additional claims about the genre: 1) its hybridity is 'a challenge to the structure of binary classification', 2) it is a 'mass cultural art', drawing on high and low art indexes, 3) it is multigeneric, composed, often ingeniously, from widely different genres and subgenres, and 4) its visual and verbal elements 'do not merely synthesize' but can tug at or tussle with each other to create meaning of unruly referents that cannot be tamed by logical or linear structures.[12] For all of these reasons—and because of a lineage of comics as a subversive, subcultural art form—graphic narratives tend to offer alternative or non-mainstream takes on the subjects they represent. The explanatory gap of neuroscience and

---

10 Rocco Versaci, *This Book Contains Graphic Language: Comics as Literature* (New York: Bloomsbury Academic, 2007), p. 6.

11 Hillary Chute and Marianne DeKoven, 'Introduction: Graphic Narrative', *Modern Fiction Studies*, 52.4 (2006), 767–82 (p. 767), https://doi.org/10.1353/mfs.2007.0002.

12 Ibid., p. 769.

the gutters of comics are a nice match in the pursuit of hypothetical knowledge about the physiology of selfhood.

In the theoretical neurosciences, the brain is routinely described as a representational organ. 'You are your synapses', writes Joseph LeDoux; the self is 'a dynamic collection of integrated neural processes, centered on the representation of the living body', writes Antonio Damasio.[13] LeDoux and Damasio disagree about quite a lot, but they agree that the brain works by representing the self—and the world—via patterns of cellular and intracellular interaction. Of course, neither of these leading neuroscientists can be sure about how the feeling of selfhood emerges from these patterns of representation. LeDoux believes the key to consciousness lies in the physiology of the cerebral cortex and Damasio believes it lies in the evolutionarily older upper brain stem. They agree that either way it will involve the multiple interactions of both these brain areas, along with a host of others. While the debates continue and research advances, the field of consciousness studies proceeds largely through two modes of investigating brain-self-world relations: the thought experiment and the brain scan. Both modes—perhaps we can call them genres—tend to obscure their representational tools. This distinguishes them from other thought experiments, about echolocating bats, color scientists locked in colorless rooms, or zombies, which are like fairytales too, but are used to wage philosophical debates and ultimately the positions they represent outshine the outlandish hypothetical stories they tell. Brain scanning technologies like PET, SPECT, and fMRI represent brain activity through a complex process that involves the collection of data about the flow of chemicals, oxygen, or blood in the brain, the algorithmic representation of that data in visual forms, and the interpretation of the images by trained human experts. As scholars in many disciplines have noted, the result is an image that appears—to the non-expert—to speak for itself; that is, to represent neural activity directly.

I am arguing that neurocomics visualize a particularly vivid version of an idea implicit in most brain memoirs and neuronovels: we need to find more effective means of communicating about how knowledge

---

13 Joseph LeDoux, *Anxious: Using the Brain to Understand and Treat Fear and Anxiety*, 1st ed. (New York: Viking, 2015), p. 324; Damasio, Antonio, *Self Comes to Mind* (New York: Pantheon, 2010) p. 8.

regarding our brains is produced. We need rhetorical techniques that account for the epistemological gaps in research, the dynamic interplay of systems proposed by the theories (including physiological, environmental, and social ones) *and* the representational tools we use to develop those theories. We need rhetorical techniques with as much appeal as thought experiments and brain scans. Graphic narratives make meaning by inviting readers into the representational process—which seldom happens in the cultural circulation of neuroscientific knowledge.

Fig. 4 David B., *L'Ascension du Haut Mal* (1999) © David B. and L'Association. All rights reserved. English translation: 'They perform gaseous encephalograms on him. They shoot gas into his brain to inflate it so they can take photos in which they hope to find traces of a lesion or tumor. When my parents tell me about it, I visualize my brother in the clutches of mad scientists'.

Images of physical brains in graphic narratives are indirectly related to the images produced by neuroimaging technologies, in the sense that the cultural pervasiveness of such images—particularly those created through fMRI—are our era's most common form of the brain image. The comics I am discussing here do not duplicate or represent brain scans, favoring images that recall comic book traditions of exposed brains or brains with agency who become characters (usually villains). But both *Neurocomic* and *Epileptic* do gesture toward a relationship between neuroimaging and the visualization of brains in comics. In *Epileptic* (which Hilary Chute calls 'the most famous of graphic illness narratives'), Jean-Christophe undergoes gaseous encephalography, an outdated and little used technology that replaces cerebrospinal fluid with gases in order to produce radiographic images.¹⁴ In David B.'s words, 'They shoot gas into his brain to inflate it so they can take photos in which they hope to find traces of a lesion or tumor. When my parents tell me about it, I visualize my brother in the clutches of mad scientists'. He is explicit here about the mediation of the image he presents for readers; it is based on third-hand knowledge. As Chute argues, *Epileptic* 'is a deeply stylized text, invested thoroughly in its own veracity but devoid of naturalism' which 'signals itself as an imaginative reconstruction of the past on every page'.¹⁵ The doctors explain the technology to his parents, they explain it to him, and he develops a terrifying fantasy about it, which he draws using stylistic features that appear throughout the memoir: the steampunk technology, the tiny doctors probing Jean-Christophe's outsized skull, the blank expression on his face, the snake-like tubes that recall the serpent of his epilepsy. If the doctors find a lesion or tumor, surgery might be able to help Jean-Christophe. But they don't. The brain scan becomes one more in a long series of attempts to find a solution to explain or eliminate Jean-Christophe's seizures, a string of epistemological failures. When David B. represents it, he is careful to cue readers to notice the mediation of the image he presents, mediation that reflects the epistemological failures of his family's quest to save his brother. The hope created by each possible cure or treatment—most of them based on false certainties—is destroying the family. As a writer and comics artist, he works in a medium that proliferates uncertainty. Even his identity is in question. David B. is a pseudonym. As a character in his memoir, the adult David B. and the child Pierre-Francois—his given name—are sometimes fused and sometimes distinct. If the memoir has a thesis, it is that he and his family need to find ways to live with such uncertainty.

By comparison, the stakes of *Neurocomic* are more academic, less personal, immediate, or visceral, but equally focused on foregrounding its representational resources—and those of the neuroscience it portrays. When Farinella and Roš explain the synapse, one of neuroscience's most basic concepts, they do it almost as a parody of conventional textbook illustrations.

---

14 Hillary Chute, 'Our Cancer Year; Janet and Me: An Illustrated Story of Love and Loss; Cancer Vixen: A True Story; Mom's Cancer; Blue Pills: A Positive Love Story; Epileptic; Black Hole (review)', *Literature and Medicine*, 26.2 (2008), 413–29 (p. 423), https://doi.org/10.1353/lm.0.0005.

15 Ibid., p. 423.

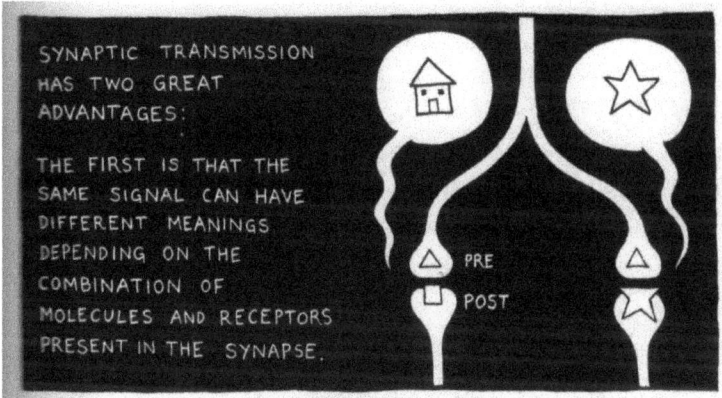

Fig. 5 Matteo Farinella and Hana Roš, *Neurocomic* (2014) © Matteo Farinella and Hana Roš. CC BY-NC-ND 4.0

Neurocomic does not make its meanings through text alone. Its meanings proliferate in their original, graphic context. They draw axons and dendrites like bones, the neurotransmitters as simple shapes (stars, squares, triangles), and the resulting signals in sperm-like thought bubbles. The representational resources of the comic—like Freud's dreams—almost demand the condensation of multiple meanings in a single image. A nerve cell is a bone, a neural signal a sperm cell. But such proliferation of meaning is not unique to comics or unheard of in neuroscience. The most conventional of textbooks describe a synapse as a gap or cleft, metaphorical language that involves similar condensation of meaning. Even the language Farinella and Roš use to explain the image closely resembles the language of textbooks: 'Synaptic transmission has two great advantages: The first is that the same signal can have different meanings depending on the combination of molecules and receptors present in the synapse'.[16] Synapses make meanings—multiple and mutable ones. Their representational resources are designed to create flexibility, to make meaning in fluid and unpredictable ways. This is a routine observation in basic neuroscience, but not one whose implications receive much attention. Whereas a conventional textbook elides the condensation of meaning in both the process of synaptic transmission and its own representation of that process, *Neurocomic* emphasizes both: a synapse is a gap, or cleft, between two cells; its

---

16 Farinella and Roš, *Neurocomic*, p. 39.

meaning is composed of parts, like the triangle and square that make a house when you combine them; the generation of meaning is like the generation of life, figured like sperm cells; the process is embodied and mobile, an idea hinted at by the joint-like depiction of the pre- and post-synaptic neurons; its meanings become functional in the context of a brain's electrical oscillations, or brainwaves, hinted at by the wavy tales of those sperm-like thought bubbles.

Because it emphasizes the representational tools of the comic as genre, *Neurocomic* suggests that synaptic gaps—or clefts—are like the gutter in a comic book or the distance between neural correlates and *qualia*, the subjective, first-person, and ineffable qualities of perception. Gaps like these are central to controversies about the power of brain imaging to explain the human mind. Graphic brain narratives and neuroimaging could not be more different in terms of their goals or the technologies and strategies involved in their representations of the brain. But they are both technologies for creating images of the brain, and both types of image make claims about understanding relations between brain physiology and selfhood.

The Tools of Neuroimaging

Brain imaging techniques—including fMRI (functional magnetic resonance imaging), PET (positron emission tomography), and SPECT (signal photon emission computed tomography)—are often described as though they offer direct images of brains at work, rather than images of brains created through a complex process of measurement, statistical analysis, and computer-aided representations. As neurobiologist Susan M. Fitzpatrick explains,

> the brain images displayed in scientific publications and in the popular media are not representations of changes in brain neuronal activity, or areas of 'activation', or the brain 'lighting up' or 'switching on'. Brain scans acquired with fMRI do not even graphically depict the magnitude of the BOLD [blood oxygen level dependent] signal. Rather, the images are computer-generated, color-coded 'maps' of statistically significant comparisons among data sets.[17]

---

17 S. M. Fitzpatrick, 'Functional Brain Imaging: Neuro-Turn or Wrong Turn?', in *The Neuroscientific Turn: Transdisciplinarity in the Age of the Brain*, ed. by Melissa M.

While scholars from multiple disciplines have made a clear-cut case that brain imaging does not provide direct access to brains, popular publications, neuro-self-help programs, and even published scientific papers promise—continuously and emphatically—that they do just that. As Dumit argues, the elision of technological complexity is bound up with assumptions about human behaviour and identity:

> Brain-imaging technologies like PET offer researchers the potential to ask a question about almost any aspect of human nature, human behavior, or human kinds and design an experiment to look for the answer in the brain. Each piece of experimental design, data generation, and data analysis, however, necessarily builds in assumptions about human nature, about how the brain works, and how person and brain are related. No researcher denies this. In fact, they constantly discuss assumptions as obstacles to be overcome as trade-offs between specificity and generalization.[18]

That trade-off between specificity and generalization is both rhetorical and methodological. Too much specialization means a smaller audience, but it also leads to more circumscribed conclusions. The potential Dumit describes is exciting, and it makes perfect sense that researchers and practitioners are interested in making the most of it. As he observes, they are well aware of the complexities involved. If the general public is not aware, it is because so much of the rhetoric about brain imaging involves misleading translations of specific research designs for the sake of emphasizing the dramatic potential of the technology.

Concrete examples of the oversimplifications Fitzpatrick laments are plentiful, and they come in a variety of forms. Many of these are well-intended translations of medical jargon designed to provide readers with accessible shorthand, though it is difficult to dismiss the dramatic effects created by the shorthand. For example, in her biography of famous neurology patient H. M., Suzanne Corkin makes a dramatic claim: 'Using MRI scans, we could look through Henry's scalp and skull to see his brain'.[19] In her biography, Corkin is aiming to create a feeling of intimacy with her subject. Looking through his scalp to see his brain, as

---

Littlefield and Jenell M. Johnson (Ann Arbor: University of Michigan Press, 2012), pp. 180–98 (p. 186), https://doi.org/10.3998/mpub.4585194.

18 Dumit, *Picturing Personhood*, p. 16.
19 Suzanne Corkin, *Permanent Present Tense: The Unforgettable Life of the Amnesiac Patient, H. M.* (New York: Basic Books, 2013), p. 80.

a character in a comic book might be able to, adds a physical dimension to that intimacy. In general, the apparent motives of such translations are more neutral, as with Cornell University's website advertising its MRI facilities:

> Neuroscientist Valerie Reyna compares functional MRI—an imaging technique that allows researchers to see the brain in action—to the microscopes and telescopes that allow scientists to peer into cells and the cosmos to explore the mysteries of life. For the first time on Cornell's Ithaca campus, she and fellow researchers can observe how the brain fires when we think and react and compare how such activity differs among age groups and populations. Such work promises to bring into focus what was once out of sight—the hidden factors that drive human behavior.[20]

Again, the writer eschews a detailed description of the technology's representational resources, describing instead a fantastical version that resembles a comic book scenario: researchers 'see the brain in action', revealing 'hidden factors that drive human behavior'. Those hidden factors—what we don't know about ourselves—appear fairly routinely in writing that makes promises about the powers of brain imaging technologies. In his book *Affective Neuroscience*, Jaak Panksepp offers a more accurate description that nonetheless is likely to be read in the tradition of rhetorical oversimplification: 'During the past decade, remarkable progress has been made in our ability to visualize what is going on inside the living human brain'.[21] The word *visualize*—as opposed to *see*—presupposes an acknowledgement of the complex representational resources entailed in brain imaging. But it is a subtle presupposition, one readers accustomed to broader and more dramatic claims are likely to miss without more explicit rhetorical cues and detailed explanations of those resources.

In a case like Panksepp's description and Cornell's account of Reyna's research, the stakes and motives of this rhetoric are relatively benign, but when they migrate from the laboratory into other spheres, they can become more troubling. For example, Dan Ariely and Gregory S. Berns published a review article entitled 'Neuromarketing: The Hope

---

20 Karene Booker, 'A Window into the Brain', *Human Ecology*, 41.2 (Fall 2013), 5–7.
21 Jaak Panksepp, *Affective Neuroscience: The Foundations of Human and Animal Emotions* (New York: Oxford University Press, 1998), p. 90.

and Hype of the Neuroimaging Business' in *Nature*'s 'Science and Society' section. As Dumit observes, the writers of these publications are usually well aware of the complexity their shorthand masks, but their shorthand has gained remarkable cultural purchase, obscuring rather the remarkable representational resources of the technologies they use to make images of the brain. In the abstract, they write, 'Although neuroimaging is unlikely to be cheaper than other tools in the near future, there is growing evidence that it may provide hidden information about the consumer experience'.[22] The title of the article indicates an agnostic stance about the potential of neuromarketing. Indeed, the authors offer detailed consideration of both its ethical and methodological pitfalls. The ethical questions they cite include the violation of 'the privacy of thought', the exploitation of 'particular neurological traits' or 'biological weakness', and the unconscious or 'peripheral' manipulation of consumers. Methodological considerations include the fact that images of 'brain activation' are not meaningful unless they are correlated with 'another behavioural measurement', that large sample sizes are necessary, that measuring responses to complex stimulus (like an ad) is not possible with current technology, and the fact that motion and time affect behavioral responses correlated with images of brain activity.[23] These lists of ethical problems and methodological obstacles are daunting. To compound matters, the authors acknowledge how little research supports the efficacy of neuromarketing as well as the considerable cost of neuroimaging versus traditional market testing. Nonetheless, the authors conclude on what they describe as an optimistic note whose implications are troubling, to say the least. Neuromarketing, they suggest, might become cheaper than current marketing methods; it 'could provide hidden information about products'; and it might 'contribute to the interface between people and businesses and in doing so foster a more human-compatible design of the products around us'.[24] Between the abstract and conclusion, the 'hidden' information described moves from the human consumers to the products they might consume. While the authors don't make the connection explicit, the move is

---

22 Dan Ariely and Gregory S. Berns, 'Neuromarketing: The Hope and Hype of Neuroimaging in Business', *Nature Reviews Neuroscience*, 11.4 (April 2010), 284–92 (p. 284), https://doi.org/10.1038/nrn2795.
23 Ibid., pp. 289–90.
24 Ibid.

dependent on an assumption that Antonio Damasio is right when he proposes that organisms and objects shape each other in the making of conscious experience.²⁵ Ariely and Berns envision a utopian future in which neuroimaging may benefit consumers by leading to the creation of products they don't realize they want, because consciousness masks their unconscious wishes. They dramatize a bizarrely Freudian capitalist fantasy that might make for a good storyline in a graphic narrative. As cultural analysis, it reveals more about the cultural neuromania involved in the circulation of ideas about brain imaging technology than it does about marketing or business.

While Ariely and Berns touch on many of the critiques of the hype around neuroimaging, they ultimately downplay them. Dumit's *Picturing Personhood* is an ethnographic study of experimental research using and cultural responses to PET scans; Fitzpatrick's 'Functional Brain Imaging: Neuro-Turn or Wrong Turn?' offers a detailed explanation of the methodologies involved in producing PET and BOLD fMRI scans, with an emphasis on 'what neuroimaging can and cannot reveal about the mind'. Hayles' 'Brain Imaging and the Epistemology of Vision: Daniel Suarez's *Daemon* and *Freedom*' offers a case study in the popular circulation of ideas in response to the ubiquity of neuroimaging. Johnson's '"How Do You Know Unless You Look': Brain Imaging, Biopower, and Popular Neuroscience' examines the representation of SPECT (single photon emission computed tomography) scans 'presented as visual evidence that is highly legible even to an untrained audience' in the neuro-self-help books by Daniel Amen. McCabe and Castel's 'Seeing Is Believing: The Effect of Brain Images on Scientific Reasoning' reports on empirical research documenting the 'persuasive power' of brain images among non-expert readers of fabricated news articles on various topics in cognitive psychology.²⁶

---

25 Antonio Damasio, *The Feeling of What Happens: Body and Emotion in the Making of Consciousness* (New York: Mariner Books, 2000).

26 Morana Alač and Edwin Hutchins, 'I See What You Are Saying: Action as Cognition in fMRI Brain Mapping Practice', *Journal of Cognition and Culture*, 4.3 ([n.d.]), 629–61 (p. 629), https://doi.org/10.1163/1568537042484977; Dumit, *Picturing Personhood*; Fitzpatrick, 'Functional Brain Imaging', p. 180; Katherine Hayles, 'Brain Imaging and the Epistemology of Vision: Daniel Suarez's *Daemon* and *Freedom*', *MFS Modern Fiction Studies*, 61.2 (2015), 320–34, https://doi.org/10.1353/mfs.2015.0025; Davi Johnson, '"How Do You Know Unless You Look?": Brain Imaging, Biopower, and Practical Neuroscience', *Journal of Medical*

The popular circulation of medical brain imaging tends to occlude its representational complexities. As Johnson observes, 'The complex averaging procedures and statistical work that go into producing such images are lost in the neat, simple-looking images presented for the readers' consumption and interpretation'.[27] In other words, the image of a brain scan is a fabrication of a complex process that requires enormous expertise both to create and to interpret, but the vivid and colorful results seem to present transparent meaning to non-experts—a fact exacerbated by a tendency in science journalism, popular neurological texts, and even textbooks to bypass the technical details involved in their production. The meaning of brain scans suggests serious implications, both medically and philosophically. As Dumit observes, 'These brain images make claims on us because they portray *kinds* of brains. As people with obviously, one or another kind of brain, we are placed among the categories that the set of images offers. To which category do I belong? What brain type do I have? Or more nervously: Am I normal? Addressing such claims requires an ability to critically analyze how these brain images come to be taken as facts about the world'.[28] Brain scans portray serious knowledge whose representational complexities are often occluded. Ironically, graphic brain narratives tend to do just the opposite, offering playful alternatives that visualize the brain with a great deal of emphasis on their own representational strategies.

N. Katherine Hayles articulates something like a critical consensus when it comes to the lack of attention to the representational tools of brain scanning technologies in popular—and many specialized—accounts of its results,

> The point is that interpretations of brain scans require careful consideration of the experimental design, knowledge of previous research linking behavior and regional brain activity, accuracy of the statistical analysis, and so forth. While the images themselves may appear seductively transparent, non-experts and even research professionals

---

Humanities, 29.3 (2008), 147–61 (p. 151), https://doi.org/10.1007/s10912-008-9062-4; David P. McCabe and Alan D. Castel, 'Seeing Is Believing: The Effect of Brain Images on Judgments of Scientific Reasoning', *Cognition*, 107.1 (2008), 343–52 (p. 343), https://doi.org/10.1016/j.cognition.2007.07.017.

27 Johnson, '"How Do You Know Unless You Look?"', p. 153.
28 Dumit, *Picturing Personhood*, p. 5.

who have not read the original article should be very cautious about deciding what the images actually show.[29]

By contrast, graphic narratives make their tools—ink strokes, interplay of words and images, frames, gutters—integral to the experience of reading them.

The authors of graphic novels render psychological experience in physical forms whose narratives are sutured with words. The juxtaposition of words and images reminds readers that representation is never transparent. Words and images translate or distort experience. When the narrative in question focuses on neurological experience, this emphasis on the representational resources of the artists becomes a vehicle for the elusiveness of the explanatory gap between physiology and subjectivity. That same explanatory gap is at play in neuroimaging, but too often it is bypassed when the meaning of brain scans appears—or is presented or received as—transparent. It is my contention that learning to read graphic brain narratives can be helpful in demystifying the representational qualities of neuroimaging, and that understanding the techniques and methods through which brain scans are created and interpreted can deepen a reader's understanding of graphic brain narratives.

A Person Surrounds This Brain

As a thought experiment, examine the image from *Neurocomic* on the facing page. Imagine the central figure isolated in negative space—minus the bird, sun, flower, and thistle or the sensory words that accompany them. You would see a human organism, with a schematic version of its brain and nervous system made visible (presumably through medical technologies). At best, the figure would appear clinical, at worst, monstrous; in either case, it would feel uncanny. The text at the top of the page is narration, offered by one of the protagonist's first guides, Santiago Ramón y Cajal, the 1906 Nobel Laureate famous for his detailed drawings of neurons emphasizing their treelike structure and proponent of the once controversial idea that brain matter is composed of distinct (rather than fused) cells we now call neurons. Ramón y Cajal

---

29 Hayles, 'Brain Imaging and the Epistemology of Vision', p. 322.

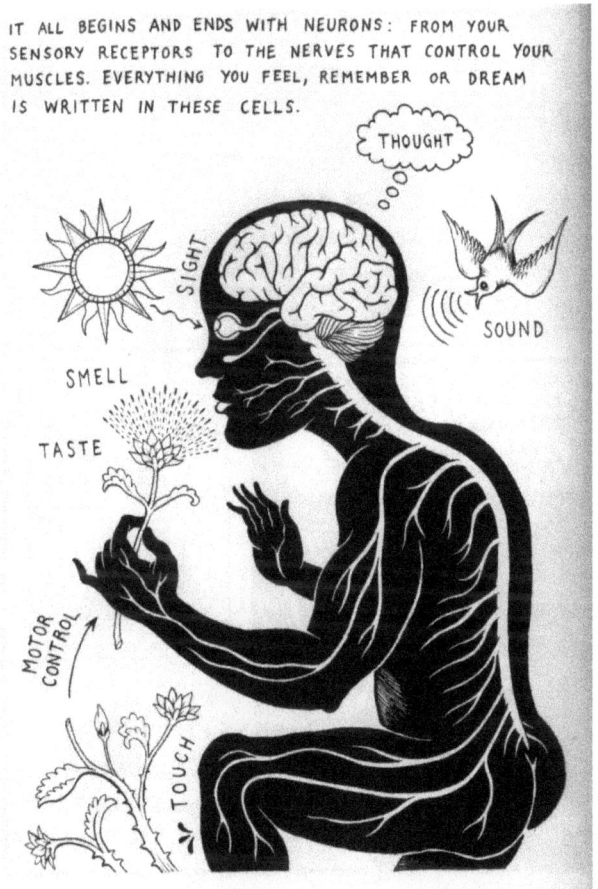

Fig. 6 Matteo Farinella and Hana Roš, *Neurocomic* (2014) © Matteo Farinella and Hana Roš. CC BY-NC-ND 4.0

explains to the tiny protagonist, 'It all begins and ends with neurons: From our sensory receptors to the nerves that control your muscles. Everything you remember, dream, or feel is written in those cells'.[30]

Like all thought experiments, the exercise I just asked you to consider replaces real world complexity with a hypothetical scenario. The image on the page represents a human being in the fullness of experience—thinking, feeling, sensing. The image emphasizes the idea that this human is an organism, stripping away the barriers of flesh,

---

30   Farinella and Roš, *Neurocomic*, p. 20.

bone, and hair to reveal the organs that enable life. Other elements in the image create a montage—a very different kind of representation—that seems to contradict Ramón y Cajal's exposition. As Scott McCloud explains in *Understanding Comics*, a montage creates an image 'where words are treated as integral parts of the picture'.[31] Farinella and Roš create tension between text and image through what McCloud calls the 'interdependent' combination, 'where words and pictures go hand in hand to convey and idea that neither could convey alone'.[32] The image adds motion, context, and feeling to the text. Text and image create tension: the words *all* and *everything*—standing in here for the fullness of experience—are misleading. A thorny thistle provides the content of touch and gives it meaning that is only possible through interconnections with other senses, with feelings, with memories: *don't touch this plant*. The same is true for the sight of the sun, the sound of a bird, or the smell of a flower. The montage of words and objects surrounding the figure gives visual form to the experience correlated with those neurons. The resulting meaning is akin to Damasio's argument that the objects of perception are integral to the making of consciousness or feeling.

Of course, tensions between text and image are central to all graphic narratives. I argue that authors of neurocomics adapt these tensions for a particular purpose, making them stand-ins for unresolved debates about the relationship between neurology and experience. The contradictions and competing ideas that proliferate from the explanatory gap between physiology and feeling make room for stories. And comics, with their fluid mixing of fantasy and realism, are well suited to exploring the contradictions.

David B. creates images that emphasize the intimate proximity and distant epistemology of the brain's relationship to the self and mind. In numerous images, David B. depicts Jean-Christophe's brain as an object probed by doctors, healers, and philosophers—all struggling to explain connections between brain physiology and the feeling of selfhood. In the words of neuroscientist Jaak Panksepp, 'All objective bodily measures [of 'interior experiences'], from facial expressions to autonomic changes, are only vague approximations of the underlying

---

31 Scott McCloud, *Understanding Comics: The Invisible Art* (New York: William Morrow Paperbacks, 1994), p. 154.
32 Ibid., p. 155.

Fig. 7 David B., *L'Ascension du Haut Mal* (1999) © David B. and L'Association. All rights reserved. English translation: 'The doctor who's treating him is stymied by my brother's epilepsy. He prescribes a new experimental therapy'.

Fig. 8 David B., *L'Ascension du Haut Mal* (1999) © David B. and L'Association. All rights reserved. English translation: 'In her mind, this sends us all back to square one. She has a vision of her son in the hospital with his head shaved. It is as if she is being pulled backwards. She reminds herself that Master N. is no longer there'.

168 Life, Re-Scaled

Fig. 9 David B., *L'Ascension du Haut Mal* (1999) © David B. and L'Association. All rights reserved. English translation: 'Unbeknownst to me, this flood of absurdities takes root in my brain. Images are born'.

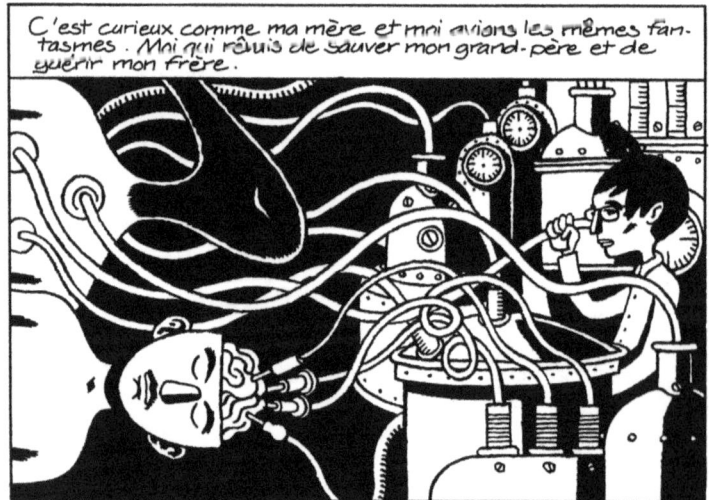

Fig. 10 David B., *L'Ascension du Haut Mal* (1999) © David B. and L'Association. All rights reserved. English translation: 'It is odd how my mother and I had the same dreams. I'd been dreaming of saving my grandfather and my brother'.

neural dynamics—like ghostly tracks in the bubble chamber detectors in particle physics'.[33] In the words of cultural critic Ann Cvetkovich,

> I tend to use affect in a generic sense, [...] as a category that encompasses affect, emotion, and feeling, and that includes impulses, desires, and feelings that get historically constructed in a range of ways. I also like to use *feeling* as a generic term that does some of the same work: namely the undifferentiated 'stuff' of feeling; spanning the distinctions between emotion and affect central to some theories; acknowledging the somatic or sensory nature of feelings as experiences that aren't just cognitive concepts or constructions.[34]

It is not surprising that the neuroscientist emphasizes the 'underlying neural dynamics' of affect and the cultural critic its historical construction. What they share—with each other and with the authors of neurocomics—is an emphasis on the elusive or ineffable quality of feelings, their subtle but immense range of expression, and the confusion they tend to create. In other words, they portray affect as a form of what McCloud calls one of graphic narrative's specialties: the interplay of 'the seen and the unseen', or the felt and the unfelt.

With his visual depictions of brains, David B. exploits the comic form's ability to mix fantastical and realist representation. Comics give form to the impossible. One key image from *Epileptic* visualizes David's fantasy that a neuroscientist could meld his brain with his brother's— one of dozens of images of physical brains David B. uses, ironically, to portray what cannot be seen or understood about his brother's illness. A related image, in which two birdlike doctors climb ladders to peer into Jean-Christophe's exposed brain, demonstrates his ironizing technique. The doctors' semi-human form casts them as fantastical hybrid creatures, belonging more to the representational world of comics (or dreams) than to medicine. The ladders give physical form to the epistemological distance between them and a cure for Jean-Christophe. The exposed brain is a reminder that in comics, you can see just about anything. In life, seeing Jean-Christophe's brain would require invasive techniques. David B. alternates images of brains with images of Jean-Christophe's skull as it is being subjected to a variety of such invasive

---

33   Panksepp, *Affective Neuroscience*, p. 9.
34   Ann Cvetkovich, *Depression: A Public Feeling* (Durham: Duke University Press Books, 2012), p. 4, https://doi.org/10.1215/9780822391852.

Fig. 11 David B., *L'Ascension du Haut Mal* (1999) © David B. and L'Association. All rights reserved. English translation: Middle Row, Left Panel: 'It'd be wonderful to let myself go'. Middle Panel: 'I could pretend to be an epileptic. I could imitate a seizure. I know how'. Right Panel: 'Anyway, I am an epileptic. These electrical discharges in my brain, like explosions, that's what they are! They are epileptic seizures!'

Fig. 12 David B., *L'Ascension du Haut Mal* (1999) © David B. and L'Association. All rights reserved. English translation Left Panel: 'I want to spill all the blood in my body'. Middle Panel: 'It would all come out at last. The anxiety, the fear, the justice, the rage'. Right Panel: 'Then I could sleep to my heart's content'.

Fig. 13 David B., *L'Ascension du Haut Mal* (1999) © David B. and L'Association. All rights reserved. English translation: Left Panel: 'It is just another way of telling stories. You cannot help yourself'. Middle Panel: 'It is a way of conjuring unhappiness. It is magic'. Right Panel: 'I've read many stories that have helped me. I want to touch people with my books in return'.

Fig. 14 David B., *L'Ascension du Haut Mal* (1999) © David B. and L'Association. All rights reserved. English translation: Banner: 'Come visit the inside of David B.'s head at the end of the 70s'.

techniques—generally figured as retro-futuristic canisters and tubes. In a typical example, Jean-Christophe's mother steadies herself atop her son's skull, while his doctors look on from a distance, poised on the head and tail of the serpent that represents his epilepsy throughout the book. Images like this collapse, condense, and distort time and space, a common technique in comics. The invasive technology belongs to a brutal history of medical experiments and to a future imagined by Jean-Christophe's doctors, one that involves the successful applications of their theoretical cures. The patient's skull, and therefore his brain, is outsized, larger than most of the other human bodies in the frame. Of course, this represents the size (or severity) of the problem, but it also represents that same epistemological distance between theoretical cures and successful applications.

A comic cannot claim to cure disease or resolve centuries of debate about the relation between brain and self. Instead, *Epileptic* offers an alternative to resolution: increasing attention to the author's identity as a writer and artist in neurological terms. David B. describes his writing as a series of 'electrical discharges' in his brain, 'like explosions' or 'tiny epileptic seizures'—through a series of frames that blur his identity with Jean-Christophe's. As the scene unfolds, he portrays a fantasy of severing his own head with macabre irony, likening himself visually to Hamlet holding Yorick's skull. He imagines he could bleed feelings: 'It would all come out at last: the anxiety, the fear, the justice, the rage'. But he revises the fantasy—and reattaches his head—within a few frames. 'Come on, admit it, you don't want to be sick [...] It is just another way of telling stories. [...] I've read many stories that have helped me. I want to touch people with my books in return'.[35] The fantasy is a personal response to the explanatory gap. Like so many theoretical accounts, it imagines that physiology and feeling are identical, and like these accounts, it undoes itself. But in this case, the undoing is an intentional bid to call attention to the tools of representation. In place of his morbid fantasy, David B. offers a revision on the fantasy of finding the immaterial in the material—that his art might *touch* other people, that the materials of his books will affect people physically via their immaterial responses to it.

---

35 B., *Epileptic*, pp. 289–90.

David B. emphasizes the materiality of reading, writing, and drawing throughout *Epileptic*. He uses facing pages to represent the complexity of his identity, or the person who surrounds his own brain (and his brother's). In a ribbon-like frame spanning the tops of both pages, he inscribes an invitation: 'Come visit the inside of David B.'s head at the end of the 70s'—words that flow backwards from a bullhorn held by a circuslike figure. This time, the inside of David B.'s head is not a brain, but a chaotic collection of fragments from stories he wrote and drew during the period. Stories that represent a decade's work, a lot of geographic wandering, and a rapidly evolving sense of identity collapse onto a single page—images from comics he has created and read, images of his brother's doctors condensed with images of the 'madmen' who populate both brothers' imaginations, and images of Jean-Christophe both healthy and sick. He represents himself directly, as writer and artist, in two bubbles, one on each page. He sits at a desk, with the tools of his medium: paper, pens, bright light. The speech bubble in the first image reads: 'There's a feverish, confused quality to these stories'. In the second: 'A pathetic bulwark, and yet it does shield me'.[36] Creating comics becomes a means of reconciling complex and apparently contradictory aspects of identity—and the flexible fluency of the form is key to ensuring that the reconciliation does not require tidy integration, that it can encompass the lumpiness of experience.

Neurocomic is a more explicitly pedagogical text than *Epileptic*, using the form of a graphic narrative to offer an accessible introduction to brain physiology and the history of neurology. But it is also a hybrid of fiction and nonfiction—and its fictional frame is by no means incidental. Farinella and Roš might have created a straightforward illustrated history of neurology, but instead they wrap it in a fictional fantasy about a shrinking man who wanders through a metaphorical forest, his own brain. Near the beginning of the story, a hypothetical human, brain exposed, examines the book page on which the protagonist is trapped. That hypothetical human is figured as the protagonist himself, in the role of reader, but also as the reader of this text. By analogy, the protagonist and reader are condensed into this hypothetical human. Near the end of the story, the woman courted by the protagonist explains: 'Our existence relies on the brain of the reader, which is able to see motion and hear

---

36   B., *Epileptic*, pp. 278–79.

Fig. 15 Matteo Farinella and Hana Roš, *Neurocomic* (2014) © Matteo Farinella and Hana Roš. CC BY-NC-ND 4.0

sounds [...] on a flat sheet of paper'.[37] This text begins on a panel featuring an unidentified character—'the reader' who is somewhat oddly, but perhaps tellingly, referred to in the second panel through the pronoun 'which' —holding a copy of *Understanding Comics*. It is a meta moment that multiplies. Readers are asked to imagine their own brains imagining these characters' brains—and to generalize the lesson to all readers, all brains. The result is a kind of recasting of the message attributed to Ramón y Cajal near the book's beginning. The 'it' in 'it all begins and ends with neurons' becomes the reader, trapped in an

---

37   Farinella and Roš, *Neurocomic*, p. 132.

epistemological loop. This hypothetical reader can only learn about its brain by using that brain, and it cannot quite know if its flesh and blood are real or products of its own ability to 'see motion and hear sounds on a flat sheet of paper'. In other words, human brains make reality, and consciousness of that reality is subjective by definition. Meaning, as a result, is always contingent, and in the case of graphic narrative, that contingency flows from the continuous interplay of text and image. Like these characters, a neural pattern is a representation of the world—one composed of neurotransmitters and electricity, whose meanings are further shaped by variables like location, the rhythms of brain waves, and the support of glial cells. Any psychology textbook will tell you that a perception is a construction—or a functional distortion—of the objects it represents.

In that sense, graphic narratives are corollaries to academic critiques of the oversimplification of brain scanning technologies. As Fitzpatrick writes, 'brain-imaging scans are highly technical and difficult to interpret without expert knowledge of the subjects participating in the studies, the tasks performed, the techniques used to acquire the data, and the complicated statistical tools used to analyze the data and create the images'.[38] I am not suggesting that brain scanning technologies and graphic narratives are equivalent. This is decidedly not the case. Meta-representational techniques are integral to the representational tools of graphic narratives, part of the reading experience. The dissection and digital reconsolidation of brains, the measurement of their electrical patterns, or the imaging of their blood flow are powerful tools for gaining knowledge of their functions. The expert knowledge necessary to make and interpret the images created through brain scanning technologies involves a great deal of meta-representation, but their cultural circulation mostly obscures this fact. Clinicians, subjects, laboratories, machines, and algorithms disappear behind appealing images.

Despite obvious and vast differences, these technologies share one significant quality with neurocomics. They are representations of brains, built not found. Simply put: the images produced by brain scans will continue to circulate as 'neurojunk' unless the people doing the circulating—including journalists, marketers, clinicians, researchers, and artists—find the rhetorical means to situate them in the representational

---

38   Fitzpatrick, 'Functional Brain Imaging', p. 194.

frameworks that make them meaningful. The sentiment in the epigraph to this chapter, from Joseph Dumit's *Picturing Personhood*, extends from the PET images he examines to brain scanning technologies in general: 'Overturning the age-old axiom that a picture is worth a thousand words, perhaps these PET images require millions of words to be understood'. Experts in neuroscience are attuned to the complexity of the materials they use to make images of brains—and to the fact that these images don't so much represent personhood as an incomplete and highly mediated set of pictures of physiology. Nonetheless, they often translate the complexities of the technology into rhetoric that makes the images in question appear to be transparent images of brains. Jaak Panksepp's description of these images as visualizations is an example of a more accurate description, but it doesn't go far enough, elaborating on the work involved in creating these visualizations.

By definition, expert knowledge belongs to specialists, but the stakes of neuroimaging belong to anybody with a brain—and that is as good a reason as I can imagine to work hard to develop a set of explanatory and rhetorical techniques that can describe the meaning of brain scans to a larger public. But neurocomics, like so many literary responses to neuroscience, demonstrate an imbalanced relationship among the arts and sciences. The writers of brain memoirs and neuronovels—including the graphic varieties of both—are highly conscious of the personal, social, and philosophical stakes of representing and circulating expert knowledge. Individual experts in the neurosciences share the awareness, but collectively, as a set of disciplines, they aren't designed to respond to concepts or tools emerging from the arts and humanities. Literary writers and critics make a vocation of working with the intricacies of representations of all varieties of human experience and knowledge. Graphic narratives demonstrate one of literature's many contributions to contemporary understandings of the brain: their emphasis on meta-representation. To imagine collective, multidisciplinary collaboration among scientists and humanists interested in the meaning of brain images—or the relationship between brain and self more generally— remains an exercise in speculative fiction. Nonetheless, graphic brain narratives offer an implicit, but concrete, suggestion to those involved in the circulation of brain scanning images. Simply to include the word *representation* in descriptions of these images would help to clarify their meanings—in no small part because the word would require some follow-up explanation, in accessible prose, of what is entailed in the representation of a brain—and, ideally, the person surrounding that brain.

Works Cited

Alač, Morana and Edwin Hutchins, 'I See What You Are Saying: Action as Cognition in fMRI Brain Mapping Practice', *Journal of Cognition and Culture*, 4.3, [n.d.], 629–61, https://doi.org/10.1163/1568537042484977

Ariely, Dan, and Gregory S. Berns, 'Neuromarketing: The Hope and Hype of Neuroimaging in Business', *Nature Reviews Neuroscience*, 11.4 (April 2010), 284–92, https://doi.org/10.1038/nrn2795

B., David, *Epileptic* (New York: Pantheon, 2006).

B., David, *L'Ascension du Haut Mal* (Paris: L'Association, 2011).

Booker, Karene, 'A Window into the Brain', *Human Ecology*, 41.2 (Fall 2013), 5–7.

Bunge, S. A. and I. Kahn, 'Cognition: An Overview of Neuroimaging Techniques', in *Encyclopedia of Neuroscience*, ed. by Larry R. Squire (Cambridge, MA: Academic Press, 2009), pp. 1063–67, https://doi.org/10.1016/b978-008045046-9.00298-9

Chute, Hillary, '*Our Cancer Year; Janet and Me: An Illustrated Story of Love and Loss; Cancer Vixen: A True Story; Mom's Cancer; Blue Pills: A Positive Love Story; Epileptic; Black Hole* (review)', *Literature and Medicine*, 26.2 (2008), 413–29, https://doi.org/10.1353/lm.0.0005

Chute, Hillary and Marianne DeKoven, 'Introduction: Graphic Narrative', *MFS Modern Fiction Studies*, 52.4 (2006), 767–82, https://doi.org/10.1353/mfs.2007.0002

Corkin, Suzanne, *Permanent Present Tense: The Unforgettable Life of the Amnesiac Patient, H. M.* (New York: Basic Books, 2013).

Cvetkovich, Ann, *Depression: A Public Feeling* (Durham: Duke University Press Books, 2012), https://doi.org/10.1515/9780822391852

Damasio, Anthony, *Self Comes to Mind: Constructing the Conscious Brain* (New York: Pantheon, 2010).

Damasio, Anthony, *The Feeling of What Happens: Body and Emotion in the Making of Consciousness* (New York: Mariner Books, 2000).

Dumit, Joseph, *Picturing Personhood: Brain Scans and Biomedical Identity* (Princeton: Princeton University Press, 2004), https://doi.org/10.1515/9780691236629

Farinella, Matteo and Hana Roš, *Neurocomic* (London: Nobrow Press, 2014).

Fitzpatrick, S. M., 'Functional Brain Imaging: Neuro-Turn or Wrong Turn?', in *The Neuroscientific Turn: Transdisciplinarity in the Age of the Brain*, ed. by Melissa M. Littlefield and Jenell M. Johnson (Ann Arbor: University of Michigan Press, 2012), pp. 180–98, https://doi.org/10.3998/mpub.4585194

Hayles, N. Katherine, 'Brain Imaging and the Epistemology of Vision: Daniel Suarez's *Daemon* and *Freedom*', *Modern Fiction Studies*, 61.2 (2015), 320–34, https://doi.org/10.1353/mfs.2015.0025

Johnson, Davi, '"How Do You Know Unless You Look?": Brain Imaging, Biopower, and Practical Neuroscience', *Journal of Medical Humanities*, 29.3 (2008), 147–61, https://doi.org/10.1007/s10912-008-9062-4

LeDoux, Joseph, *Anxious: Using the Brain to Understand and Treat Fear and Anxiety*, 1st ed. (New York: Viking, 2015).

Levine, Joseph, 'Materialism and Qualia: The Explanatory Gap', *Pacific Philosophical Quarterly*, 64 (October 1983), 354–61, https://doi.org/10.1111/j.1468-0114.1983.tb00207.x

Malabou, Catherine, *What Should We Do with Our Brain?*, trans. by Sebastian Rand (New York: Fordham University Press, 2008).

McCabe, David P. and Alan D. Castel, 'Seeing Is Believing: The Effect of Brain Images on Judgments of Scientific Reasoning', *Cognition*, 107.1 (2008), 343–52, https://doi.org/10.1016/j.cognition.2007.07.017

McCloud, Scott, *Understanding Comics: The Invisible Art* (New York: William Morrow Paperbacks, 1994).

Panksepp, Jaak, *Affective Neuroscience: The Foundations of Human and Animal Emotions* (New York: Oxford University Press, 1998).

Tougaw, Jason, *The Elusive Brain: Literary Experiments in the Age of Neuroscience* (New Haven: Yale University Press, 2018), https://doi.org/10.12987/yale/9780300221176.001.0001

Versaci, Rocco, *This Book Contains Graphic Language: Comics as Literature* (New York: Bloomsbury Academic, 2007).

# III. PANDEMIC IMAGINARIES

7. The Fiction of the Empty Pandemic City

Race and Diaspora in Ling Ma's *Severance*

Rishi Goyal

> 'It takes hard work *not* to see'.
> —Toni Morrison, *Playing in the Dark*

During a zoom conversation in May of 2020, my younger brother, a conservationist living in Kathmandu, mentioned how clean and sweet the air seemed. During that year, we were flooded with colour coded satellite images of improved air quality and lower levels of pollution, from the megacities of the global South to the Northern highlands, as people fled infected cities and the world economy came to a screeching halt. As humans supposedly abandoned Rome, London, Mumbai and Tel Aviv, reports suggested that wildlife were repopulating urban spaces and that countless animals and vegetation were rewilding the world's cities. But many of these stories, both about the cities emptying of people and about animals and plant life resurging, are fictions. The photos showing the return of animals or of emptied cities have often been revealed as fakes: doctored images or historically unrelated. And while certain people definitely fled the city, a new twenty-first-century white flight, many more people could not. In New York City, the quiet of the streets was punctuated by both an increase in bird song *and* an increase in ambulance sirens.

Fig. 16 Sourav Chatterjee, *49th St. between 7th and 8th Ave* (2020) © Sourav Chatterjee. Editorial use.

Fig. 17 Reena Rupani, *Times Square* (2020) © Reena Rupani. CC BY-NC.

Ambulance call volumes nearly doubled from late March to early April. But they only increased in lower socioeconomic status (SES) areas which are predominantly populated by black and brown people, while the call volumes decreased in higher income areas where many residents fled the city.[1] The emptying of the city is a pandemic fiction that masks the uneven class and race effects of biological catastrophes.[2] The 'empty city', I suggest, focuses our attention on the racialized nature of biopolitical existence while also magnifying the uneven effects of restrictions or regulations on movement: immigration bans, cordon sanitaire, quarantine, shelter-in-place orders.

As an Emergency Medicine physician, I treated hundreds of patients with severe pneumonia from COVID-19 when New York City was the epicentre of the pandemic. Many, if not most of them, died. The overwhelming majority of those that were sick and those that died were black and brown. The neighbourhood that I live in, Harlem, and the neighbourhood that I work in, Washington Heights, were full of people, people who often got COVID-19. While other neighbourhoods, SoHo, the West Village, the Upper East Side, felt like ghost towns as their primarily white and/or wealthy residents fled the city. The spatial logic of COVID-19 was overwhelming even as it revealed class and race differences.

The COVID-19 pandemic has highlighted the biopolitical assemblage: biology and politics have grown ever increasingly intertwined in the last two centuries as populations become the site of political power. Public Health concerns and policies now influence all aspects of civic life ranging from urban planning and transportation services to housing

---

[1] Zip code level data about covid death rates, infections and ambulance call volumes is available from the NYC Department of Health (https://www1.nyc.gov/site/doh/covid/covid-19-data.page#maps); *The New York Times* and other mainstream media reported on the disparity between income, infection rate and urban flight, e.g. Kevin Quealy, 'The Richest Neighborhoods Emptied Out Most as Coronavirus Hit New York City', *The New York Times* (15 May 2020), https://www.nytimes.com/interactive/2020/05/15/upshot/who-left-new-york-coronavirus.html. Carrión et al. developed a NYC zip code level index to demonstrate the association between neighborhood social disadvantage and rate of covid infections, and mortality in Spring 2020. See: Daniel Carrión et al., 'Neighborhood-level disparities and subway utilization during the COVID-19 pandemic in New York City', *Nature Communications*, 12.3692 (2021), https://doi.org/10.1038/s41467-021-24088-7.

[2] Editors' note: This masking is also discussed by Pieter Vermeulen (chapter 9) in his study of elitist and racist fantasies of a depopulated Earth, typical of what he calls (after Brian Aldiss) 'cosy catastrophe fictions'.

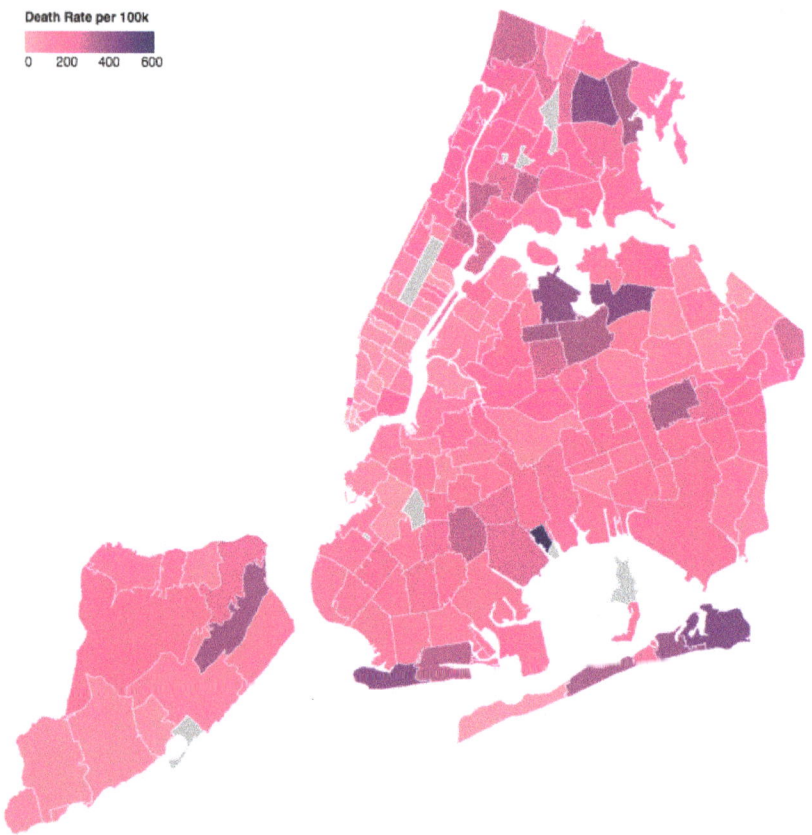

Fig. 18 NYC Health Department, *Coronavirus death rate by zip code* (2020) © Courtesy of NYC Health Department. The interactive map reveals that there was a striking increased mortality rate among people of colour and those that lived in poor neighbourhoods.

and environmental protection. Pandemic fictions, an increasingly popular genre, help mould and imagine this constellation. Unlike other dystopic or catastrophic imaginaries (like ecological disaster, alien invasion, supernatural apocalypse, AI or technology) pandemic fiction often portrays a city left standing but empty of its humans. The physical infrastructure of highways and buildings remain mostly undisturbed, but the human infrastructure has been decimated. People are imagined leaving the city behind because there is something particularly infectious about the city itself. In films like Francis Lawrence's *I am Legend* (2007), in video games like *The Last of Us* (2013), and in novels like Emily St.

John Mandel's *Station Eleven* (2014), and Ling Ma's *Severance* (2018), buildings, highways and roads remain undamaged, but cities are rapidly depopulated as people are either killed by the infection or flee to the countryside. Even Daniel Defoe's 1722 *Journal of the Plague Year*, a fictionalized account of the 1665 great plague of London, depicts the flight from the plague city: '[O]ne would have thought the very City itself was running out of the Gates, and that there would be no Body left behind'.[3]

Pandemic fiction and the fiction of the pandemic both imagine an emptied city as a response to or an effect of biological catastrophe. But why does the catastrophic imaginary of pandemic fiction envision an empty city when the reality is more likely that the poor and brown and black people will not and cannot leave? In *White Flights*, Jess Row writes, 'In a country shaped by colonization, enslavement, and racialized capitalism—practices of exclusion and exploitation very much alive in the early twenty-first century—few spaces remain empty by accident'.[4] This is true for fictional and counterfeit emptiness as well. Row then observes that, '[t]here are implicit and explicit barriers that govern the visible world, and these lines are themselves a design, or authorship, that deserves to be read'. The empty city of the pandemic is empty of some people but *full* of others, even if they are not visible. In what follows, I want to explore the fiction of the empty city of the pandemic. I want to repopulate the fictional city with the many essential workers, precarious peoples, everyday inhabitants, street homeless and other individuals who lived through New York City in particular as it was struck by the COVID-19 pandemic, even as essential services—the subways, street cleaning, garbage pickup, and park maintenance—were cut in real time.[5] These cuts and austerity measures follow a long tradition of racialized public health and urban planning like ghettoization, gentrification, and quarantine. While exposing the dangerous fiction of the empty city, I

---

3 Daniel Defoe, *Journal of the Plague Year* (New York: Norton, 1992), p. 79.
4 Jess Row, *White Flights: Race, Fiction, and the American Imagination* (Minneapolis: Graywolf Press, 2019), p. 105.
5 Stacey Sager, 'Coronavirus Update NYC: Neighborhoods hit hardest by COVID now buried under garbage', *ABC7* (15 September 2021), https://abc7ny.com/covid-garbage-hard-hit-nyc-neighborhoods-buried-by-trash-new-funding/11022921/; Caroline Spivack, 'Here's how the coronavirus pandemic is affecting public transit', *Curbed* (21 May 2020), https://ny.curbed.com/2020/3/24/21192454/coronavirus-nyc-transportation-subway-citi-bike-covid-19.

will develop a reading of Ling Ma's debut novel *Severance* as it entangles these questions with global immigration patterns, racial identity formation and the haunted cityscape.

'White flight', the rapid post World War II migration of white Americans out of cities and into the suburbs and exurbs, was not a random or self-originating event, somehow representing the spirit of the people. It was a function of racist urban planning and housing policies in response to the new (albeit temporary) mobility of black Americans. As black people migrated to cities like Chicago and New York, some whites, threatened perhaps by the loss of what W. E. B. Dubois called the 'psychological wage' of being white (whiteness as a psychological compensation for low actual wages)[6] first resorted to terrorist fire bombings and harassment,[7] and then ultimately left the cities for segregated planned communities like Levittown. As Ta-Nehsi Coates writes, '[W]hite flight was a triumph of social engineering, orchestrated by the shared racist presumptions of America's public and private sectors'.[8] The urban blight resulting from redlining and white flight during the 1950s and 1960s was further promoted by diminished funding for the inner cities, the crack wars and a presumed threat of violence during the 1970s and 1980s.

That urban breakdown occurring in the context of the intermingling of blacks and whites is often described in terms of biological metaphors

---

6   'It must be remembered that the white group of labourers, while they received a low wage, were compensated in part by a sort of public and psychological wage. They were given public deference and titles of courtesy because they were white. They were admitted freely with all classes of white people to public functions, public parks, and the best schools. The police were drawn from their ranks, and the courts, dependent on their votes, treated them with such leniency as to encourage lawlessness. Their vote selected public officials, and while this had small effect upon the economic situation, it had great effect upon their personal treatment and the deference shown them'. W. E. B. Du Bois, *Black Reconstruction* (New York: Harcourt, Brace & Co., 1935), pp. 700–01.

7   On 11 July 1951, a young black family, Henry Clark Jr., his wife Johnetta, and their two children, had to flee their new home in Cicero, a suburb of Chicago, on the very same day they moved in, because a mob of 4,000 white people stormed their apartment, destroyed their property and firebombed their building. Like other young black families, the Clarks had moved to a Northern city (Chicago) from the South (Mississippi). While this was one of the most egregious instances of Northern white hostility to the arrival of Black people during The Great Migration, it was one of countless examples. See: Isabel Wilkerson, *The Warmth of Other Suns: The Epic Story of America's Great Migration* (New York: Random House, 2010).

8   Ta-Nehisi Coates, 'The Case for Reparation', *The Atlantic* (15 June 2014), https://www.theatlantic.com/magazine/archive/2014/06/the-case-for-reparations/361631/.

like *blight* (a disease in plants caused by other species like fungi) is no accident. There is a long history of understanding urban degradation in biological terms, especially in relation to infectious diseases. But the effects of real infectious diseases can provide the optics for a social analysis of the urban. Historian Charles Rosenberg derived a narrative structure of a society's and individual's response to epidemics: 'Epidemics start at a moment in time, proceed on a stage limited in space and duration, follow a plot line of increasing revelatory tension, move to a crisis of individual and collective character, then drift toward closure'.[9] In this three-act play, the epidemic reaches an individual and societal crescendo that is marked by a crisis which puts pressure on the social fabric. Epidemics perform a kind of social analysis, revealing who and what is valued. Continuing the theatrical analogy, if the epidemic is a tragedy, then this moment is an *anagnorisis*, what Aristotle in his poetics referred to as 'a change from ignorance to knowledge' or recognition of a hidden truth,[10] the truth here being that some people are valued more highly than others. This lifting of the veil exceeds the scale of local or national politics, and that of a well-defined period. As we engage in a history of the present, we might ask whether our current pandemic is an exceptional event or an everyday event. Malaria kills more than 400,000 people a year, AIDS killed 770,000 in 2018 and Tuberculosis kills more than 1.5 million annually. 800,000 children die from diarrheal illness every year.[11] These deaths, like the COVID-19 deaths, are not distributed evenly but follow the geospatial patterns of colonial history. The vast majority of all of these deaths from treatable and curable infectious disease are in Africa and Southeast Asia. Even during the AIDS epidemic, leading health experts suggested that the era of infectious disease was over and had been replaced by chronic diseases of lifestyle.

Every year in the winter months our urban emergency rooms see a surge of patients with fever and cough. The number of emergency visits doubles in the winter, especially in areas that are servicing poor,

---

9   Charles E. Rosenberg, 'What is an epidemic? AIDS in historical perspective', *Daedalus*, 188 (1989), 1–40 (pp. 1–17).
10 Aristotle, *Poetics*, trans. by S. H. Butcher (New York: Cosimo Classics, 2008), p. 20.
11 *Global Health Estimates 2015: Deaths by Cause, Age, Sex, by Country and by Region, 2000–2015*, World Health Organization, section 'Mortality data' (2016); Katherine F. Smith et al., 'Global rise in human infectious disease outbreaks', *Journal of the Royal Society Interface*, 11.101 (2014), https://doi.org/10.1098/rsif.2014.0950.

minority, and otherwise marginalized communities. Because of antigenic shift (the recombination of genes from different strains of a virus) and antigenic drift (the accumulation of minor genetic mutations), influenza is rendered a 'constantly emerging disease'. Flu kills at least 50,000 people in the US per year and perhaps a million worldwide. Our seasonal flu epidemic is the norm every winter. Epidemics and pandemics are catastrophic but singular events that have demarcated beginnings and ends. But in the contemporary world, they have become regularized, ongoing, and without end. Epidemics have become endemic. How else can we still refer to AIDS as an epidemic if it has persisted for forty years? Infectious diseases, modelled and theorized as epidemics, resist the rhetorical transformation to endemics. The colonial language of epidemics links it to early nineteenth-century problems of swamps or miasma or nonhuman spaces; but the modern flu and COVID-19 pandemics are specifically the results of human practices and intervention: global urbanization, agribusiness, industrial livestock production, and deforestation.

To begin to understand and theorize the uneven distribution of deaths and the contradictions of care they entail in endemic and epidemic diseases, we may find some provisional approaches in Michel Foucault's concepts of biopower and biopolitics. While the notion is sometimes more suggestive than precise, biopower is comprised of a series of regulatory policies, disciplinary practices, and epistemic shifts centred on the body and population that we can all recognize. Distinguishing between the rights of the sovereign and those under biopower, Foucault writes: 'If genocide is indeed the dream of modern powers, this is not because of a recent return of the ancient right to kill; it is because power is situated and exercised at the level of life, the species, the race, and the large-scale phenomena of population'.[12] Foucault marks a shift from a sovereign power with its right to kill to a late eighteenth-century biopower which circulates between the disciplinary order of the body and the biopolitical optimization of the population. But Foucault asks, if the end of biopower is to increase life, prolong its duration, improve its chances, how is it possible for this political power to kill? It is here that we see the birth of state racism. Race enters directly into the basic

---

12   Michel Foucault, *The History of Sexuality, Volume 1: An Introduction*, trans. by Robert Hurley (New York: Vintage, 1990), p. 137.

mechanisms of state power. Racism is the 'indispensable precondition' that authorizes the biopolitical state's right to kill.[13] The optimization of life under biopower relies on a distinction and hierarchy among races to produce groups within the population—on the one hand, fit and vital, and on the other, endemically unfit, inferior, and unhealthy: '[R]acism justifies the death function in the economy of biopower by appealing to the principle that the death of others makes one biologically stronger insofar as one is a member of a race or a population, insofar as one is an element in a unitary living plurality'.[14] The disproportionate mortality rates among minority populations during the COVID-19 (and also the Spanish Flu) pandemic seems to reflect this state. Social distancing is an example of biopolitical regulation/regulatory biopower. Only the 'fit and vital' can engage in social distancing while the unfit populations of racialized essential workers—who are more likely to take public transportation, to not have second homes to escape the cities, to work in jobs that cannot be performed at home, to live in multigenerational families, to suffer the effects of pollution and overcrowding—die.

Ling Ma's 2018 debut novel, *Severance*, a love affair with New York City as well as a dystopian pandemic fiction, hints at this contradiction. A fungal pandemic originating in China disrupts global supply chains and ultimately wipes out most of the world's population. While the novel is conventionally realistic in many aspects, the symptoms of the infection include becoming stuck in an endless loop of everyday tasks, eventually dying of something like malnutrition. The novel proceeds from the protagonist Candace Chen's perspective in alternating recursive chapters that narrate her flight to and away from the city. In the post-flight narrative, Candace and a small band flee the city together to make a new home at a commune ironically set in a suburban shopping mall. But the pre-flight narrative that details her parent's emigration from Fungzhou, China and her own immigrant or diasporic experience sustains the interest in the story. In narrating her parents' achingly banal attempts but heartbreaking failures to assimilate into American life,

---

13 Michel Foucault, *Society Must Be Defended*, trans. by David Macey (New York: Picador, 1997), p. 257. For extended discussions of race and biopower, see: Ann Laura Stoler, *Race and the Education of Desire: Foucault's History of Sexuality and the Colonial Order of Things* (Durham: Duke University Press, 1995), and Etienne Balibar, *Masses, Classes, Ideas* (London: Routledge, 1994).

14 Foucault, *Society*, p. 258.

Candace prefigures the effects of the pandemic. Shen fever, caused by a fungal infection originating in Shenzen, China, ultimately decimates and destroys the human population, but not before transforming the victims into zombie like automatons, endlessly repeating everyday routinized tasks like setting a table or folding clothes in a department store: 'It is a fever of repetition, of routine'.[15] As the post-colonial scholar Anjali Raza Kolb notes, *Severance* is about 'race, waste, ennui, fevers, cults, suburbia and apocalypse'.[16] But *Severance* is also very much about cities and New York City in particular. The interest in urban theory is signalled by Candace and her boyfriend's courtship which happens over a copy of Jane Jacobs' *The Death and Life of Great American Cities*. And the image of the empty city serves as an *ekphrastic* figure chronicled in Candace's blog *NY Ghost* in which, as one of the last living non-fevered humans in New York City, she photo-documents the empty buildings, places, spaces, and icons: subway stops, the Flatiron Building, Times Square. Near the end of the novel, as the city becomes increasingly emptied and devoid of infrastructure, as the subways stop running, Candace takes a taxicab from her apartment in gentrified Bushwick across the iconic Brooklyn Bridge to her now empty office in Times Square. She strikes up a conversation with her 'middle-aged Hispanic' driver, both somewhat surprised that the other is still in New York City. When Candace directly asks why he chose to remain, he responds, '[N]ow that all the white people have finally left New York, you think I'm leaving?' (261). What is often experienced as necessity, contingency, and exploitation, can at least momentarily be reclaimed as choice, or at least a moment of recognition. The city is basically unliveable, but for these two minority immigrants, there is a compensation. With the white majority gone, their experiences are no longer defined by the indignities of racism. The mutual recognition of the two minority characters suggests that they have a shared experience as immigrants or racialized persons. They find comfort in each other's existence as they both identify as non-white. The presence of a white majority had left them marginalized, diminished

---

15 Ling Ma, *Severance* (New York: Farrar, Straus and Giroux, 2018), p. 62. Further quotations from the novel will be cited internally.
16 Kali Handelman and Anjuli Raza Kolb, 'Epidemic Empire: A Conversation with Anjuli Raza Kolb', *The Revealer* (6 May 2020), https://therevealer.org/epidemic-empire-a-conversation-with-anjuli-raza-kolb/.

their life chances, and rendered them invisible. For a moment, they can see and recognize each other after the majority white populations have either fled or been fevered. But their bonds are still tethered to whiteness since their only affiliation is as non-white, while their differences are further cemented by their class positions. Candace is constituted in and through the myth of the model minority while the cab driver belongs to the working class.

In Defoe's *Journal of the Plague Year*, the fictional narrator HF, like Candace, stays in the fevered city even after many of its inhabitants have fled. It is not clear what impels him to stay—curiosity, individual will, God's design (both Candace and HF practice the medieval art of *sortes biblicae* or bibliomancy where a reader opens a page and verse of the Bible at random and is guided by it)—but it provides him with an opportunity to observe the effects of the plague on the denizens of London. While he notices and describes the centrifugal flight of the rich ('Those that had mony always fled farthest'), he is sympathetic to those that remained or were left behind. Even as he describes the emptying of the city ('there would be no Body left behind'), he is acutely aware not only that the city is in fact still full of people, but that it is certain kinds of people, the poor, the vagrants, the disenfranchised and immiserated masses: 'It must be confest, that tho' the plague was chiefly among the poor; yet, were the Poor the most Venturous and Fearless of it, and went about their employment, with a sort of brutal Courage'.[17] The flight of the rich is contrasted with the often strict restriction in movement experienced by the vast majority of London's population. Houses suspected of harbouring plague were shut up, literally locked from the outside, and presided over by watchmen, day, and night. The shutting up of houses under quarantine was complemented by strictly enforced curfews and shelter in place orders. But alongside the roll calls of the dead, HF imbues the London poor with pathos *and* agency as they outwit their jailors and continue the essential work of city life. Ultimately as much a love letter to London as *Severance* is to New York, both pandemic fictions inscribe and then trouble the fiction of the empty pandemic city.

Severance is unique and relevant because it is both dystopic pandemic science fiction *and* a coming-of-age diaspora novel by an Asian American.

---

17 Defoe, *Journal*, p. 75.

It makes the link between emerging global pandemics and emerging immigrant identity formations explicit. Pandemics are often coded as threats to national identity from an Other. The Spanish Flu. The English Malady. The China Virus. The earliest response to pandemics has been the closing of borders and ports, followed by limitations on internal mobility. In America, hate crimes and violence against Asian Americans have noticeably increased since the beginning of the pandemic.[18]

Fig. 19 George Hirose, *Protesters against Anti-Asian Hate*, Union Square Park, March 21 (2021) © George Hirose. All rights reserved.

This follows pre-existing circuits of hate and racialization. From the Chinese Exclusion Act of 1882 to the eugenicist Immigration act of 1924 which banned all immigration from Asia to the forced internment of Japanese Americans in camps during WWII, the United States periodically restricts immigration from Asia, while symbolically and

---

18 Associated Press, 'More than 9,000 anti-Asian incidents since pandemic began', *POLITICO* (12 August 2021), https://www.politico.com/news/2021/08/12/more-than-9-000-anti-asian-incidents-since-pandemic-began-504146.

juridically casting Asian Americans as internal foreign threats ('the foreigner within') against which society must be defended.

Fig. 20 Philip Timms Studio, *Boarded-up businesses after race riots in Chinatown at northwest corner of Carrall Street at Pender* (1907) © Courtesy of Philip Timms Studio/Vancouver Public Library/VPL 940.

In *The Melancholy of Race*, Anne Cheng explores the ambivalent psychodynamic process by which racial identity is formed in America through a study of art by Asian and African Americans. Cheng forcefully asserts that '[a]s both the targeted, racialized group in United States immigration policy and yet the least 'coloured' group, Asian Americans offer a charged site where American nationhood invests much of its contradictions, desires, and anxieties'.[19] By specifically invoking the Freudian concept of melancholy, Cheng reconstitutes racial identity formation as radically ambivalent:

> Racialization in America may be said to operate through the institutional process of producing a dominant, standard, white national ideal, which is sustained by the exclusion-yet-retention of racialized others. The national topography of centrality and marginality legitimizes itself by retroactively positing the racial other as always Other and lost to the

---

19 Anne Cheng, *The Melancholy of Race: Psychoanalysis, Assimilation, and Hidden Grief* (New York: Oxford University Press, 2001), p. 21.

heart of the nation. Legal exclusion naturalizes the more complicated 'loss' of the unassimilable racial other.[20]

Just as the melancholic subject constitutes its ego through the internalization of the lost object, racial identity is constituted as loss, exclusion, *and* incorporation. Asian Americans are both American and not American. But melancholy is constitutive of both subjects and objects. Dominant white identity is itself melancholic because it is caught in the bind of incorporation and rejection. This excluded presence haunts the American experiment. In *Playing in the Dark*, Toni Morrison argues that African American presence, the ghost in the machine, is the excluded centre of the American literary production. They are invisible by design. This canonical exclusion is a purposeful form of invisibility that is similar to the fiction of the empty pandemic city that I have been discussing. The inability to see allows for ongoing police brutality, the squalor of migrant camps and the continued marginalization of racial minorities.

While previous commentators on Freud have noted that there might not be a self without melancholy,[21] I would suggest that there might not be an immigrant self without nostalgia. Melancholy and nostalgia are certainly related, and the title of Ma's novel, *Severance*, is an example of the 'vocabulary of grievance' which allows Cheng to associate the legal work of grievance with the affective work of grief in identity formation. But nostalgia may be an even more important affect in immigrant racial identity formation than melancholy. Nostalgia derives from the Greek *nostos* ('homecoming') and *algos* ('pain, grief'). In contemporary usage it most often refers to a wistful yearning for the past that has recently become even more abstract. But in its original formation, nostalgia was specifically a longing for a place ('home'), and it was understood primarily as a medical condition: 'severe homesickness considered a disease'.[22] Initially applied in the context of war, nostalgia was often the

---

20   Ibid., p. 10.
21   In their readings of 'Mourning and Melancholia', Judith Butler, Elin Diamond, and Anne Cheng have all suggested that melancholia, rather than being only a pathological later life formation, may actually be constitutive of the ego in early development. See: Judith Butler, *The Psychic Life of Power: Theories in Subjection* (Palo Alto: Stanford University Press, 1997); Elin Diamond, 'The Violence of "We": Politicizing Identification', in *Critical Theory and Performance*, ed. by Janelle G. Reinelt and Joseph R. Roach (Ann Arbor: University of Michigan Press, 1992); Anne Cheng, *The Melancholy of Race*.
22   Marcos Piason Natali, 'History and the Politics of Nostalgia', *Iowa Journal of Cultural Studies*, 5 (2004), pp. 10–25, https://doi.org/10.17077/2168-569X.1113. During the

purported cause of death for soldiers whose death had no immediate organic reference. Ma seems to have both ideas in mind as she works out the central role which nostalgia plays in her novel. Nostalgia is the most non-realistic element of Shen fever. It is a symptom *and* a cause of a mortal medical condition. As Candace notes after Ashley becomes fevered, 'Don't you think it's strange Ashley became fevered in her childhood house? It's like nostalgia had something to do with it' (143). And a few pages later, the narrative voice becomes even more affirmative and didactic, 'Shen Fever being a disease of remembering, the fevered are trapped indefinitely in their memories' (160). The narrative transformation from subjective introspection to clinical objectivity is a feature that unsettles the reading experience, by manipulating the expected conventions of realism and the fiction of the psychologically whole and reliable narrator.[23]

Nostalgia is also central to the novel's intertwined narrative, where it is the critical affect in immigrant racial identity formation. Candace's parents' flight from China is steeped in the tropes of immigrant literature: escape from an oppressive family and state structure, the desire for educational and economic opportunities for self and child, and the liberating conveniences of American life. But immigrant life is equally marked by the centripetal pull of the homeland: 'My mother wanted to return to China', Candace writes, 'if not today then eventually' (42). The competing pushes and pulls of nostalgia and assimilation situate the trope of immigrant identity in America. Nostalgia is a longing (perhaps in bad faith) to return home, but to a home that no longer (or perhaps never even) existed. The spatial logic of immigrant nostalgia is bad politics and bad history. Watching live footage of the Tiananmen Square protests on TV from their Salt Lake City basement apartment, surrounded by 'tchotchkes that did not belong to them, that they did not know or understand within any cultural context and did not find beautiful', Candace's father says to his wife, 'We are never going back' (176). Nostalgia links the two anti-parallel narrative strands like base

---

American Civil War, there were 74 deaths from nostalgia on the Union side, and more than 5,200 cases in the Surgeon General's records. Susan Matt, *Homesickness: An American History* (New York: Oxford University Press, 2011).

23 One of my students, Junjie Ren, even suggested that the entire novel could be read as a retrospective account of Candace's own fevered nostalgia as she, like the other characters, has fallen victim to Shen fever.

pairs in DNA, connecting its own historical life first as disease and now as affect. The novel actually reverses this history of nostalgia. Nostalgia is a disease that became an affect but in *Severance*, nostalgia is an affect that becomes a disease.

Candace's approach to nostalgia and the past is to keep moving forward: 'The past is a black hole, cut into the present day like a wound, and if you come too close, you can get sucked in. You have to keep moving' (120). But you cannot escape a black hole. The double metaphor of black hole and wound serve to locate the past in physical space and in the body. The past as wound and as unescapable are explicitly linked to her immigrant identity formation. In her reconstruction, before she immigrated to America, as a young child in Fuzhou, she 'seemed to lack all neurosis or anxiety' (183). It was only after moving to America to be with her parents that she describes herself as 'angry, chronically dissatisfied, bratty' (184). Her tantrums lasted well into her teen years and are a way of expressing the body's knowledge.

Personal meaning in the novel is revealed through time, through history, memory, and nostalgia, but it is also registered in terms of space, place, and geography. In *Urban Underworlds*, Thomas Heise explores the process by which immigrant and other marginalized communities and the geographies they inhabit become associated with miasma and disease as they are quarantined to the underworld in literature and space. He cites journalist Edgar Saltus' view in 1905 from atop the newly built world's first skyscraper, the Flatiron Building, also photo-chronicled by Candace for her blog. From up high, Saltus imagines a vision of frenzied capitalism 'erupting gold'. But Saltus' vertical vision grasps the uneven social, geographic, and racial space. Turning from the northern upper Fifth Avenue with its gods of industry, Saltus describes the southward Chinese quarter as a 'sewer'.[24] The North-South Modernist grid of the city seen from high above is coded as moral, economic, and racial hierarchy all at once. The view from a height both highlights and

---

24 Quoted in Thomas Heise, *Urban Underworlds: A Geography of Twentieth-Century American Literature and Culture* (New Jersey: Rutgers University Press, 2011), p. 2. Similarly, in *Contagious Divides: Epidemic and Race in San Francisco's Chinatown* (Berkeley: University of California Press, 2001), Nayan Shah recounts how nineteenth-century public health officials described San Francisco's Chinatown as a '"plague spot", a "cesspool", and the source of epidemic disease and physical ailments', p. 1.

obscures the racial, economic, and colonial effects of capitalism on the new geography of the city. That paradox of perspective is also reflected in the glass-box architecture Candace works in and comes to inhabit. Candace's contemporary New York City is a postmodern kaleidoscope where the ethnic ghettos of Brooklyn have gentrified and rebranded on a global scale. The vice districts and sexual cruising of Times Square seedy theatres and sex shops have been replaced by multi-national brands, made for Instagram neon signs and family entertainment (for upper-class tourists).

Candace's New York office, which also becomes her home briefly, takes up the thirty-second and thirty-third floors of a Times Square building. And it is from this vantage that she initially describes the effects of the pandemic:

> I looked out the windows. For the first time, I noticed that Times Square was completely deserted. There were no tourists, no street vendors, no patrol carts. There was no one. It was eerily quiet, as if it were Christmas morning. Had it become this way without my noticing? I walked around the perimeter of the office, trying to spot a fire truck or police car pulling up outside, trying to discern a siren in the distance, something. It wasn't just the emptiness. In the absence of maintenance crews, vegetation was already taking over; the most prodigious were the fernlike ghetto palms, so-called because they exploded in prolific waves across urban areas, seemingly growing from concrete, on rooftops, parking lots, and all sidewalk cracks. (252)

We find here all of the traps and tropes of pandemic fictions. The emptying of people and the resurgence of prehistoric wildlife, as if the city were being swallowed up into its past. But height and distance create a false sense of emptiness and totality. Describing a view from a similar vantage point atop the 107th floor of the World Trade Center, the philosopher Michel de Certeau notes that the height falsely transforms the 'extremes of defiance and poverty, the contrasts between races and styles' into a unified surface. For de Certeau, the totalizing panoramic view erases difference and context, obscures life and 'creates the fiction of knowledge'.[25] Like the fiction of the empty city, this panoramic 'fiction of knowledge' is the result of optical transformations and desires. The empty

---

25 Michel de Certeau, 'Practices of Space', in *On Signs*, ed. by Marshall Blonsky (Baltimore: Johns Hopkins University Press, 1985), pp. 122–23.

city of pandemic fiction displaces cultural anxieties over immigration, changes in racial makeup, and altered standards of behaviour onto imagined contagion. It obscures real social and geographic deprivation. When Candace comes down from the heights and begins to document New York City, she not only falls in love with the 'real' city, she also begins to notice signs of human life. Like HF, Defoe's fictional narrator, Candace chronicles places and the remaining people: a fruit vendor, an old lady in a nightgown, a homeless teen couple. But New York City is also home to a significant non-fevered population, Sentinel security guards: 'The Institutions here were still being maintained, as if someone expected everyone to come back eventually. They were guarded by security guards [...] The city had contracted Sentinel to protect public institutions from looters. Private owners also hired the company to guard evacuated homes' (255). Even as the pandemic rages, the conjunction of capital and state power maintain the value of specific geographic spaces. Lights blazing without any tourists, and guards to protect when there are no threats. As Candace asks, 'If a horse rides through Times Square, and no one is there to see it, did it actually happen?' (253–54).

Race and racism are legislated and buttressed through a logic of spatial politics predicated on exclusion. Racialized bodies and identities are constituted through acts of immigration restriction that limit or block entry. But just as travel bans fail to prevent the spread of infectious disease in a globalized era, immigration restrictions only highlight the real permeability of borders. And like the ancient single cell organisms that became incorporated into other cells as mitochondria, the immigrant is always already the foreigner within. The racialized immigrant body is further distinguished by spatial limitations that confine them to specific neighbourhoods, ghettos and underworlds. But, as if that is not enough, in states of emergency and crisis, they are further relegated to the interstices of the visible. The trope of the empty city, both in pandemic fictions and fictions of the pandemic, inflicts psychological and physical harm on immigrant and minority bodies. During the pandemic, the fiction of the empty city supported an austerity reduction in city services that augmented the already disproportionate death toll in black and brown populations while alleviating the guilt of a dominant white majority that escaped the cities. Federal, state, and local governments contained the virus by allowing it to multiply and thrive in low resource settings amongst the most marginalized and vulnerable populations. Cordon sanitaire prevented the virus from getting out, but it also prevented the people from getting out or the healthcare workers from getting in. While literature can support and reflect these obfuscations, it can also lay bare the contradictions in the naturalized order of things. *Severance* engages with nostalgia as both a symptom of a global infectious disease and a process of racial formation, to question and highlight the spatial logic of exclusion and invisibility at the heart of the American Experiment. This novel works hard to make us see what hard work had made us not see.

Works Cited

Anne Cheng, *The Melancholy of Race: Psychoanalysis, Assimilation, and Hidden Grief* (New York: Oxford University Press, 2001).

Aristotle, *Poetics*, trans. by S. H. Butcher (New York: Cosimo Classics, 2008).

Associated Press, 'More than 9,000 anti-Asian incidents since pandemic began', *POLITICO* (12 August 2021), https://www.politico.com/news/2021/08/12/more-than-9-000-anti-asian-incidents-since-pandemic-began-504146

Balibar, Etienne, *Masses, Classes, Ideas* (London: Routledge, 1994).

Butler, Judith, *The Psychic Life of Power: Theories in Subjection* (Palo Alto: Stanford University Press, 1997).

Carrión, Daniel et al., 'Neighborhood-level disparities and subway utilization during the COVID-19 pandemic in New York City', *Nature Communications*, 12.3692 (2021), https://doi.org/10.1038/s41467-021-24088-7

Cheng, Anne, *The Melancholy of Race: Psychoanalysis, Assimilation, and Hidden Grief* (New York: Oxford University Press, 2001).

Coates, Ta-Nehisi, 'The Case for Reparations', *The Atlantic* (15 June 2014), https://www.theatlantic.com/magazine/archive/2014/06/the-case-for-reparations/361631/

De Certeau, Michel, 'Practices of Space', in *On Signs*, ed. by Marshall Blonsky (Baltimore: Johns Hopkins University Press, 1985), pp. 325–38.

Defoe, Daniel, *Journal of the Plague Year* (New York: Norton, 1992).

Diamond, Elin, 'The Violence of "We": Politicizing Identification', in *Critical Theory and Performance*, ed. by Janelle G. Reinelt and Joseph R. Roach (Ann Arbor: University of Michigan Press, 1992), pp. 390–98.

Du Bois, W.E.B., *Black Reconstruction* (New York: Harcourt, Brace & Co., 1935).

Du Bois, *The History of Sexuality, Volume 1: An Introduction*, trans. by Robert Hurley (New York: Vintage, 1990).

Foucault, Michel, *Society Must Be Defended*, trans. by David Macey (New York: Picador, 1997).

Handelman, Kali and Anjuli Raza Kolb, 'Epidemic Empire: A Conversation with Anjuli Raza Kolb', *The Revealer* (6 May 2020), https://therevealer.org/epidemic-empire-a-conversation-with-anjuli-raza-kolb/

Heise, Thomas, *Urban Underworlds: A Geography of Twentieth-Century American Literature and Culture* (New Jersey: Rutgers University Press, 2011).

Ma, Ling, *Severance* (New York: Farrar, Straus and Giroux, 2018).

Matt, Susan, *Homesickness: An American History* (New York: Oxford University Press, 2011).

Natali, Marcos Piason, 'History and the Politics of Nostalgia', *Iowa Journal of Cultural Studies*, 5 (2004), 10–25, https://doi.org/10.17077/2168-569X.1113

Quealy, Kevin, 'The Richest Neighborhoods Emptied Out Most as Coronavirus Hit New York City', *The New York Times* (15 May 2020), https://www.nytimes.com/interactive/2020/05/15/upshot/who-left-new-york-coronavirus.html

Rosenberg, Charles E., 'What is an epidemic? AIDS in historical perspective', *Daedalus*, 188. 2 (1989), 1–40.

Row, Jess, *White Flights: Race, Fiction, and the American Imagination* (Minneapolis: Graywolf Press, 2019).

Sager, Stacey, 'Coronavirus Update NYC: Neighborhoods hit hardest by COVID now buried under garbage', *ABC7* (15 September 2021), https://abc7ny.com/covid-garbage-hard-hit-nyc-neighborhoods-buried-by-trash-new-funding/11022921/

Shah, Nayan, *Contagious Divides: Epidemic and Race in San Francisco's Chinatown* (Berkeley: University of California Press, 2001).

Smith, Katherine F. et al., 'Global rise in human infectious disease outbreaks', *Journal of the Royal Society Interface*, 11.101 (2014), 1–6.

Spivack, Caroline, 'Here's how the coronavirus pandemic is affecting public transit', *Curbed* (21 May 2020), https://ny.curbed.com/2020/3/24/21192454/coronavirus-nyc-transportation-subway-citi-bike-covid-19

Stoler, Ann Laura, *Race and the Education of Desire: Foucault's History of Sexuality and the Colonial Order of Things* (Durham: Duke University Press, 1995).

Wilkerson, Isabel, *The Warmth of Other Suns: The Epic Story of America's Great Migration* (New York: Random House, 2010).

World Health Organization, 'Global Health Estimates 2015: Deaths by Cause, Age, Sex, by Country and by Region, 2000–2015', World Health Organization (2016).

8. Dead Gods and Geontopower
An Ecocritical Reading of Jeff Lemire's *Sweet Tooth*[1]

Kristin M. Ferebee

Let me begin with a proposition: Jeff Lemire's 2009–2013 comic title *Sweet Tooth*, adapted by Netflix as a science fiction series that began airing in 2021, is a story about extractive resource exploitation. This claim will surprise those familiar with the comic, who know it as a post-apocalyptic narrative about the half-animal hybrid children who rise to found their own post-human world after a pandemic destroys human civilization and, simultaneously, a narrative about the journey across America undertaken by an ex-hockey bruiser, Jepperd, and a small half-deer boy named Gus. Indeed, very little that is even tangentially evocative of extractive industries appears in the first twenty-five issues of the comic, which conform to the post-apocalyptic fiction genre and to the genre of frontier captivity narrative popularized by American Westerns in which a vulnerable child must be returned to human civilization from the savage wilderness. The arc that begins in *Sweet Tooth* #26 does not overtly broach the topic of extraction, either. This arc, an historical flashback to an ill-fated Arctic voyage in 1911, presents the experiences of a surgeon who is traveling to a missionary settlement in what is now Alaska in search of his sister's vanished fiancé, a man named Simpson. When the surgeon, Thacker, reaches the settlement, he finds all the missionaries dead of a plague save one, Simpson, who has been assimilated into a local Inuit

---

[1] The research for this chapter was carried out with the support of the European Research Council.

© 2022 Kristin M. Ferebee, CC BY-NC 4.0   https://doi.org/10.11647/OBP.0303.08

community. Simpson relates an extraordinary tale of how, venturing into a passage deep below the ice one day, he trespassed upon the skeletal animal-human bodies of what he later learned were Inuit gods. He explains that a shaman told him that this place was where the gods rested after their earthly bodies died—that 'their *spirits* remained free, but their bodies would rest there until they returned one day'.[2] There are two consequences of Simpson disturbing the tomb of the hunting god Tekkeitsertok:[3] viral sickness wipes out the mission, and Simpson's Inuit wife gives birth to a deer-human hybrid son whom Simpson believes to be an incarnation of Tekkeitsertok 'come to reclaim this land'.[4] However, the repulsed Thacker massacres the entire Inuit community before perishing of the sickness alongside all of his shipmates, still locked in the ice.

The exact relationship of this historical anecdote to the central narrative of the comic is not entirely transparent. It is clear, based on later scenes, that Thacker's journal (containing the account of his search for Simpson) has been recovered from the icebound wreck of his ship, and that this either led to or occurred contemporaneously with the rediscovery of the passage below the ice by American scientists who attempted to clone a living creature (Gus) from the bones of Tekkeitsertok's 'earthly body'. Their violation of the Inuit sacred space and of the gods' dead bodies caused the pandemic that ravaged the world and triggered the birth of the hybrid children, some of whom are reborn gods. Further detail is never offered, nor is there an attempt to explain how the Inuit beliefs described in Thacker's journal might 'translate' to Western understanding in an acceptable way.

Here, I argue that the *Sweet Tooth*'s refusal to suture Inuit ontology to Western ontology by providing 'reasonable' answers to these questions situates the comic within the collapse of what anthropologist Elizabeth Povinelli has termed geontopower: the ongoing effort on the part of settler capitalism to maintain essential boundaries between Life and Nonlife and regulate who or what is considered capable of 'being'. At the same time, *Sweet Tooth* tries to productively imagine ways and scales of biological and

---

2   Jeff Lemire, *Sweet Tooth* #26 (New York: Vertigo Comics, 5 October 2011). As *Sweet Tooth* is inconsistently paginated, I will cite by issue rather than by page.
3   The misspelling 'Tekkietsertok' and the accurate spelling 'Tekkeitsertok' are both used in the comic, but I have opted to use the more accurate transliteration here.
4   Jeff Lemire, *Sweet Tooth* #26.

quasi-biological being that geontopower has largely worked to invisibilize, utilizing the technics of narrative to engage the reader in new knowledge-practices that chart a course out of the fly-bottle of the Anthropocene age.

The term 'fossil fuel' is a moderately accurate evocation of the source from which such resources come. Though natural gas and crude oil may not be derived from the corpses of dinosaurs, as both the popular imaginary and Timothy Morton (in his evocation of 'liquefied dinosaur bones burst[ing] into flame') suggest, they are created by the decomposition of plants and organisms over millennia.[5] In other words: petroleum is the remains of the nonhuman dead. This element of the carbon imaginary is rarely emphasized in ecocritical accounts, which tend to focus on the impact and imagined *futures* of petroleum products—as Kathryn Yusoff puts it, the 'future fossilization of humanity' or the 'human-as-fossil-to-come'.[6] Even where Yusoff suggests that the slogan 'Welcome to the Anthropocene!' might be replaced with 'The Carboniferous Lives Again!',[7] she opts not to engage with the notion of fossil fuels as possessed of a history and (arguably) a life cycle. Amanda Boetzkes and Andrew Pendakis describe oil as 'time materialized by sediment' and as 'the energy made possible by eons of fossilized death',[8] but choose not to treat the organic life that underlies this death.

Boetzkes and Pendakis also describe oil as 'an oddly feral god', and *Sweet Tooth* literalizes this description in its representation of dead ancient animal gods whose corpses—buried beneath the Arctic ice—are 'extracted' as resources by modern technology. The mythology that is mobilized in *Sweet Tooth* reads these corpses as simultaneously non-living things (the skeletons that first Simpson and later the scientists disturb) and as merely one non-living aspect (the dead 'earthly bodies') of beings that are capable of what Povinelli calls 'chang[ing] states'.[9] These are dead bodies whose deadness does not divorce them from

---

5 Timothy Morton, *Hyperobjects: Philosophy and Ecology after the End of the World* (Minneapolis: University of Minnesota Press, 2013), p. 58.
6 Kathryn Yusoff, 'Anthropogenesis: Origins and Endings in the Anthropocene', *Theory, Culture & Society*, 33.2 (2016), 3–28, https://doi.org/10.1177/0263276415581021.
7 Ibid.
8 Amanda Boetzkes and Andrew Pendakis, 'Visions of Eternity: Plastic and the Ontology of Oil', *e-flux*, 47 (September 2013), [n.p.].
9 Elizabeth Povinelli, *Geontologies: A Requiem to Late Liberalism* (Durham: Duke University Press, 2016), p. 28, https://doi.org/10.1215/9780822373810.

their relationship to the living world, but is rather part of a larger being-ness that constitutes the intentional 'thing' that is a god. Povinelli analyses how rock and mineral formations, in indigenous Australian ontology, possess such an intentionality. This is an intentionality that allows them to 'intend, desire, [and] seek'[10]—in the titular examples that Povinelli gives,[11] to listen or die—as part of an identity as *durlg* or type of Dreaming that is capable of, among other things, engendering human life through a conception that is both nonbiological (in that it is separate from human biological conception) and material (in that it is related to an understanding of shared biological material—in the example that Povinelli discusses, sweat).[12] This ontology points towards a profoundly different 'slicing' of the world than that practiced by settler capitalism, as Povinelli details—not into human and nonhuman, or into life and nonlife, but rather into affiliations of being and kinship that do not differentiate on these bases. (That this categorization dissolves the problems associated with the ontological status of a virus, which we struggle to characterize as alive or dead, seems evocative in the context of *Sweet Tooth*'s plague.)

Tekkeitsertok and his fellow Inuit gods emerge, when viewed through the possibility of such alternative ontologies, as the type of bodies that literary theorist Monique Allewaert has described as 'disaggregated' or 'decomposed': 'pulled into parts' in a way that does not 'vanish' the bodies or beings in question, but rather changes them into another state. Allewaert suggests that at the centre of disaggregate being is relation, which 'describes an enmeshment that is not a merging and that forecloses the possibility of exchange'.[13] In the titular example of Allewaert's *Ariel's Ecology*, the Shakespearean spirit Ariel describes a drowned traveller as transformed into coral and pearl through a 'sea-change' that, Allewaert argues, pulls the body 'into parts' and renders it

---

10 Ibid., p. 46.
11 Elizabeth Povinelli, 'Do Rocks Listen? The Cultural Politics of Apprehending Australian Aboriginal Labor', *American Anthropologist* New Series, 97.3 (September 1995), 505–18, https://doi.org/10.1525/aa.1995.97.3.02a00090; Povinelli, *Geontologies*, p. 30.
12 Elizabeth Povinelli, *The Cunning of Recognition: Indigenous Alterities and the Making of Australian Multiculturalism* (Durham: Duke University Press, 2002), p. 219, 241.
13 Monique Allewaert, *Ariel's Ecology: Plantations, Personhood, and Colonialism in the American Tropics* (Minneapolis: University of Minnesota Press, 2013), p. 8.

ecological, yet preserves 'the apparently paradoxical possibility that the personhood is not vanished by the disaggregation but instead changed'.[14] Applying this model, we might see that the skeleton in the underground cavern that Simpson discovers is a part that exists in a certain relation to the living body of the hybrid Inuit infant whom Thacker murders, both of which exist in a relation to the living body of Gus that we might call 'Tekkeitsertok'. Thus, when Gus is close to death in *Sweet Tooth* #25 and a vision of Tekkeitsertok appears to him and guides him to the slaughtered Inuit village where the hybrid child and its mother lie dead, Tekkeitsertok is revealing himself to himself: awakening Gus to the relationality that characterizes his/their own being. This is significant because at no other point in the narrative or its world is there any contact between Gus and the slaughtered Inuit. Indeed, Thacker's narrative, confined within its hundred-year-old journal, is isolated from the comic's central storyline. The Thacker narrative unfolds over three issues of *Sweet Tooth* (#26–28) and makes a central appearance in another (#17), meaning that it constitutes a full tenth of the comic—yet it exists almost entirely in parallel with the narrative present of the twenty-first-century pandemic, an enmeshment of the type (never quite separate or quite merging) that Allewaert describes. Simpson and Thacker's violation of the Inuit gods does not seem to *cause* the pandemic, nor does it *solve* the pandemic; it barely even *explains* the pandemic through its thirdhand account of Inuit lore. Only the presence of this journal in the hands of a scientist at the remote Alaskan base that extracted the gods' skeletons, which the half-mad doctor Singh discovers, gestures towards any place where these two narratives diegetically intersect. The Tekkeitsertok relation, however, offers a schema according to which we can understand the two story 'parts' as related.

This emphasis on partedness (a condition of being *part of* a larger whole yet also *parted from* in a way that implies independent wholeness) seems congruent with a reading of the gods as a form of extractive resource. Extractive resources occupy several simultaneous scales: they are microscopic and macroscopic in ways that draw attention to them as always already incomplete and that act to obscure meaningful perception of their substance. After all, oil may be an 'oddly feral god', but it is rarely

---

14 Ibid., p. 2.

imagined as such. As Amitav Ghosh famously points out in an early treatment of 'petrofiction', oil is scarcely imagined at all. Ghosh suggests a number of reasons why this might be the case: oil is aesthetically unpleasant (it looks and smells bad). At the time when Ghosh was writing (1992), oil 'reek[ed] of unavoidable overseas entanglements'; it 'smell[ed] of pollution of environmental hazards'.[15] Even now, when scholars seek to visualize oil, they most often do so in terms of oil-as-infrastructure and oil-as-pollution. The collection *Petrocultures: Oil, Politics, Culture* (2017) contains twenty-two chapters examining aspects of the titular petrocultures, from plastics to automobiles to cosmetics to dead ducks to the Gulf War. Appel, Mason, and Watts suggest that oil functions as a metonym for other forces—'authoritarianism, corruption, violence, misallocation of money'[16]—but their point is obscured by the extent to which *oil itself* is never quite what is being discussed. Indeed, Jenny Kerber assumes their intent is to suggest that 'the difficulty of grasping "oil" writ large means that we often turn to stand-ins'.[17] When we imagine oil, we do not imagine oil. We imagine the thing that oil is inside, the thing that oil has made, the thing that oil has ruined, failing to apprehend it as part or partial being.

All of this seems to define what Timothy Morton has termed a 'hyperobject': so 'massively distributed' across spacetime relative to human existence that it cannot be completely or directly perceived. Morton's notion of hyperobjects as being *massive* relative to human existence, however, overlooks the possibility that the problems associated with such objects also accrue to objects that are *hypo*: in other words, massively small. Furthermore, Morton spends little time dwelling on the fact that many of the phenomena he classifies as hyperobjects are actually at once massively large *and* massively small—most notably all of the examples involving radiation (Chernobyl, Fukushima, the Trinity test), and arguably those involving weather as well. In the same way, these phenomena distort our conception of time. They are extremely long-lasting, but also inconceivably instant (*vide*

---

15 Amitav Ghosh, 'Petrofiction: The Oil Encounter and the novel', *The New Republic* (2 March 1992), p. 30.
16 Hannah Appel, Arthur Mason and Michael Watts (eds), 'Introduction: Oil Talk', in *Subterranean Estates: Life Worlds of Oil and Gas* (Ithaca: Cornell University Press, 2015), p. 10, https://doi.org/10.7591/9780801455407-002.
17 Jenny Kerber, 'Up from the Ground: Living with/in Petrocultures in the US and Canadian Wests', *Western American Literature*, 51.4 (Winter 2017), 383–9 (p. 386), https://doi.org/10.1353/wal.2017.0001.

the discrepancy between the lightning-quick timescale of radioactive exposure and the long half-lives of radioactive isotopes); they are also often simultaneously past and future. Astrid Schrader has explored the characterization of the microscopic as evolutionarily primitive and inherently *prior* to complex life,[18] while macroscopic complexity is often associated with modernity or with advancing futures, yet both orders or structures of life are invoked in hyperobjects such as oil. Clearly the central obstacle presented by these phenomena is not that they are spatiotemporally massive, since they are also spatiotemporally miniscule, or even that they exist on a nonhuman scale (since they exist on more than one scale). I suggest that the obstacle is that we struggle to attribute meaningful being to objects that we perceive as 'parted'—that is to say, either: 1) parts of other objects or, 2) made up of parts in a way that does not form a coherent unity.

In my analysis of plural subjectivity in science fiction,[19] I have discussed the threat that partedness poses to the human fantasy of bodily completeness and coherence—in short, the microscopic threatens to remind us of the microscopic scale at which 'we' dissolve into vital systems, organs, and microorganisms (what Stefan Helmreich describes as *Homo microbis*, 'the microbial human'),[20] while the macroscopic threatens to remind us that we, too, are only parts of material systems that exist at a macroscopic scale. Schrader suggests that '[e]mpathy requires the unity of an "I"' and that identification with the Other must begin with 'auto-affection, the becoming-present to self',[21] rendering parted beings outside of the possibility of compassion insofar as they not only lack a unified self that can be identified with, but also trouble the plausibility of a unified 'I' from which the human can identify and therefore empathize.[22] This is particularly significant insofar as the

---

18 Astrid Schrader, 'The Time of Slime: Anthropocentrism in Harmful Algal Research', *Environmental Philosophy*, 9.1 (2012), 71–94 (p. 79), https://doi.org/10.5840/envirophil2012915.
19 K. M. Ferebee, 'Pain in Someone Else's Body: Plural Subjectivity in TV's *Stargate: SG-1*', *LLIDS: Language, Literature, and Interdisciplinary Studies*, 4.3 (Summer 2020).
20 Stefan Helmreich, *Sounding the Limits of Life: Essays in the Anthropology of Biology and Beyond* (Princeton: Princeton University Press, 2016), p. 62, https://doi.org/10.1515/9781400873869.
21 Astrid Schrader, 'Abyssal intimacies and temporalities of care: How (not) to care about deformed leaf bugs in the aftermath of Chernobyl', *Social Studies of Science*, 45.5 (2015), 665–90 (p. 679), https://doi.org/10.1177/0306312715603249.
22 The fragmented being might arguably be particularly difficult for humans to empathize with if one accepts the argument (emerging from Lacan) that Lennard

most common human engagement with both the microscopic and the macroscopic takes the form of inter-assimilation of human and machine parts—the usage of cameras, telescopes, microscopes, etcetera, in a way that Helmreich has described as becoming 'part of a compound eye'.[23]

There are obvious implications in this compound or 'parted eye' for responsibility in the practical sense that Karen Barad describes: the 'ability to respond to the other' that they argue 'cannot be restricted to human-human encounters',[24] but that is often restricted to a certain subdivision of Life—the subdivision that is possessed of a soul, or, as Povinelli picks out of science studies and the Aristotelian tradition, the 'carbon-based metabolism [that] provides the inner vitality (potentiality) that defines Life as absolutely separate from Nonlife'.[25] Barad asserts that this responsibility/response-ability is '"the essential, primary and fundamental mode" of objectivity as well as subjectivity', but acknowledges the difficulty inherent in a situation where 'the "face" of the other that is "looking" back at me is all eyes, or has no eyes, or is otherwise unrecognizable in human terms'[26]—a point that they explore in more depth through their study of the 'all eyes' brittlestar that dwells on the sea bottom.[27] However, the brittlestar, as an animate creature, is relatively recognizable as responsive in human terms in comparison to Povinelli's rock formations, and both examples escape the problems caused by the being that cannot be seen except by the compound, parted eye that disallows any fantasy of the human as other-than-assemblage. Barad addresses this issue in their discussion of Ian Hacking's account of microscopic 'seeing', writing that what the use of microscopes permits is not a practice of seeing so much as it is a practice of bringing-into-being certain kinds of phenomena that we are prone to calling 'objects'.[28] Of course, the assemblage human-plus-microscope is not

---

      Davis makes regarding the discomfort that humans feel when confronted with fragmented bodies. Davis argues that such bodies uncomfortably call attention to the always-already-fragmented nature of the human body, in which wholeness is only ever hallucinated.

23 Stefan Helmreich, *Alien Ocean: Anthropological Voyages in Microbial Seas* (Berkeley: University of California Press, 2009), p. 43.
24 Karen Barad, *Meeting the Universe Halfway* (Durham: Duke University Press, 2007), p. 392, https://doi.org/10.1215/9780822388128. Barad uses nonbinary pronouns.
25 Povinelli, *Geontologies*, p. 49.
26 Barad, *Meeting*, p. 392.
27 Ibid., pp. 369–84.
28 Ibid., pp. 50–54.

inherently different from the assemblage human, which also operates to bring-into-being certain kinds of phenomena (sensory perceptions/ organizations of the world) in certain ways—and it is this destabilizing fact that we must reckon with when our consideration of the compound eye brings it to the fore.

Barad suggests that Hacking's failure lies in his inability to resist 'one of representationalism's fundamental metaphysical assumptions: the view that the world is composed of individual entities with separately determinate properties'.[29] Rather than emphasize the individuality or separateness in this assumption, as I would argue Barad does, I wish to critique the implicit wholeness of the 'entity' to which Barad refers. In their mobilization of Niels Bohr's quantum philosophy, Barad writes of a 'wholeness' that Bohr attributes to phenomena marked by their inseparability, a wholeness that is made possible in their philosophy by an experimental arrangement that performs a 'cut'. Though Barad understands this wholeness as always contingent, it nevertheless seems like one of the less developed parts of their work. Further development would perhaps require our contemplation of what the relationship is between the inseparable inner workings of the phenomena and everything that is excluded from it, which we might also frame as the relationship between the *actual* (or the actual*ized*) and the *potential* (all other actualizations that are possible). Take, for instance, the case of oil: how can a study of petroleum-based plastics take these objects not *merely* as symptomatic of or stand-ins for (that is, parts of) the larger oil phenomenon, while also acknowledging their relatedness to microorganisms that lived and died and were compressed into petroleum? How can the lives of those microorganisms be attended to as lives *qua* lives, while they are simultaneously viewed as part of an oil 'thing' that encompasses dead ducks and smog over Los Angeles? If, as Schrader notes, much philosophical development of empathy for the nonhuman takes mortality and its corollary experience of vulnerability to be its basis,[30] and even the 'hetero-affection' of her proposed human-nonhuman intimacy 'inscribes mortality within life', then how, in the case of the microorganisms who have become oil, can we share the experience of mortal vulnerability with something that is already dead? Even further: if we understand mortality, in a less material sense, to

---

29 Ibid., p. 55.
30 Schrader, *Intimacies*, p. 672.

refer to a cessation of being-as-something, and argue therefore that it is possible for certain rock formations and certain creeks to be mortal through their vulnerability to being transformed into some other kind of thing, then the microorganisms at the base of oil have already undergone such a transformation and constantly threaten to undergo another. How can we feel compassion for these things and, thereby, be moved to respond to and be responsible for them in the ways that seem so necessary in the Anthropocene?

Something about oil, in its ooziness and its infiltration, its microscopic (*ci devant* organic) and macroscopic (infrastructural, Anthropocene) qualities, its already-deadness and its not-yet-aliveness (its potential to fuel or be made into things) frustrates any effort to understand it as a thing that deserves empathy, or even to understand it as an *oil thing*. So how can we reach towards the actualization of the oil thing without making the mistake of thinking that by doing so we will somehow grasp a whole (unparted) object in a way that will force it to cohere and thereby yield itself as visible—knowable—to our eyes?

It has become a trope of Anthropocene scholarship to note the destructive effects wrought by new forms of knowledge production birthed in the Enlightenment era—an era 'shaped by human beings' preoccupation', Allewaert writes, 'with uncovering, mapping, measuring, and (in most cases) instrumentalizing the natural world'.[31] Jason Moore describes these forms of knowledge-production as 'premised on a new quantitativism whose motto was: reduce reality to what can be counted, and then "count the quanta"'.[32] This epistemological practice is premised on a Cartesian dualism in which an 'objective' subject (the counter) who is, to reverse engineer Donna Haraway's words, 'disengaged' and 'from everywhere and so nowhere... free from interpretation, from being represented [...] fully self-contained [and] fully formalizable',[33] counts the quanta of an 'individually determinate entit[y] with inherent properties',[34] wholly external and therefore capable of being

---

31 Allewaert, *Ecology*, p. 9.
32 Jason W. Moore, *Capitalism in the Web of Life: Ecology and the Accumulation of Capital* (London: Verso, 2015), p. 211.
33 Donna Haraway, 'Situated Knowledges: The Science Question in Feminism and the Privilege of Partial Perspective', *Feminist Studies*, 14.3 (Autumn 1988), 575–99 (p. 590), https://doi.org/10.2307/3178066.
34 Barad, *Meeting*, p. 137.

anatomized. Moore suggests that this conception of the object as entirely external works to render objects better capable of being 'subordinated and rationalized, [their] bounty extracted, in service to capital and empire'.[35] In other words, the concept of the individual discrete object as something separate from the subject-observer is at the heart of settler-capitalist world-ecology.

Crucially, *Sweet Tooth* frames its colonial history in terms of this knowledge-production. The three-issue arc in which Thacker's narrative is related is entitled 'The Taxidermist', and a large panel early in #26 (Fig. 21) reveals that the titular taxidermist is Thacker himself, who occupies himself with taxidermy on the long sea voyage to the Arctic.

He is, he writes, 'a seasoned naturalist', whose 'hunger for adventure' and eagerness to catalogue 'exciting new species' motivates his journey as much as the need to find his lost brother-in-law. A panel depicts Thacker's tool stitching the pieces of a preserved bird together; the next reveals the ship's cabin that he has filled with other such birds and fish. Lemire's choice to title the arc after this single activity, which is never again seen or mentioned, suggests his awareness that it speaks to some larger theme of the episode, and indeed the comic as a whole. Certainly, a parallel exists between the fragmented, reassembled bodies of the animals that Thacker preserves and the skeletal remains of the Inuit gods' earthly bodies, even leaving aside the ways in which the disturbance and extraction of the latter evokes controversies surrounding the colonial-scientific misappropriation of indigenous remains. Where indigenous cultural prohibition forbids the disturbance of the gods' bodies while the gods are 'resting', in spite of the fact that these bodies are dead, Thacker's taxidermy speaks to a belief that animal bodies are fundamentally divorced from any former or future being—that they can be disassembled and rearranged according to his whim without any effect on other beings. Defining Thacker as 'taxidermist' centers this belief as his primary characteristic and positions his anatomization and reconstruction of animal bodies as representative of a larger onto-epistemology in which his deft hand with a needle sewing up fur and feathers does not really count as a touch; or rather in which a touch does not engineer a relation between two beings. This is an 'unsticky' world where division upon division can be smoothly and unproblematically

---

35   Moore, *Capitalism*, p. 18.

Fig. 21 Jeff Lemire, *Sweet Tooth* #26 (2011) © Jeff Lemire and Vertigo Comics. All rights reserved.

enacted or *un*enacted: a cell from an organ, an organ from a body, a body from its environment. It is a quality that is mirrored in the work of *Sweet Tooth*'s modern-day scientists, who indifferently dissect hybrid children in their quest for a plague cure—a resemblance that is highlighted by Lemire's cover art for the comic, which features, in #5, Gus' head mounted like a taxidermized hunting trophy and, in #8, his body preserved in a giant specimen jar.

To say that Thacker is the titular taxidermist is not to overlook the extent to which Louis Simpson also plays this role in the story. Simpson is superficially the less sinister of the two characters—a missionary who finds himself 'more and more enamored with [Inuit] ways' and comes to consider the Inuit 'so much more enlightened than [colonizers] are'.[36] Abandoning his mission, he marries into an Inuit community and happily adopts their ways. Yet his willingness to ignore his own instinct of boundaries and prohibitions—'Deep down I knew I wasn't meant to be here', he narrates, 'I knew that as much as I'd come to be accepted by these people, I was still *an outsider*'[37]—suggests the extent to which his 'going native' is part of a tradition whose basis reinscribes white male agency. Sara Ahmed has detailed how fantasies of becoming-other (and particularly, as in her reading of *Dances with Wolves*, becoming-indigenous) are enabled by a white male ability to make and unmake boundaries at will: 'the border between self and other, between natives and strangers'.[38] Moreover, Ahmed connects these fantasies to epistemology: 'the Western subject can *have* the difference and thus *knows* the difference'.[39] Knowledge here indicates an occupation that is also assimilation, or vice versa—a type of mastery that hearkens to Wittgenstein's questions regarding the nature of the connection between 'knowing' and 'knowing-how-to-go-on'.[40] Colonial knowledge practices elide the gap between these two zones, collapsing description and agency. The agency that Ahmed attributes to becoming, which she diagnoses as the 'ability to transform oneself', is of a piece with the capacity to anatomize, and therefore 'master', the natural world.

---

36 Jeff Lemire, *Sweet Tooth* #26.
37 Ibid.
38 Sara Ahmed, *Strange Encounters: Embodied Others in Post-Coloniality* (London: Routledge, 2000), p. 124, https://doi.org/10.4324/9780203349700.
39 Ibid.
40 Ludwig Wittgenstein, *Philosophical Investigations* (Oxford: Blackwell, 1953).

Simpson's perception of himself as the super-agential knower who can touch the other without being touched by it (becoming and *un*becoming the other as convenient) is akin to that of Thacker: the taxidermist. (It is also relevant that the two characters are said to have met in medical school.)

It is simple enough to read Simpson's crime, in this way, as one of incorrect relation—he did not behave towards the gods in the way that is correct. Yet, I want to press harder on this point. Simpson more specifically recounts the Inuit shaman as saying that the plague was 'the price of [his] betrayal. Mankind's price for disturbing the gods'. The syntactical apposition of 'betrayal' and 'disturbing the gods' suggests an equation: Simpson's crime is not violation of religious prohibitions, nor transgression into a sacred space, but his 'disturbance' of the bones in the cave, which are dead/inanimate and yet characterized by a form of being that is capable of being disturbed. I emphasize this in part because Povinelli[41] has compellingly explored how often indigenous tradition is enjoyed as an authentic difference until it offers substantial challenge to universalized liberal settler beliefs—humoured according to the 'profound asymmetry' that Wendy Brown notes as characterizing the 'culturalization of politics', wherein 'liberalism's conceit about the universality of its basic principles' demands that these principles and the culture of which they are representatives be perceived as not-culture, and liberalism therefore as cultureless.[42] An indigenous belief that a certain cavern is sacred, or that specific kinds of behaviour are forbidden within that cavern, might be enjoyed as harmlessly cultural; however, Simpson's actions are not described in terms of belief or culture. The Inuit, indeed, are absent from this issue, which is about what Simpson has done to the gods. What we are asked to accept is not a *belief* that Simpson has done something, but rather the more challenging assertion that *he has*—and that his actions are mirrored by the extractive efforts of scientists in our own era.

When the twenty-first-century Dr. Singh arrives in the cavern of the gods, which has been transformed into a scientific laboratory, he

---

41 Povinelli, *The Cunning of Recognition*.
42 Wendy Brown, *Regulating Aversion: Tolerance in the Age of Identity and Empire* (Princeton: Princeton University Press, 2006), p. 20–21, https://doi.org/10.1515/9781400827473.

discovers that the dead gods are still present, and they are still dead: their bones lie, exposed to the air, on slabs. Broken industrial incubation tanks mark where DNA was extracted from the bones and used to create new clones of the gods. Where Gus was successfully 'born' from his tank and raised by a runaway janitor (the only survivor when the plague swept through Fort Smith) the other new gods were left to smash their way out and become feral animal-children. Singh encounters these feral gods in a scene that sees him surrounded by the gods as both dead *and* living bodies: two different materializations of the same source. The simultaneous visibility of these materializations draws attention to their relation, foregrounding not only the historicity and potentiality of the separate-yet-inseparable Tekkeitsertok parts, but also the very real impact that disturbance of the one can have upon the other. Though the ancient bones may be inanimate, they are not *inert* insofar as, for example, Tekkeitsertok's bones participate in a meaningful and ongoing lineage of Tekkeitsertok relation that connects Gus to the slaughtered Inuit and beyond them to the dead bodies of the gods. The Inuit manner of describing this—that the bodies 'rest' while waiting for the gods to return—emphasizes the characterization of earlier material incarnations as dissolved-but-involved, dead-yet-responsive, past in a way that does not dissolve obligations on either side of the encounter. The disturbance of the gods, in this sense, manifests as an intrusion into this lineage: a disruption of the process of becoming, and one for which no acknowledgement is given or responsibility taken. This is not dissimilar to the intrusion into and disruption of indigenous cultures that colonialism causes, another parallel highlighted by the relationality of Thacker's and Gus' tales.

This resonance between human and nonhuman histories[43] is heightened by the comic's use of several panels (and one piece of cover art) in which Gus is pictured against a backdrop or atop a pile of dead human bodies (Fig. 22).

These bodies are not literally present, but represent the millions of humans who have died of the 'Sick' in the course of Gus/Tekkeitsertok's inadvertent 'cleansing' of the world. The bodies, unindividuated and compressed into the ground, might suggest the victims of colonialism

---

43 Sylvia Wynter has explored how human and nonhuman histories are linked together by their othering under a settler-colonial regime intent on easy capitalization.

Fig. 22 Jeff Lemire, *Sweet Tooth* #36 (2012) © Jeff Lemire and Vertigo Comics. All rights reserved.

on top of whom much of modern civilization has been constructed — or, equally, the masses of microscopic dead whose bodies have literally fuelled that same civilization. Meanwhile, the images, with their striking juxtaposition of childhood innocence and mass death, also confront the reader with the question of what the exact relation is between Gus and these dead. After all, the plague is entwined with Tekkeitsertok-being: described, in the 'taxidermist' interlude, as the 'breath' of Louis Simpson's hybrid son, and in its modern incarnation as something to which hybrid children are immune at a genetic level. ('Something fucked-up in yer DNA', Jepperd tells Gus.[44]) In #37, Dr. Singh insists that Gus was 'the carrier' of the plague, 'sent to kill all of us', and Gus seems to accept this idea, later repeating to his pig-girl friend, Wendy, that he 'carried the Sick, and whatever it was that made all the hybrids, in [him]', and that he 'killed everyone. [He] killed [his] daddy and [hers]'.[45] Even Jepperd, Gus' father-figure, admits that Gus' creation probably did cause the plague.[46]

Yet Jepperd, Wendy, and their human ally Becky are also reluctant to attribute culpability to Gus, perhaps not only because Gus is a child, but it is, on the face of it, absurd to view Gus as responsible for a phenomenon that he is only one part of. In the face of their inability to assign blame and, correlatively, agency in this context, these characters instead repeatedly assert the impossibility of any objective answers to the conundrum. Becky at first shifts responsibility to humans, before saying that she simply 'can't believe' that God sent Gus to kill people;[47] Jepperd renounces quantitativism, telling Gus, 'Truth is we ain't ever gonna understand how it happened... what the hell they were doing here in the lab... none of it. There is no big secret. At least not one that you or I or Singh or your daddy are ever going to be able to explain'.[48] Perhaps most interestingly, *Sweet Tooth*'s villain, the psychopathic Abbot, is left crazed by his inability to pry forth the answers from any text—notably from the flesh of hybrid bodies, written in the code of their 'fucked-up' DNA. 'What did Singh find?' Abbot demands when he reaches the Alaskan cavern. 'Where did *the boy* come from? *The plague?*'

---

44   Jeff Lemire, *Sweet Tooth* #2 (New York: Vertigo Comics, 7 October 2009).
45   Jeff Lemire, *Sweet Tooth* #38 (New York: Vertigo Comics, 3 October 2012).
46   Ibid.
47   Ibid.
48   Jeff Lemire, *Sweet Tooth* #37 (New York: Vertigo Comics, 5 September 2012).

Singh has already told him, however: 'There are no answers. At least not the ones you want'.⁴⁹ And, ultimately, Abbot is slain by Gus himself: the vexation of all of these questions made flesh insofar as he both *is* Tekkeitsertok and is *not* Tekkeitsertok, is not *all* Tekkeitsertok and is not *all of* Tekkeitsertok: a paradox of agency and wholeness that can find no resolution in the text.

It is easy to say that this paradox offers a critique of settler capitalist knowledge practices and their inability to encompass certain kinds of problems, and to argue that this is why the text itself mirrors the partedness it takes as topic. The metaconundrum of the relation between the Thacker and Gus parts of the narrative thus reflects the conundrum of the relation between Gus, the Sick, the dead bones of the gods, and the Inuit child: a problem of understanding the relation between parts. Yet, as I have previously referenced, critiques of settler capitalist knowledge practice are not novel—they are a trope of Anthropocene scholarship. What distinguishes the critique that *Sweet Tooth* offers is the ways in which its focalization through (principally) the character of Gus highlights a very specific problem that we are confronted with when we utilize our current knowledge practices to perceive the hypo-/hyperphenomena that are defined by what I have termed their 'partedness'. I have discussed the simultaneously macro- and micro-qualities of such objects, and the temporal distortions that render them simultaneously past and future, instantaneous and prolonged. Missing from this discussion, however, is attention to the question of why oil seems absent from our conversations about oil. In other words: between microscopic and macroscopic, between past and future, there ought surely to be a *here-now* of parted objects that never seems to appear. It is precisely when we try to fix our gaze on such a thing that it recedes into very small or very large spatiotemporal scales—becoming particulate or massively made up of parts. When we work to 'bring into focus' (gesturing back towards Hacking's work on microscopy) the *thing itself*, we cannot bring a thing into focus at our own scale. We are confronted by an absence that, often, we take as an essential characteristic, a quality of weirdness that defines (in the case of Morton) a special quality of object, or that, at the very least, becomes constitutive of a quasi-horror that births what Gry Ulstein describes as 'Anthropocene monsters', situated

---

49   Jeff Lemire, *Sweet Tooth* #39 (New York: Vertigo Comics, 7 November 2012).

in a spatiotemporal landscape (the Anthropocene) that is distinguished by 'disorientation and chaos[,] overwhelming confusion and terrifying realizations'.[50] Perhaps it is no surprise that this 'New Weird' would so often be identified as or with a kind of horror, when the absent presence of the *thing* functions as a kind of specter in the Derridean sense, agential in spite of its displacement in both space and time.

In fact, the specific genre in which *Sweet Tooth* most clearly participates is a kind of Arctic horror that clearly works in the sense Ulstein[51] suggests, as a metaphor for ecological issues. This genre addresses itself to the hyper-systems of resource extraction and climate change, focusing on the massive, world-destroying horrors that might be birthed from such systems, while simultaneously materializing these horrors as microscopically viral or parasitic. The 2015–2018 TV series *Fortitude*, for example, builds its central horror on the premise that an ancient parasite might be preserved in thawing permafrost, capitalizing on a popular news narrative that has spawned headlines about possible 'frankenviruses'[52] or 'zombie viruses',[53] reanimated after millennia, emerging due to anthropogenic global warming. *Sweet Tooth's* virus, released from beneath the Alaskan ice, clearly echoes these fears of the alien agent that is both too small and too large for us to engage with directly, too ancient and yet too futural (in its ability to define a coming apocalypse). Like *Fortitude*, *Sweet Tooth* also builds on media narratives about the emergence of animal bones and mummies from permafrost, themselves now a precious resource that 'prospectors' in the tundra extract.[54] These bones, parasites, viruses, oil, and rare earth minerals all emerge from the Arctic as haunting parts of a spectral whole that we cannot visibilize but sense must be there. The absence of this projected whole impels us to treat all of these parted objects as, well, *parts*: partial beings to which we need not attribute the kind of responsiveness

---

50 Gry Ulstein, 'Brave New Weird: Anthropocene Monsters in Jeff VanderMeer's *The Southern Reach*', *Concentric: Literary and Cultural Studies*, 43.1 (March 2017), 71–96 (p. 78), https://doi.org/10.6240/concentric.lit.2017.43.1.05.

51 Ibid., p. 74.

52 Chris Mooney, 'Why you shouldn't freak out about ancient "Frankenviruses" emerging from Arctic permafrost', *The Washington Post* (11 September 2015).

53 Michaeleen Doucleff, 'Are There Zombie Viruses—Like the 1918 Flu—Thawing in the Permafrost?', *NPR.org* (19 May 2020).

54 Andrew Roth, 'Permafrost thaw sparks fear of "gold rush" for mammoth ivory', *The Guardian* (14 July 2019).

that is characteristic of a whole being, though we trouble ourselves with supernatural visions of how this hallucinatory whole being, the Anthropocene monster, might respond were it to awaken.

Our failure to engage with parted objects on their own terms is evident in the kind of language we use for them. Ulstein notes how prominent Anthropocene theorists use language to describe the Anthropocene and its parts that is 'strikingly horror-evocative and apocalyptic', marked by evocations of 'malevolence', 'trauma', 'annihilations', 'intrusion', and 'terrors'.[55] Within the realm of petrohumanities, the language used to specifically describe oil is no more neutral: oil is 'dirty', 'toxic'; in the words of Juan Pablo Pérez Alfonso, 'the devil's excrement'. Scholars, as much as anyone, fall prey to this embedded moralizing: 'To unveil mounds of petrochemical debris and the fungus of derricks everywhere', Georgiana Banita writes,

> is akin to an autopsy on a body whose death, we are made to feel, could have been avoided. The images convey a surgical violence in their attention to malignant sprawl... The film of oil mingled with the earth's surface has the unfinished, un-chewed quality of something our bodies secretly consume or excrete.[56]

Stephanie LeMenager suggests that something of this expressive revulsion may result from the fact that oil's 'biophysical properties have caused it to be associated with the comic "lower bodily stratum"'.[57] Andrew Pendakis argues that oil is 'arguably the dirtiest of liquids [...] not just on the level of its (highly racialized) material properties (its blackness, its stickiness, its opacity, etc.), but on the terrain of its social and political usage'.[58] This discourse seems to subsume oil itself within a dread of the objects that it is read as part of—appalled and revolted by these beings in a way that denies and forecloses their potential to *have*

---

55 Ulstein, 'New Weird', p. 78.
56 Georgiana Banita, 'Sensing Oil: Sublime Art and Politics in Canada', in *Petrocultures: Oil, Politics, Culture*, ed. by Sheena Wilson, Adam Carlson, and Imre Szeman (Montreal: McGill-Queen's University Press: 2017), pp. 431–57 (p. 446).
57 Stephanie LeMenager, *Living Oil: Petroleum Culture in the American Century* (Oxford: Oxford University Press, 2014), p. 92, https://doi.org/10.1093/jahist/jav188.
58 Andrew Pendakis, 'This Is Not a Pipeline: Thoughts on the Politico-Aesthetics of Oil', in *Energy Humanities: An Anthology*, ed. by Imre Szeman and Dominic Boyer (Baltimore: Johns Hopkins University Press, 2017), pp. 504–11 (p. 506), https://dx.doi.org/10.17742/IMAGE.sightoil.3–2.2.

been and to *become* other things, the very quality that Povinelli identifies (rejecting the alternative of 'vitality') as intrinsic to being itself.

It is therefore the very partedness of parted things, their quality of binding together things in relations of unwholeness, that marks them as response-able. This is the aspect that the focalization of *Sweet Tooth*'s narrative through Gus most productively allows to emerge: not only is Gus, as a part of Tekkeitsertok, bound to the massacred Inuit community, but the comic also shows us his identification with the animal nonhuman (Fig. 23), the immanent sacred (Fig. 24), and the Alaskan landscape (Fig. 25)—the last through the skeletal face of Tekkeitsertok and, in one case, Gus himself shown haunting the clouds of the Arctic sky.

Fig. 23 Jeff Lemire, *Sweet Tooth* #1 (2009) © Jeff Lemire and Vertigo Comics. All rights reserved.

Fig. 24 Jeff Lemire, *Sweet Tooth* #20 (2011) © Jeff Lemire and Vertigo Comics. All rights reserved.

In one striking vision that Gus experiences, he encounters an ambiguous figure who appears to be Tekkeitsertok, but who looks extremely similar to the adult Gus we later see in *Sweet Tooth* #40. This figure's answer of 'Not yet' when Gus asks, 'Who?'[59] (*Sweet Tooth* #13) amplifies the sense that it gestures towards some potential that exists within Gus—yet another form, like the murdered Inuit child, like the deer he encounters, like the metaphysical form of Tekkeitsertok and the land, that he could have been/could be/could become. In many ways, *Sweet Tooth* is the story of how Gus—whose creation was also his disruption, an act of parturition and partition that brought him into being as a being separate (yet also inseparable) from another being—explores and reconciles himself to the paradoxes of his parted existence.

Perhaps, even beyond its mischievous invitation to imagine what oil or a virus might be like if it were a little boy with antlers growing out

---

59 Jeff Lemire, *Sweet Tooth* #13 (New York: Vertigo Comics, 1 September 2010).

Fig. 25 Jeff Lemire, *Sweet Tooth* #26 (2011) © Jeff Lemire and Vertigo Comics. All rights reserved.

of his head, *Sweet Tooth* asks us to reconcile ourselves to the paradoxes of *our* existence. After all, as Matthew Zantingh has observed, the comic 'calls on readers to witness the suffering of Indigenous lives at the hands of colonialism and to imagine a different future',[60] just as

---

60 Zantingh, Matthew, 'Tekkietsertok's Anger: Colonial Violence, Post-Apocalypse, and the Inuit in Jeff Lemire's *Sweet Tooth* Series', *Studies in Canadian Literature*, 45.1 (2020), p. 7.

Tekkeitsertok poses the same implicit demand to Gus—and, too, asks us to imagine the possibility of affinity with nonhuman others from the position of someone who looks (mostly) like us. Furthermore, as genetic and microbial humans embedded in Earthly ecological systems, we are also confronted by the tensions of partedness. Like Gus, we must ultimately ask ourselves: how can we conceive of ourselves as beings defined by affinities that cross species, scale, and animacy and interpret ourselves as existing in relation with a past and a future that we have responsibilities to?

Works Cited

Ahmed, Sara, *Strange Encounters: Embodied Others in Post-Coloniality* (London: Routledge, 2000), https://doi.org/10.4324/9780203349700

Allewaert, Monique, *Ariel's Ecology: Plantations, Personhood, and Colonialism in the American Tropics* (Minneapolis: University of Minnesota Press, 2013).

Appel, Hannah, Arthur Mason, and Michael Watts (eds), *Subterranean Estates: Life Worlds of Oil and Gas* (Ithaca: Cornell University Press, 2015), https://doi.org/10.7591/9780801455407-002

Banita, Georgiana, 'Sensing Oil: Sublime Art and Politics in Canada', in *Petrocultures: Oil, Politics, Culture*, ed. by Sheena Wilson, Adam Carlson, and Imre Szeman (Montreal: McGill-Queen's University Press, 2017), pp. 431–57.

Barad, Karen, *Meeting the Universe Halfway* (Durham: Duke University Press, 2007), https://doi.org/10.1215/9780822388128

Boetzkes, Amanda and Andrew Pendakis, 'Visions of Eternity: Plastic and the Ontology of Oil', *e-flux*, 47 (September 2013), [n.p.].

Brown, Wendy, *Regulating Aversion: Tolerance in the Age of Identity and Empire* (Princeton: Princeton University Press, 2006), https://doi.org/10.1515/9781400827473

Davis. Lennard J, *Enforcing Normalcy: Disability, Deafness, and the Body* (New York: Verso Books, 1995).

Doucleff, Michaeleen, 'Are There Zombie Viruses—Like the 1918 Flu—Thawing in the Permafrost?', *NPR.org* (19 May 2020).

Ferebee, K.M., 'Pain in Someone Else's Body: Plural Subjectivity in TV's *Stargate: SG-1*', *LLIDS: Language, Literature, and Interdisciplinary Studies*, 4.3 (Summer 2020), 26–48.

Fox, A.C. Heron, and M.Q. Sutton, 'Characterization of natural products on Native American archaeological and ethnographic materials from the Great

Basin region, USA: A preliminary study', *Archaeometry*, 37.2 (August 1995), 363–75.

Ghosh, Amitav, 'Petrofiction: The Oil Encounter and the novel', *The New Republic* (2 March 1992), 29–34.

Haraway, Donna, 'Situated Knowledges: The Science Question in Feminism and the Privilege of Partial Perspective', in *Feminist Studies*, 14.3 (Autumn 1988), 575–99, https://doi.org/10.2307/3178066

Harrell, J.A. and M.D. Lewan, 'Sources of mummy bitumen in ancient Egypt and Palestine', *Archaeometry*, 44.2 (May 2002), 285–93.

Helmreich, Stefan, *Alien Ocean: Anthropological Voyages in Microbial Seas* (Berkeley: University of California Press, 2009).

Helmreich, Stefan, *Sounding the Limits of Life: Essays in the Anthropology of Biology and Beyond* (Princeton: Princeton University Press, 2016), https://doi.org/10.1515/9781400873869

Kerber, Jenny, 'Up from the Ground: Living with/in Petrocultures in the US and Canadian Wests', in *Western American Literature*, 51.4 (Winter 2017), 383–9, https://doi.org/10.1353/wal.2017.0001

Kurlansky, Mark, *Salt: A World History* (New York: Walker & Co., 2002).

LeMenager, Stephanie, *Living Oil: Petroleum Culture in the American Century* (Oxford: Oxford University Press, 2014), https://doi.org/10.1093/jahist/jav188

Lemire, Jeff, *Sweet Tooth* (New York: Vertigo Comics, 2009–2013).

McDonald, Grantley, 'Georgius Agricola and the Invention of Petroleum', *Bibliothèque d'Humanisme et Renaissance*, 73.2 (2011), 351–63.

Mooney, Chris, 'Why you shouldn't freak out about ancient "Frankenviruses" emerging from Arctic permafrost', *The Washington Post* (11 September 2015).

Moore, Jason W., *Capitalism in the Web of Life: Ecology and the Accumulation of Capital* (London: Verso, 2015).

Morton, Timothy, *Hyperobjects: Philosophy and Ecology after the End of the World* (Minneapolis: University of Minnesota Press, 2013).

Pendakis, Andrew and Ursula Biemann, 'This Is Not a Pipeline: Thoughts on the Politico-Aesthetics of Oil', in *Energy Humanities: An Anthology*, ed. by Imre Szeman and Dominic Boyer (Baltimore: Johns Hopkins University Press, 2017), pp. 504–11, https://dx.doi.org/10.17742/IMAGE.sightoil.3-2.2

Povinelli, Elizabeth, *Geontologies: A Requiem to Late Liberalism* (Durham: Duke University Press, 2016), https://doi.org/10.1215/9780822373810

Povinelli, Elizabeth, *The Cunning of Recognition: Indigenous Alterities and the Making of Australian Multiculturalism* (Durham: Duke University Press, 2002).

Povinelli, Elizabeth, 'Do Rocks Listen? The Cultural Politics of Apprehending Australian Aboriginal Labor', *American Anthropologist* New Series, 97.3 (September 1995), 505–18, https://doi.org/10.1525/aa.1995.97.3.02a00090

Roth, Andrew, 'Permafrost thaw sparks fear of "gold rush" for mammoth ivory', *The Guardian* (14 July 2019).

Schrader, Astrid, 'The Time of Slime: Anthropocentrism in Harmful Algal Research', *Environmental Philosophy*, 9.1 (2012), 71–94, https://doi.org/10.5840/envirophil2012915

Schrader, Astrid, 'Abyssal intimacies and temporalities of care: How (not) to care about deformed leaf bugs in the aftermath of Chernobyl', *Social Studies of Science*, 45.5 (2015), 665–90, https://doi.org/10.1177/0306312715603249

Ulstein, Gry, 'Brave New Weird: Anthropocene Monsters in Jeff VanderMeer's *The Southern Reach*', *Concentric: Literary and Cultural Studies*, 43.1 (March 2017), 71–96, https://doi.org/10.6240/concentric.lit.2017.43.1.05

Wilson, Sheena, Adam Carlson, and Imre Szeman (eds), *Petrocultures: Oil, Politics, Culture* (Montreal: McGill-Queen's University Press, 2017).

Wittgenstein, Ludwig, *Philosophical Investigations* (Oxford: Blackwell, 1953).

Wynter, Sylvia, 'Unsettling the Coloniality of Being/Truth/Power/Freedom: Towards the Human, After Man, Its Overrepresentation—An Argument', *CR: The New Centennial Review*, 3.3 (Fall 2003), 257–337.

Yusoff, Kathryn, 'Anthropogenesis: Origins and Endings in the Anthropocene', *Theory, Culture & Society*, 33.2 (2016), 3–28, https://doi.org/10.1177/0263276415581021

Zantingh, Matthew, 'Tekkietsertok's Anger: Colonial Violence, Post-Apocalypse, and the Inuit in Jeff Lemire's *Sweet Tooth* Series', *Studies in Canadian Literature*, 45.1 (2020), 5–28.

9. Depopulating the Novel

Post-Catastrophe Fiction, Scale, and the Population Unconscious

Pieter Vermeulen

… je ne peux pas, personnellement, avoir d'espoir pour un monde trop plein.
—Claude Lévi-Strauss

The Population Unconscious

Having Kids was an organization dedicated to, as the 2019 version of its homepage had it, 'universal child-centric family planning policies that promote smaller families cooperatively investing more in every child'.[1] This baseline vertiginously traversed a wide array of scales: from the singular ('every child') to the universal, from the downscaled family to the aggrandized child. Still, the visual rhetoric of the website consistently privileged a smaller bandwidth of scales, as it prominently featured relentlessly cheerful images of families with one or two children and testimonials by families who have adopted the organization's model of delayed, reflexive, and constrained reproduction. Under the rubric 'The Facts Supporting Fair Start Family Planning', the website featured a colourful illustration demonstrating the scalar satisfactions of smaller families (see Fig. 26). The image promises easy scalar mobility (and here I am scanning the image counter-clockwise): between the (consciously small) nuclear family, a network of similarly sized families, and an eminently manageable (not too much) larger community. Not

---

1 'Homepage', *Having Kids.org* (7 June 2019), https://havingkids.org.

only does this scale of life promise plenty of room for date nights where participants can celebrate what the image calls 'well-being' and 'gender equality', it even boasts well-deserved me-time where individuals can withdraw and enjoy their intimacy with nature, away from their loved ones.

Fig. 26 Having Kids, *The Benefits of Smaller Families* (2018) © Fair Start Movement. All rights reserved. https://havingkids.org/wp-content/uploads/2018/06/New-2100-graphic-rearranged.jpg

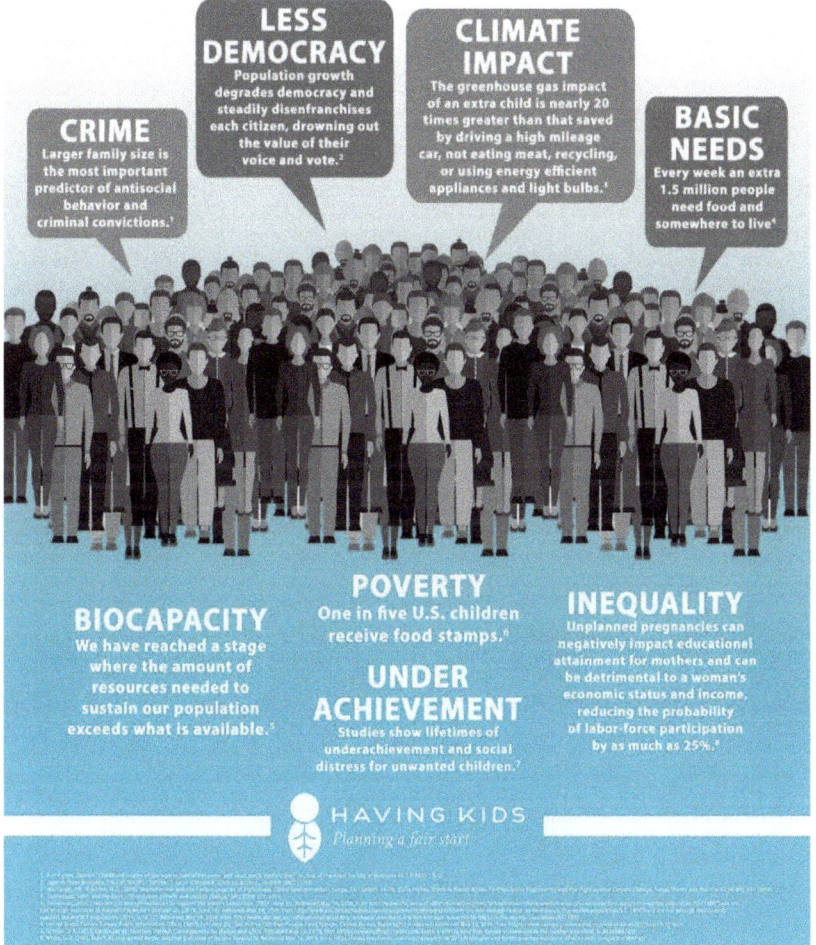

Fig. 27 Having Kids, *The Consequences of Poor Family Planning* (2018) © Fair Start Movement. All rights reserved. https://havingkids.org/wp-content/uploads/2018/07/New-2100-graphic-part-2-1.jpg

All this is only possible, the website implied, because these loved ones are not too numerous. Remarkably, the website did not really consider the plight of the unloved many—the people who will have to disappear (or stop reproducing) for the lifestyle promoted here to (literally)

find a place within the lifetime of the website's audience. Indeed, the reality of demographic excess only surfaced in one image that contrasts with the previous one: illustrating 'The Consequences of Poor Family Planning', the visual shows a greyscale drawing of an undifferentiated crowd surrounded by slogans expressing the miseries of overcrowded life (see Fig. 27). Most of the website was dedicated to the affirmation of a particularly scaled mode of life: small and manageable, egalitarian, with constant face-to-face interaction. The reality that this solution would be hard to scale to a planet populated by 7.7 billion people was not made explicit. The attractive lifestyle the website advanced depends on a disavowed population politics—what I will call its 'population unconscious': it leaves implicit the inconvenient fact that implementing its vision of the good life within the lifetime of the audience it addresses would require a spectacularly rapid demographic decrease.

This chapter examines a much broader—and, I want to argue, deeply problematic—tendency to articulate a vision of the good life while obfuscating the need to get rid of the majority of the world population for that vision to materialize. It is not that such initiatives *obscure* the question of scale: their blueprints for thriving are emphatically *scaled*— they belong to a particular scale domain (ranging from a few dozen to a few hundred individuals)—but foreground the scale of their particular version of the good life only to sideline the challenge of demographic surplus and the fact that their templates for human thriving do not seem amenable to scaling up. What I call the underlying 'population unconscious' is as ineluctable as it is unspeakable, and has, as I will argue below, existed at least since Malthus.

What makes this relevant for literary scholars is that this combination of affirmation (of a particular scale of flourishing) and disavowal (of a population politics based on radical decrease) also marks the popular genre of post-catastrophe fiction. While some of the most iconic instances of the genre—such as Cormac McCarthy's relentlessly grim *The Road* (2006) or Alfonso Cuarón's gruelling *Children of Men* (2006)—are clearly not committed to an imagining of the good life, other examples of post-catastrophe fiction (or post-, neo-, crypto-, or ana-apocalyptic fiction, a terminological confusion which testifies to the genre's affective and temporal complexity), are much less depressing and bleak, and instead present post-catastrophe life as somehow desirable.² In novels like Emily St. John Mandel's *Station Eleven* (2014), which will be my main touchstone in this essay, post-catastrophe life is presented as pleasant and attractive.³ These novels sideline the fact that this life is only attractive for a limited group of survivors, and only *because* the group of survivors is limited. Through their engrained survival bias, which draws attention away from the mass extinction that powers them, these novels invite readers to invest in their visions of the good life while participating in their disavowal of demographic catastrophe. By outsourcing that catastrophe to a nonhuman, biological force—a pandemic in the three novels studied here—they distract from the realization that their templates for survival *depend* on demographic elimination.

But is it really plausible that these novels' imaginings of post-catastrophe life have an encrypted normative force—that the lives they represent are meant to be desirable? Surely there is a difference between a nakedly programmatic family planning initiative like Having Kids and the fairly sophisticated works of upper middlebrow fiction? The rest of this essay argues that, while the normative force of these fictions is vague enough to afford them plausible deniability, the difference is not as great as one might be inclined to think. I make this point by situating the genre of post-catastrophe fiction in three different generic genealogies—all of which participate in the longer history of literary engagements with population: utopian fiction (which, as I explain in detail below, has always imagined the good life on a particular, limited scale), science fiction (where, as I show, anxieties about over- or de-population are traditionally confronted through the genre's defining spatiotemporal displacements), and the realist novel (which recent scholarship has shown to negotiate the pressures that growing populations put on political and aesthetic forms). To the extent that post-catastrophe fiction recombines elements of all three trajectories in a conjuncture that also generates initiatives like Having Kids, it borrows their normative and affective dimensions in a way that make their avoidance of population issues problematic at best. As we will see, at a time when there is no

---

2   Cormac McCarthy, *The Road* (New York: Vintage, 2006); Alfonso Cuarón, *Children of Men* (Universal Picture, 2006), DVD; for the variety of terms, see Heather Hicks, *The Post-Apocalyptic Novel in the Twenty-First Century: Modernity beyond Salvage* (New York: Palgrave Macmillan, 2016), p. 6, https://doi.org/10.1057/9781137545848.

3   Emily St. John Mandel, *Station Eleven* (Basingstoke: Picador, 2015).

longer any scientific evidence that a 'population bomb' is looming (as research published in such venues as *The Lancet* makes clear, and as the widely reported falling fertility rates during the COVID-19 crisis in both Asia and the West have made inescapable), the fantasy of a less overcrowded planet is displaced, and hides behind the pleasing images of decluttered life that post-catastrophe fiction and certain family planning initiatives project. These works and initiatives provide what Fredric Jameson has called 'blueprints for bourgeois comfort' that offer little in the way of a solution to planetary challenges.[4]

Cosy Catastrophe

Post-catastrophe fiction's promotion of particular versions of the good life is not a recent phenomenon. Already in 1973, British science fiction writer Brian Aldiss identified the 'cosy catastrophe' subgenre of apocalyptic fiction: a spate of fictions from the 1950s and 1960s, the most famous of which is John Wyndham's *The Day of the Triffids*, where 'global disaster is survived by a typically prosperous remnant that adapts, with reasonable aptitude and plenty of common sense, to the new conditions of a post-collapse world'.[5] Aldiss' term ('cosy') underlines the decidedly tame and diminutive natures of post-catastrophic comfort—his most enthusiastic description of Wyndham's works calls them 'urbane and pleasing'. His most direct definition of the subgenre directly links this affective downscaling to a more radical diminishment: 'The essence of cosy catastrophe is that the hero should have a pretty good time (a girl, free suites at the Savoy, automobiles for the taking) while everyone else is dying off'.[6] Post-catastrophe's diminished pleasures are only available to reduced numbers of the human race; in most cosy catastrophe fictions, the imagining of the good life is intimately connected to an eliminative population politics—a connection that Aldiss' formulation intimates ('while') even while the works themselves bury it underneath the narrative pleasures they afford.

---

4   Fredric Jameson, *Archaeologies of the Future: The Desire Called Utopia and Other Science Fictions* (London: Verso, 2007), p. 12.
5   Andrew Tate, *Apocalyptic Fiction* (London: Bloomsbury, 2017), p. 8, https://doi.org/10.5040/9781474233545-004.
6   Brian Aldiss, *Billion Year Spree: The History of Science Fiction* (London: Weidenfeld and Nicolson, 1973), pp. 315–16.

Emily St. John Mandel's *Station Eleven* (2014) is a case in point. The gentle and subdued nature of the world the novel imagines is already apparent in its (at least) double generic affiliation: *Station Eleven* gained critical acclaim both as a work of science fiction—it won an Arthur C. Clarke Award—and as a work of literary fiction—being shortlisted for the National Book Award and the PEN/Faulkner Award. This double consecration is enabled by the novel's decision to organize itself around a genre element—a pandemic that wipes out 99.6% of the world population—without dwelling on the grim horrors and deprivations such events typically unleash in works like *World War Z* or *The Walking Dead*:[7] the novel studiously skips 'the first ten or twelve years after the collapse' and shifts to the 'calmer age' that follows it.[8] Indeed, the most conspicuous aspect of the life of the survivors is how cosy and pleasant it is. There is a travelling group of actors who perform Shakespeare for the communities they encounter; there is a character living the life with his wife, his son, his newborn, and a puppy, 'baking bread in an outdoor oven'; there is, most notably, a small community establishing itself in an airport terminal and organizing itself around an improvised 'Museum of Civilization' containing defunct iPhones, stiletto heels, and Nintendo consoles.[9] Life here, as Caroline Edwards remarks, is set in 'a pastoral world of slowed-down time that asserts the mundane and the domestic over the catastrophic and the dramatic'.[10]

In *Station Eleven*, life after collapse is nothing if not desirable—at least, for people who are not part of the 99.6%, because for them, there are no outdoor ovens, only the burning pits where contaminated corpses

---

7   These works, significantly, feature images of mindless and uncontrollable zombie hordes, as if to underscore that the horror of the end of times essentially has to do with surplus populations; the demographic excess figured in the zombie apocalypse, which is typically presented as a teeming horde of mindless bodies invading all residual comfort zones, is the flip-side of a novel like *Station Eleven*'s minimalism; if a novel like *Station Eleven* disavows the fear of overpopulation, the zombie apocalypse hysterically acts it out. Andreu Domingo has argued that contemporary zombie 'demodystopias' function as technologies for elaborating the 'radical split between [...] "redundant" and the "resilient"' populations'. Andreu Domingo, 'Resilient Evil: Neoliberal Technologies of the Self and Population in Zombie "Demodystopia"', *Utopian Studies*, 30 (2019), 444–61 (p. 446), https://doi.org/10.5325/utopianstudies.30.3.0444.

8   Mandel, *Station*, p. 145.

9   Ibid., pp. 312, 255, 258.

10 Caroline Edwards, *Utopia and the Contemporary British Novel* (Cambridge: Cambridge University Press, 2019), p. 161, https://doi.org/10.1017/9781108595568.

end up. Or so readers can presume, because the novel does not disclose any details, as its economy of attention conspicuously skips the realities of mass extinction and focuses readerly attention on the surviving few, not the superfluous many. If the novel kills off 99.6% of the world population, that means that 0.4% survive—which is to say (although the novel does not tell us), 28 million people. Yet the novel's imagination is set at an even smaller scale: we read that '[t]here were 320 people living in the Severn City Airport that year, one of the largest settlements Kirsten had seen'.[11] *Station Eleven*'s imagining of a more or less pleasant post-pandemic life—which leads Caroline Edwards to define it as an example of 'pastoral post-apocalypse'—invites readers to forget what Aldiss' definition of the cosy catastrophe subgenre foregrounds: that the pleasures of post-pandemic domesticity depend on the vast majority of the world population 'dying off'.[12] The novel's refrain is *'Because survival is not sufficient'*—which is cast as an answer to the implicit question of why we need a meaningful and comfortable life. What does not enter the novel is the perspective of the 99.6% for whom survival would, if not sufficient, at least have been something.

This logic is not exclusive to *Station Eleven*. In Margaret Atwood's tonally very different *MaddAddam* trilogy, for instance, which outsources its demographic solution to the workings of an evil genius who creates a lethal virus, the final volume presents a small community of human survivors living a simple but nurturing life (together with the Crakers, an improved posthuman species) in a fortified cobb house. The novel intercuts the story of the simple life in this 'makeshift community of sustenance and care', sustained by a simple vegetable garden and beehives, with analepses to the brutal and violent life of some of the characters in the pre-catastrophe world.[13] This narrative organization underscores that the life at the cobb house is a desirable retreat 'away from the urban rubble'.[14] And in the post-pandemic world, the large numbers of city life are no longer a threat: life feels like 'a vacation of sorts', but '[t]hey aren't escaping from daily life. This is where

---

11   Mandel, *Station*, p. 306.
12   Edwards, *Utopia*, p. 161.
13   Shelley Boyd, 'Ustopian Breakfasts: Margaret Atwood's *MaddAddam*', *Utopian Studies*, 26 (2015), 160–81 (p. 161), https://doi.org/10.5325/utopianstudies.26.1.0160.
14   Margaret Atwood, *MaddAddam* (London: Virago, 2014), p. 6.

they live now'.¹⁵ The former world was divided between secured elite Compounds on the one hand and the slums, suburbs, and malls that make up the 'pleeblands' on the other. If this old world was riven by violence and hate, the post-pandemic world overcomes violence (a large part of the plot tells the story of how the survivors unite to attempt to kill off their last enemies). This neat opposition between the violent and overcrowded past and the downscaled and peaceful present only holds up, however, if we forget that the survivors' pleasures depend on the death of most of the population. As in Mandel's novel, the comforts of simplicity and scarcity serve to obfuscate a fairly sinister population politics.

In a disgruntled review of a spate of post-catastrophe fictions, Ursula Heise has remarked on the genre's disavowed demographic politics. In these works' depiction of a return to a simpler life, '[w]hat really counts is that the characters, in their break from the corruptions of the past, no longer have to deal with things like crowded cities, cumbersome democracies, and complex technologies. Whatever the hardships of their lives may be, they are better off without the world of corporations, biotech, and the Internet—even, apparently, at the price of genocide'.¹⁶ Such an insight into what I call the genre's population unconscious has not been an explicit focus in the scholarly books about the genre that have begun to appear in recent years: Heather Hicks' *The Post-Apocalyptic Novel in the Twenty-First Century*, Andrew Tate's *Apocalyptic Fiction*, which, in spite of the different genre named in the title, deals with many of the same texts, and Diletta De Cristofaro's *The Contemporary Post-Apocalyptic Novel*, which again covers the usual suspects (*Station Eleven*, *The Road*, Will Self's *The Book of Dave*, the novels of David Mitchell).¹⁷ While these books offer sophisticated accounts of the forms and functions of such fictions, they do not explore to what extent they serve to channel particular disavowed population fantasies. In order to fill in this gap, the next sections situate the genre in different generic genealogies that, I argue, cumulatively endow their scaled depictions of the good life with

---

15 Ibid., p. 184.
16 Ursula Heise, 'What's the Matter with Dystopia', *Public Books* (2 January 2015), https://www.publicbooks.org/whats-the-matter-with-dystopia/.
17 Hicks, *Post-Apocalyptic*; Tate, *Apocalyptic*; Diletta De Cristofaro, *The Contemporary Post-Apocalyptic Novel* (London: Bloomsbury, 2020).

a peculiar affective and normative force: science fiction (which I situate in the context of popular scientific concerns over demographic growth), utopian fiction, and the realist novel.

Population between Science and Speculation in Science Fiction

Contemporary fiction is not the only place where questions of population scale figure only indirectly. Even while questions over the impact of human action on the ecological and chemical makeup of the planet have taken an increasingly prominent place in environmental thought, it has been remarkably reluctant to confront the question of population head-on. Environmental ethicist Patrick Curry has pointed to a taboo on talking about 'the P-word' and uses Sandy Irvine's term 'overpopulation denial syndrome (ODS)'.[18] Ecocritic Greg Garrard has noted the failure of environmental literary studies to talk about the rescaling of the human population. For Garrard, art and literature are regrettably better at imagining an empty world without humans than a 'world with *far fewer* of us'.[19] Political theorist Diana Coole talks about a vast 'disavowal of the population question', and identifies 'five categories of silencing discourse' that make it much more difficult for people to raise the issue of demographic growth.[20]

Taboo, denial, disavowal, silence, shame: these terms strongly indicate that population discourse is never merely scientific, but always overdetermined by fantasies and affective investments. This is all the more obvious because contemporary science no longer supports the idea that overpopulation is a major planetary problem. Given that the ecological footprint of affluent Westerners is vastly larger than that of the regions where the population is actually growing (most current and projected growth being situated in Africa), pointing to population numbers as an index of environmental impact reveals a disregard for global

---

18 Patrick Curry, *Ecological Ethics: An Introduction* (Cambridge: Polity, 2006), pp. 123–24.
19 Greg Garrard, 'Worlds Without Us: Some Types of Disanthropy', *SubStance*, 41 (2012): 40–60 (p. 59), https://doi.org/10.1353/sub.2012.0001.
20 Diane Coole, 'Too Many Bodies? The Return and Disavowal of the Population Question', *Environmental Politics*, 22 (2013), 195–215 (p. 197), https://doi.org/10.1 080/09644016.2012.730268.

wealth inequalities. As Diana Coole notes, what makes problematizing population almost shameful is 'a pervasive suspicion that limiting population actually means limiting certain categories of people who are deemed redundant or undesirable'.[21] Contemporary population discourse also suffers from its association with the disputable political track record of earlier reflections on the environmental ramifications of population growth—not only in the eugenic tradition, but also in what Michelle Murphy has analyzed as the 'historically specific regime of valuation' she terms 'the economization of life', which held population to be adjustable and manageable in relation to the macroeconomy of the nation-state.[22] Discussions on the impact of demographic growth cannot help but resonate with the earlier interventions of Thomas Malthus' *Essay on the Principle of Population* (1798) and Paul and Anne Ehrlichs' *Population Bomb* (1968), even if these earlier works' emphasis on food security has been replaced with a focus on ecological footprints, carbon emissions, and the threat of global pandemics. Such discussions inevitably slip from the domains of ecology and biology into more ethically and politically fraught terrain, where those concerns end up taking on more sinister biopolitical connotations.

Population increase only became a critical concern with the work of Thomas Malthus. As Emily Steinlight underlines, demographic increase was an unambiguous marker of social wellbeing until the end of the eighteenth century.[23] Malthus famously made the case that the planet would not be able to sustain population growth: while food production can increase at only arithmetic rates, populations tend to grow exponentially at geometric rates, and the result is a mismatch that will threaten the lives of many. Malthus' fears of overpopulation are often traced back to his concerns over the French revolution (whose danger might be located in 'the sheer numbers—the swarm-like quality of the poor');[24] they were formulated before demographic data about England

---

21 Coole, 'Too Many', p. 199.
22 Michelle Murphy, *The Economization of Life* (Durham: Duke University Press, 2017), pp. 5–6.
23 Emily Steinlight, *Populating the Novel: Literary Form and the Politics of Surplus Life* (Ithaca: Cornell University Press, 2018), p. 4.
24 Alex McCauley, 'The Promise of Disaster: Specters of Malthus in Marxist Dreams', *Ecozon@*, 9 (2018), 53–65 (p. 63), https://doi.org/10.37536/ECOZONA.2018.9.1.1649.

were even available and were rooted less in empirical observation and ecological awareness and more in principles of macroeconomic thought. The spectre of overpopulation, from its Malthusian origins despite its claim to scientific credentials, is a matter of speculation and projection— as Frances Ferguson has it, less 'a response to the pressure of too many bodies' than to 'the felt pressure of too many consciousnesses'.[25] Malthus' writings evoke the Kantian figure of the *mathematical sublime:* a plurality that is immeasurable and uncontainable and overwhelms the human's rational capacities. Crucially, overpopulation is here *not* an empirical observation—the data was simply not available—but a projection powered by a fantasy of withdrawal, control, and containment—a downscaling operation that is still at work in contemporary post-catastrophe fiction.

The return of the population question in the 1960s and 1970s *was* supported by biological knowledge and real science. Garrett Hardin, who wrote 'The Tragedy of the Commons' in 1968, was a professor of human ecology (as well as a eugenicist and a white supremacist), while the crucial notion of 'carrying capacity' made its way from biology into demography and human ecology.[26] Paul and Anne Ehrlich, who wrote the bestselling *The Population Bomb* in the same year, were conservation biologists, and this allowed them to present a detailed vision of the planetary deterioration that would befall an exhausted planet unable to feed its human population. The very title of the book obviously resonated with contemporaneous fears over a coming nuclear winter, as the book balanced the Malthusian fear of overcrowding with an ecological fear of environmental exhaustion. The book's tagline underlines this double movement, simultaneously threatening exhaustion and overcrowding: 'While you are reading these words four people, most of them children, will die of starvation—and twenty-four more babies will have been born'.[27] The dash between the two declarations and the future anterior ('will have been') suggests that Malthus' fear was materializing *right now*; the affect it inspired was dread and panic.

---

25 Frances Ferguson, *Solitude and the Sublime: The Romantic Aesthetics of Individuation* (London: Routledge, 1992), p. 114.
26 Sabine Höhler, '"Carrying Capacity": The Moral Economy of the "Coming Spaceship Earth"', *Atenea*, 26 (2006), 59–74 (p. 70).
27 Paul Ehrlich, *The Population Bomb*, revised and expanded edition (New York: Ballantine Books, 1971).

The scientifically grounded fears of the Ehrlichs and Hardin found their way into '60s and '70s overpopulation dystopias, where they intersected with concerns over eroding class privilege. As Ursula Heise has shown, these works tended to cast overpopulation as an urban phenomenon marked by 'nightmarish crowding and the erosion of individual privacy'. Observations of global demographic excess beyond the planet's carrying capacity were reflected in 'quite class-specific paranoias' and concerns over inconveniences afflicting metropolitan middle-class life.[28] The abiding fear in these dystopias, as Fredric Jameson notes, 'is that of proletarianization [...] of losing a comfort and a set of privileges which we tend increasingly to think of in spatial terms: privacy, empty rooms, silence, walling other people out, protection against crowds and other bodies'.[29] Eva Horn has noted that both *The Population Bomb* and the 1973 ecological dystopian thriller *Soylent Green* (a film based on Harry Harrison's 1966 novel *Make Room! Make Room!*) begin with the image of a third-world overcrowded city.[30] In these works, the reality of demographic growth is represented, not disavowed under the guise of a pandemic or natural disaster. The support for the diagnosis of rampant population growth in ecology (Hardin), biology (the Ehrlichs), and interdisciplinary environmental research (as in the Club of Rome's famous 1972 report on *The Limits of Growth*, drafted by a team of 17 researchers) gave population politics a prominent place in these 'demodystopias'.[31]

One reason for the more recent backgrounding of population politics is that scientific evidence has reliably failed to provide evidence for unchecked population growth. Works of popular science such as *Peoplequake* and *Too Many People?* downplay the dangers of overpopulation, and a 2009 forum in *The New Scientist* on 'The Population Delusion' is similarly unimpressed.[32] A study published in *The Lancet* in the summer

---

28   Ursula Heise, 'The Virtual Crowd: Overpopulation, Space, and Speciesism', *ISLE*, 8 (2001), 1–29 (p. 2), https://doi.org/10.1093/isle/8.1.1.
29   Fredric Jameson, *Postmodernism, Or, The Cultural Logic of Late Capitalism* (Durham: Duke University Press, 1989), p. 286.
30   Eva Horn, *The Future as Catastrophe: Imagining Disaster in the Modern Age* (New York: Columbia University Press, 2018), pp. 117–18.
31   Andreu Domingo, '"Demodystopias": Prospects of Demographic Hell', *Population and Development Review*, 34 (2008), 725–45, https://doi.org/10.1111/j.1728-4457.2008.00248.x.
32   Fred Pearce, *Peoplequake: Mass Migration, Ageing Nations and the Coming Population Crash* (New York: Eden Project, 2011); Ian Angus and Simon Butler, *Too Many People?*

of 2020 predicts that the world population will not exceed ten billion in this century and begin to decline as early as 2064—a position echoed in a work of popular science like *Empty Planet: The Shock of Global Population Decline*.[33] Already in 2001, Ursula Heise noted that '[o]verpopulation has lost its terror for the Western imagination' and that 'so many of the Western overpopulation dystopias of the 1960s and 1970s seem dated by now'.[34] This lack of scientific alarmism makes apparent the 'dubious cultural politics' of earlier fictions engaging with demographic growth, which retain traces of a discredited eugenics, elitism, and racism— nowhere more clearly, perhaps, than in Jean Raspail's racist classic *The Camp of the Saints* (1973), which depicts a future France overtaken by nameless Third World hordes.[35] It also makes clear that population fantasies will require updated literary templates. Contemporary post-catastrophe fiction is one such template where population fantasies and paranoias survive the erosion of their evidentiary basis.[36]

Survival at Scale in Post-Catastrophe Science Fiction

Contemporary post-catastrophe fiction no longer draws on the tension between the multitude and the individual that has dominated demographic speculation since Malthus; instead, it resolutely focuses on minor scales that it presents as desirable while removing all too palpable images of overcrowding.[37] Post-catastrophe fiction imagines an alternative to overcrowding, not the reality of crowdedness. The vision

---

Population, Immigration, and the Environmental Crisis (Chicago: Haymarket Books, 2011); 'The Population Delusion', *New Scientist* (23 September 2009), https://www.newscientist.com/round-up/population/.

33 Stein Emil Vollset, et al., 'Fertility, Mortality, Migration, and Population Scenarios for 195 Countries and Territories from 2017 to 2100: A Forecasting Analysis for the Global Burden of Disease Study', *The Lancet*, 396 (2020), 1285–306, https://doi.org/10.1016/S0140-6736(20)30677-2; Darrell Bricker and John Ibbitson, *Empty Planet: The Shock of Global Population Decline* (New York: Crown, 2019).

34 Heise, 'Virtual', pp. 1–2.

35 Lionel Shriver, 'Population in Literature', *Population and Development Review*, 29 (2003), 153–62 (p. 158), https://doi.org/10.1111/j.1728-4457.2003.00153.x.

36 Editors' note: In chapter 7, Rishi Goyal studies how comparable 'population fantasies and paranoias' shape the imagining of the pandemic city in Ling Ma's *Severance* (2018), notably through its use of uniformising descriptions 'from above'.

37 See Timothy Clark, '"But the real problem is....": The Chameleonic Insidiousness of "Overpopulation" in the Environmental Humanities', *Oxford Literary Review*, 38 (2016), 7–26 (pp. 16–18), https://doi.org/10.3366/olr.2016.0177.

of the good life that novels like *Station Eleven* and *MaddAddam* project is not imagined as scalable. Instead, the scale on which they operate (the small community) is intrinsic to their visions; they promote a Marie Kondo-style minimalism rather than indulgent excess. As one of the characters in Atwood's cobb house muses, '[o]nce, there were too many people and not enough stuff; now it's the other way round'. In the cobb house, there is enough stuff, but not too much, as '[n]ow that history is over, [they]'re living in luxury, as far as goods and chattels go'.[38] The cobb house is developed according to a hipster aesthetic marked by vintage quaintness: a hand pump that used to be 'a retro decoration' becomes both 'the source of their drinking' and a source of delight for the children; the structure of the house survives as 'ersatz antiquity, like a dinosaur made of cement'.[39] Even at the serene end of the story, DIY projects continue apace, as extensions, a nursery, and extra solars are being developed. While these initiatives point forward to a better future, there are no indications that this future will be scaled differently and that life will return to pre-pandemic numbers.[40]

The intrinsic connections between the good life and scarcity are nowhere clearer than in the way *Station Eleven* affirms the value of commodities that make up the 'Museum of Civilization'. These objects are repeatedly said to be 'beautiful', and their beauty is directly linked to their finitude. The characters 'had always been fond of beautiful objects, and in [their] present state of mind, all objects were beautiful'.[41] When one of the child actors in the play that opens the novel, Kirsten, who will survive and play a prominent role in the narrative's post-apocalyptic strand, receives a paperweight on the eve of destruction, it is described as 'the most beautiful, the most wonderful, the strangest thing'; after the collapse, she still finds it 'nothing but dead weight' yet 'beautiful'.[42] After the collapse, the world is rendered more lovely by the prospect that '[p]erhaps soon humanity would simply flicker out', which releases '[t]he beauty of this world where almost everyone was gone'.[43] Even an abandoned Toronto strikes one survivor with '[a] stark and unexpected

---

38   Atwood, *MaddAddam*, p. 45.
39   Ibid., pp. 55, 116.
40   Ibid., p. 458.
41   Mandel, *Station Eleven*, p. 255.
42   Ibid., pp. 15, 66.
43   Ibid., p. 148.

beauty, silent metropolis, no movement'.[44] It is only when cars have stopped driving and planes have stopped flying that people 'recognize the beauty of flight'.[45] The removal of excess populations, in other words, creates room for the value of objects and experiences to emerge. The removal of the boredom and waste of contemporary life generates beauty. The inverse observation, that it is the overabundance and excess of our current overpopulated lives that robs the world of its beauty, is left implicit—tangible enough for readers to indulge in, but vague enough that it affords both the novel and its readers plausible deniability.

The association between fantasy and scale is so pervasive that it also appears in Ling Ma's post-apocalyptic zombie novel *Severance*, a work that strongly rejects *Station Eleven*'s and *MaddAddam*'s belief in the promise of small pockets of survivors, since it shows a small commune fall prey to authoritarianism and violence. The novel's protagonist flees a collapsing New York to join a group of survivors; neither of these scales suit her, and she ultimately leaves for an emptied Chicago. The crew's stalking missions, in which they indulge the lure of commodities in abandoned houses, offers an aesthetic experience not that different from what *Station Eleven*'s museum affords: 'Room by room, we amassed boxes', in a looting project involving 'empty boxes and garbage bags' and 'supply vans'.[46] Scavenging is figured as 'an aesthetic experience', in which Candace, very much like a post-pandemic Marie Kondo, 'would get lost in the taking of inventory, with the categorizing and gathering, the packing of everything into space-efficient arrangements'.[47] Again, it is the emptiness of the houses they visit (and the demographic reduction on which it depends) that enables experiences of beauty and that phantasmagorically converts scarcity into value, converts survival into a vision of the good life.

This logic also accounts for the peculiar organization of the novel: while Candace's life story up to the pandemic and her stay with the survivors are narrated non-chronologically in (mostly) alternating chapters, the novel builds toward the period of Candace's stay in an abandoned New York as a somehow almost benign period. When the city is deserted, 'everything seemed to take longer. The city was

---

44 Ibid., p. 182.
45 Ibid., p. 247.
46 Ling Ma, *Severance* (London: Picador, 2019), p. 65.
47 Ibid., pp. 58, 65.

operating on a different kind of time', an alteration she takes it upon herself to document—'documenting the deserted city', 'deserted but not abandoned'.[48] As a cab driver tells her, one reason not to abandon New York is that '[i]t's too beautiful not to enjoy', especially when '[t]he fevered stumbled around New York in ever-diminishing numbers'.[49] As in *Station Eleven* and *MaddAddam*, the intricate relation between diminishment and desire obscures the horror of demographic reduction by substituting it with a palpable image of the enjoyment of small numbers.

Utopian and Realist Fictions

If the genre of post-catastrophe fiction is informed by deep affective investments and an eschewal of scientific insight, as I have argued, an appreciation of its intrinsic normative claims requires an exploration of its affiliation with utopian and realist fiction. Indeed, the idea that literature's imagining of life has normative value is nowhere more apparent than in the genre of utopia—and post-catastrophe fiction's reliance on utopian devices transports this normativity to its imagining of community. Crucially, scale is intrinsic to the make-up of the forms of community that utopia imagines. Lyman Tower Sargent, one of the leading thinkers of utopia, notes that nineteenth- and twentieth-century imaginings of utopia typically posit so-called 'intentional communities', which he defines as 'a group of five or more adults and their children, if any, who come from more than one nuclear family and who have chosen to live together to enhance their shared values or for some other mutually agreed upon purpose'.[50] Size is often central to the definition of an intentional community: the scale of utopian thought is typically that of the 'communal experiment' or the 'commune'.[51] As Fredric Jameson has it: the 'insistence on *the small group itself* [...] is the libidinal fountainhead of all Utopian imagination'.[52]

---

48 Ibid., pp. 248, 255.
49 Ibid., pp. 260, 258.
50 Lyman Tower Sargent, 'The Three Faces of Utopianism Revisited', *Utopian Studies*, 5 (1994), 1–37 (p. 15). Italics removed.
51 Lucy Sargisson, 'Utopia and Intentional Communities', unpublished manuscript (Uppsala: ECPR Conference, 2004), p. 4, https://perma.cc/HB77-N2QG.
52 Fredric Jameson, *The Seeds of Time* (New York: Columbia University Press, 1996), p. 67.

In what is customarily considered the first modern utopia, Thomas More's book that coined the term 500 years ago, the issue of scale is confronted directly—and immediately solved: in utopia, cities consist of no more than 6,000 households of some 10 members each; when there are too many people, they can simply go and settle somewhere else—until they are needed again. Already in More (or, indeed, already in Plato's Republic, which was supposed to be smaller than a typical village), an ideal society is a community operating on a particular scale; it can be cloned but not scaled up.[53] When post-catastrophe fiction then borrows formal and thematic devices from utopian fiction—the meticulous description of the minutiae of everyday life (which explains, for instance, why *Robinson Crusoe* is such an important intertext for post-catastrophe fiction); the fairly static nature of the world (which allows *Station Eleven* to simply skip an eventful decade); the rejection of some aspects of contemporaneous culture (overcrowding, in the case of *Station Eleven*); and, not least, the eminently manageable scale of life—it also imports the normative dimension of the genre, even if it does not make it explicit.

Although they are organized around an imagined cataclysm, the fairly conventional textures of novels like *MaddAddam*, *Severance*, and *Station Eleven* are also clearly indebted to the tradition of the realist novel. Recent scholarship has unearthed the intricate relationship between realist fiction's commitment to particular scale domains, its normative claims, and its implicit population unconscious. Anna Kornbluh's *The Order of Forms* sees a measure of utopia's 'social dreaming' as constitutive of realist form. For Kornbluh, realist fiction's worldbuilding amounts to 'a speculative projection of hypothetical social space'.[54] Realist novels intrinsically project a particular normative ordering of society—they are essentially 'about making spaces and order deliberately and justly'.[55] If we bring the vocabularies of utopian and realist fiction together, we can say that the community that a realist novel imagines is always potentially also an intentional community gathered around a normative conception of the good life: 'Every project *of* is a project *for*'.[56] For Kornbluh, there is

---

53   Edwards, *Utopia*, p. 47.
54   Anna Kornbluh, *The Order of Forms: Realism, Formalism, and Social Space* (Chicago: University of Chicago Press, 2019), p. 30.
55   Ibid., p. 4.
56   Ibid., p. 10.

no opposition between realism's commitment to limits and its utopian force: realism's recognition of limits is not simply a conservative affirmation of the status quo; rather, the 'encounter with limits [that] is built into realism's architecture' amounts to 'realism's utopianism'.[57]

This 'recognition of constrainedness' sets realism apart from romance and traditional science fiction. The realist embrace of diminishment and limitation provides a good gloss for the kind of writing we find in novels like *Station Eleven*, *MaddAddam*, and *Severance*: while they are all organized around a science fictional device, their 'realist' attention to all-too-human concerns is in line with the peculiar utopianism Kornbluh observes in realism. In an essay on the work of Kim Stanley Robinson, Kornbluh has noted that a number of contemporary works that officially qualify as science fiction abandon science fiction's customary imaginative excess and replace it with an acceptance of 'constraints of finitude [and] mortality' and a 'constriction of the environment'.[58] This 'diminution of science fiction to realism', I argue, also pertains in the way contemporary post-catastrophe fiction projects intended communities—in abandoned airport terminals, in repurposed cobb houses, and in Candace's community of one.[59] Realism's constitutive self-limitation provides it with an alibi to focus on a small-scale reality and leave demographic multiplicity unaddressed—an alibi that post-catastrophe fiction borrows in the form of an excuse for the removal of surplus populations.

Realism's constrainedness has arguably always implied a sinister population politics. In her recent book *Populating the Novel*, Emily Steinlight has foregrounded the genre's population unconscious. Steinlight finds 'a systematic emplotment of superfluity' through which novels aim to come to terms with the reality of 'demographic excess'.[60] *Station Eleven*'s refrain that '*survival is insufficient*' (which implies that only particular forms of life are worthy to be recognized as life) inscribes it in a longer novelistic tradition in which, in Steinlight's words, 'some qualitative human essence' works hard to escape subsumption by 'the sheer quantitative excess of human beings pressing against each

---

57 Ibid., p. 42.
58 Anna Kornbluh, 'Climate Realism, Capitalist and Otherwise', *Mediations*, 33 (2020), 99–118 (pp. 102, 104).
59 Ibid., p. 103.
60 Steinlight, *Populating*, pp. 21, 3.

other'—the 'sublime' scenario we observed in Malthus and in the 1970s population panic.[61] For Steinlight, this means that the realist novel is not only a form that negotiates the relation between subject and society, between individual and aggregate (as influential accounts by the likes of Nancy Armstrong and Franco Moretti have it), but also a form that metabolizes 'demographic excess' and engages in 'a recalibration of the relationship between society and species, polity and population, power and life'.[62] What a critical tradition has investigated as the novel's imagining of 'a coherent body politic' is often also a way of managing demographic surplus.[63]

In the face of the reality of demographic excess, fiction is faced with the challenge of attending to an overcrowded world. One possible strategy for managing this impossible imaginative feat is exercising what Steinlight calls 'relative indifference to most of what [the work] sees': most of what the world contains is waste and superfluity, mere matter that cannot figure in designs of the good life.[64] Historically, one way of 'managing the human aggregate' has been the 'intensified psychologism' typical of modernist fiction, a strategy that *Station Eleven* and *Severance*, in their minute attention to the moods, thoughts, and feelings of their character, also adopt.[65] As Steinlight makes clear, this cultivation of interiority is always also a training in accepting the superfluity of most of the lives that make up the exterior world. I submit that post-catastrophe fiction's displacement of demographic solutions to viruses and other natural disasters is another strategy for giving shape to such indifference; the lack of sentimentality in the way *MaddAddam* and *Station Eleven* deal with excess populations is indicative of this. This is one way in which Atwood's trilogy, as Steinlight notes, 'reveal[s] the specter of disposability lurking behind the longstanding fantasy of optimizing health and longevity'.[66]

The one death that *Station Eleven* mourns extensively is that of Arthur Leander in the novel's extremely powerful and moving opening scene. While performing the lead in *King Lear*, Leander dies on stage—not

---

61 Ibid., p. 2.
62 Ibid., pp. 3, 6.
63 Ibid., p. 10.
64 Ibid., p. 14.
65 Ibid., p. 212.
66 Ibid., p. 222.

from the virus, but from a heart attack. Characteristically, *Severance* is more self-conscious in the way it stages indifference as a hard-won survival strategy that involves a painful excision of sympathy. When Candace suspects that the cab driver she has befriended has become infected, she escapes from New York by dragging him out of his car and stealing it. Candace pauses to consider that '[i]t's possible that there is another true story', a version in which 'he wasn't fevered': 'It's possible. I can't be sure. Because I wasn't really all that careful. All I thought about was myself. It got me where I needed to go'.[67] The following final two chapters consolidate this resoluteness as they describe Candace (in an open-ended present tense) escaping from her confinement and going her own way. The novel does not imagine a clear plan or resolution—it leaves us with the protagonist 'step[ping] out and start[ing] walking'.[68] Indifference, it seems, is the key achievement—an indifference to excess populations that the post-catastrophe genre enacts through pandemic or catastrophic solutions.

Conclusion: Downscaling Survival

This essay has argued that post-catastrophe fiction functions as a refuge for demographic fantasies that no longer find as much scientific support for the looming threat of global overpopulation. Lacking an evidentiary basis that was—or seemed to be—much more solid a few decades ago, those fantasies no longer inform the stark oppositions between the multitude and the individual that science fiction demodystopias inherited from Malthus, but rather take shape as positive blueprints for comfortable living that leave their reliance on mass-extinction implicit. It is the genre's peculiar blend of science fiction elements, utopian devices, and realist notation that makes it a site for the projection of a normative vision of the good life that does not need to avow what I have called its population unconscious.

As my opening discussion of a certain type of family planning initiative already indicated, post-catastrophe fiction is not the only place where these fantasies and imaginative traditions surface. When we analyse the visual and textual rhetoric on the websites of projects such

---

67 Ma, *Severance*, p. 278.
68 Ibid., p. 291.

as Uncrowded (whose tagline in 2019 read 'Smaller Families for a Better World')[69] or the organization Population Matters, at least two features stand out. First, while the 'sublime' opposition between an overcrowded multiplicity and a more cosily scaled community persists (especially in the infographics), there is an undeniable shift toward affirmative images of the blessings of a less crowded life—images in which the reality of overcrowding, just as in post-catastrophe fictions, is magically resolved. One of these blessings of decluttered life, to judge from these websites, is the capacity to enjoy nature. This is a second key feature: the natural world is not, as in earlier population discourse, mainly a source of food provision, but a resource to restore a qualitatively superior form of human life; this resonates with Matthew Hart's argument that post-catastrophe fiction offers a world marked by 'archipelagic insularity': dotted by small, populated territories surrounded by unclaimed wilderness.[70]

The gradual shift from a rhetoric of sublime opposition between the many and the few to an affirmation of the joys of small-scale life is most apparent on the website of Uncrowded.[71] In 2021, the home page provided images of traffic congestion, busy walkways, and other urban inconveniences of what it called 'Our Crowded World'; scrolling down, we could see these make way for sunset images of an individual woman and a one-child family (a shift that characteristically moves us from the city to the countryside, as it overwhelmingly does in post-catastrophe fiction, especially in the strand Caroline Edwards analyses as 'pastoral post-apocalypticism').[72] The images on the organization's 'The Facts' page repeated this visual rhetoric that opposes a scary and confusing multiplicity to a clear and wholesome individuality: an infographic visualizing the CO_2-equivalent for different actions foregrounded the self-evident simplicity of 'hav[ing] one fewer child' against the confusing background of a multiplicity of comparatively insignificant options (see Fig. 28); another image juxtaposed a human figure to an overkill of cars, appliances, and light bulbs to underline the self-evident

---

69 Uncrowded, *Our Crowded World* (7 June 2019), https://havingkids.org/uncrowded-home/.
70 Matthew Hart, *Extraterritorial: A Political Geography of Contemporary Fiction* (New York: Columbia University Press, 2020), p. 134.
71 Uncrowded, *Our Crowded World*.
72 Edwards, *Utopia*, p. 160.

superiority of family planning as a strategy for reducing environmental impact (see Fig. 29). The Global Footprint Network, an organization that does not explicitly promote family planning, adopts a similar visual rhetoric of number: an infographic that shows how many earths would be necessary if everyone adopted particular lifestyles is fairly useless in terms of information value—letters are illegibly small, and there is simply too much information—but the sheer multiplicity of planets and the dynamic generated by their escalating increase successfully conveys a sense of a world spinning out of control by the force of number (see Fig. 30).[73]

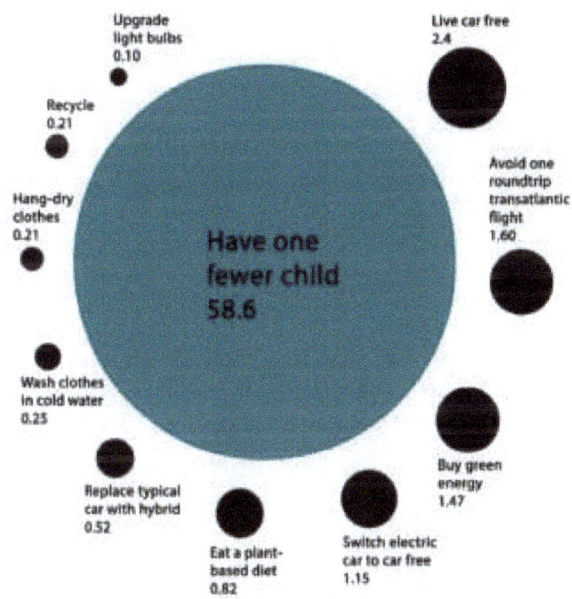

Fig. 28 Having Kids, *Having one fewer child will save 58.6 tonnes of CO2-equivalent per year* (2018) © Fair Start Movement. All rights reserved. https://havingkids.org/wp-content/uploads/2018/12/child-impact-guardian-graphic-1-358x400.jpg

---

73 Global Footprint Network, *How Many Earths?* (2019), https://www.footprintnetwork.org/.

Fig. 29 Having Kids, *The carbon legacy of just one child* (2018) © Fair Start Movement. All rights reserved. https://havingkids.org/wp-content/uploads/2018/10/27972800_1575734402545365_7111663051996650685_n-400x400.png

It is significant that, as of 2020, the Uncrowded website was nested inside the more affirmative Having Kids website—a change that reflected the shift toward more affirmative visions of what it called 'a world of smaller and truly democratic communities, surrounded by nature'. This vision also dominates the website of the UK-based Population Matters initiative.[74] Population Matters has a more scientific orientation: there are more graphs and numbers, as well as quotes from David Attenborough. Like Having Kids, it foregrounds the pleasant realities of downscaled families under rubrics such as 'Having a Smaller Family' or 'Life in a Smaller Family'. There is a remarkable consistency in the images used to advertise small families across these websites: a predilection for sunsets, seas, and faceless figures; children are without exception small children, never bored teenagers; cities have disappeared and made way for

---

74 Population Matters, *Homepage* (2021), https://populationmatters.org/.

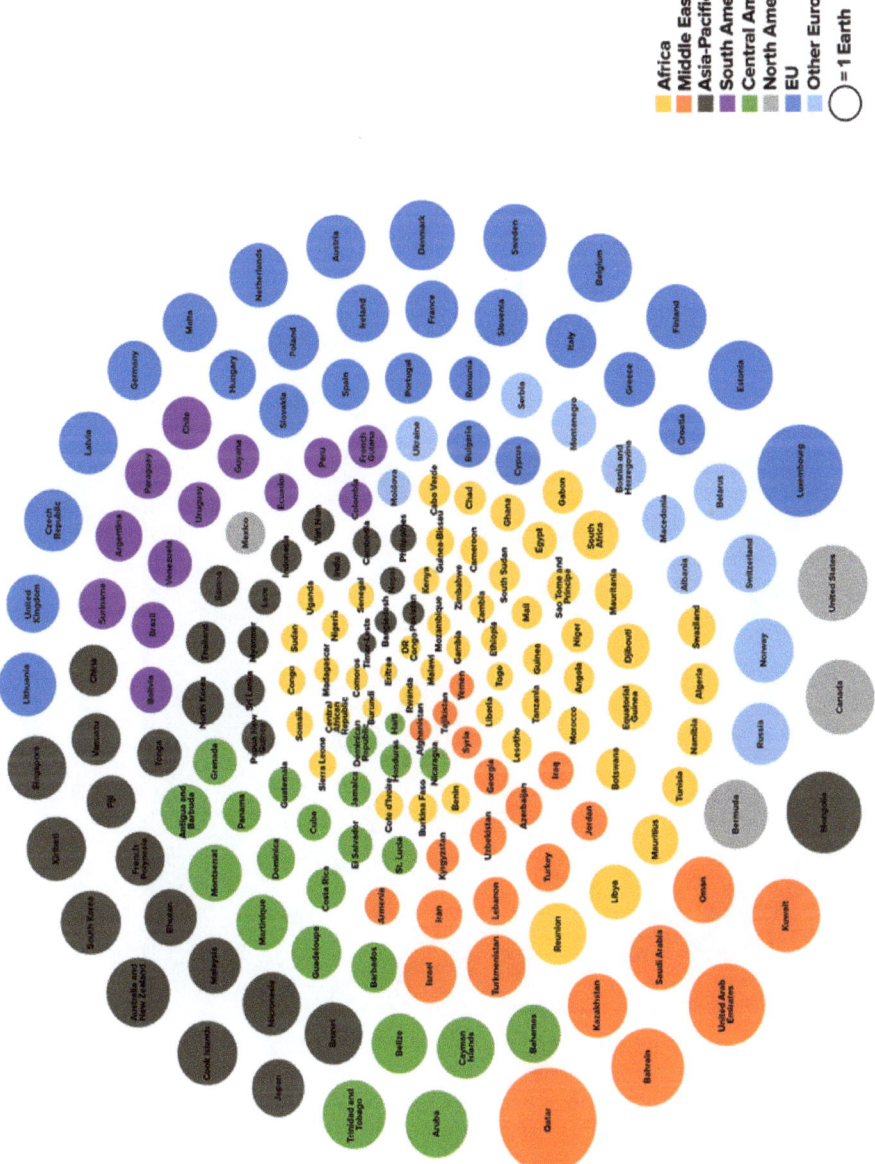

Fig. 30 Global Footprint Network, *How Many Earths?* (2019) © Global Footprint Network. All rights reserved. https://www.footprintnetwork.org/content/uploads/2019/01/Infogrpahic-Pub-Data-Circle-v3.png

nature and suburbs. The issue of where all this free space comes from is not addressed: it is as if the rest of the world had *also* adopted this agenda and decided to take up less room (and, given the ostentatious contemporaneity of the images, done so overnight), or as if the problem of surplus populations had magically resolved itself. It is hard not to see these images as works of post-catastrophe fiction in their own right.

Fig. 31 Population Matters, *Having a Smaller Family* (2021) © Shutterstock. Free to use. https://populationmatters.org/having-smaller-family

Fig. 32 Alberto Casetta, *Family Walking in the Rain* (2021) © Alberto Casetta. Free to use under Unsplash licence. https://populationmatters.org/sites/default/files/styles/full_width_image/public/alberto-casetta-349138-unsplash_0.jpg?itok=UzNpoVfv

Fig. 33 Uncrowded, *Homepage* (2021) © Fair Start Movement. All rights reserved.
https://fairstartmovement.org/uncrowded-home/

With remarkable frequency, these images of nature-adjacent, downscaled life are set in fairly moderate weather: it is often windy, cloudy, rainy, and grey (See Fig. 31–33). It is tempting to see this as a complement to the population fantasy informing these images: they suggest that a world with smaller and fewer families will be a world without global warming and rampant climate change—a world where the moderation and predictability of the Holocene will be magically restored, as if putting a break on population growth will achieve nothing less than a solution to climate change. This fantasy points to a desire to preserve current inequalities and to continue to enjoy privileges that many of us still take for granted. This desire is all the more apparent in the figures (typically with their faces turned away or cropped out of the images) who inhabit these mild landscapes—figures who profit from expensive rain clothes, hiking gear, bikes, campers, and the like. The good life that these images project is *not* a life without the commodities affluent audiences currently enjoy; it is a life where there will be more space and less intruders to enjoy these things.

A common imaginary of the down-scaled good life links those initiatives to the post-catastrophe fiction examined in this study. The novels I have analysed show a commitment to current inequalities and an inability to think of planetary challenges at a proper scale—which is to say, for 7.7 billion people rather than some 320. In *Station Eleven*, global warming is pre-empted by a drastic reduction of the human impact on the planet. As the novel notes, 'automobile gas goes stale after two or three years'—which is what happens in the years following the collapse.[75] In *MaddAddam*, the survivors simply decide to adopt a

---

75 Mandel, *Station Eleven*, p. 31.

'hipster DIY and maker culture in their attention to walking, traditional medicine, mushroom foraging, and roasting of local wild roots for coffee'—even if, as Ursula Heise has remarked, there is nothing in the world's fictional givens that compels them to do so.[76] *Severance*, as I have already noted, concludes with the end of automobility when Candace arrives in Chicago, where traffic has come to a standstill as cars have been abandoned across town. All there is left to do for Candace (as for Atwood's characters) is to 'start walking'.[77] Walking, we know, is good for you, but it is not a solution for environmental crisis. Nor is it an activity that cannot be shared with 7.7 billion others, for that matter.

Works Cited

Aldiss, Brian, *Billion Year Spree: The History of Science Fiction* (London: Weidenfeld and Nicolson, 1973).

Angus, Ian, and Simon Butler, *Too Many People?: Population, Immigration, and the Environmental Crisis* (Chicago: Haymarket Books, 2011).

Atwood, Margaret, *MaddAddam* (London: Virago, 2014).

Boyd, Shelley, 'Ustopian Breakfasts: Margaret Atwood's MaddAddam', *Utopian Studies*, 26 (2015), 160–81, https://doi.org/10.5325/utopianstudies.26.1.0160

Bricker, Darrell, and John Ibbitson, *Empty Planet: The Shock of Global Population Decline* (New York: Crown, 2019).

Clark, Timothy, '"But the real problem is....": The Chameleonic Insidiousness of "Overpopulation" in the Environmental Humanities', *Oxford Literary Review*, 38 (2016), 7–26, https://doi.org/10.3366/olr.2016.0177

Coole, Diane, 'Too Many Bodies? The Return and Disavowal of the Population Question', *Environmental Politics*, 22 (2013), 195–215. https://doi.org/10.1080/09644016.2012.730268

Cuarón, Alfonso, *Children of Men* (Universal Pictures, 2006), DVD.

Curry, Patrick, *Ecological Ethics: An Introduction* (Cambridge: Polity, 2006).

De Cristofaro, Diletta, *The Contemporary Post-Apocalyptic Novel* (London: Bloomsbury, 2020).

Domingo, Andreu, '"Demodystopias": Prospects of Demographic Hell', *Population and Development Review*, 34 (2008), 725–45, https://doi.org/10.1111/j.1728-4457.2008.00248.x

---

76 Heise, 'What's the Matter', [n.p.].
77 Ma, *Severance*, p. 291.

Domingo, Andreu, 'Resilient Evil: Neoliberal Technologies of the Self and Population in Zombie "Demodystopia"', *Utopian Studies*, 30 (2019), 444–61, https://doi.org/10.5325/utopianstudies.30.3.0444

Edwards, Caroline, *Utopia and the Contemporary British Novel* (Cambridge: Cambridge University Press, 2019), https://doi.org/10.1017/9781108595568

Ehrlich, Paul, *The Population Bomb*, revised and expanded edition (New York: Ballantine Books, 1971).

Ferguson, Frances, *Solitude and the Sublime: The Romantic Aesthetics of Individuation* (London: Routledge, 1992).

Garrard, Greg, 'Worlds Without Us: Some Types of Disanthropy', *SubStance*, 41 (2012), 40–60, https://doi.org/10.1353/sub.2012.0001

Global Footprint Network, *How Many Earths?* (2019), https://www.footprintnetwork.org/.

Hart, Matthew, *Extraterritorial: A Political Geography of Contemporary Fiction* (New York: Columbia University Press, 2020).

Having Kids, *Homepage* (Fair Start Movement, 2021), https://havingkids.org.

Heise, Ursula, 'The Virtual Crowd: Overpopulation, Space, and Speciesism', *ISLE*, 8.1 (2001), 1–29, https://doi.org/10.1093/isle/8.1.1

Heise, Ursula, 'What's the Matter with Dystopia', *Public Books* (2 January 2015), https://www.publicbooks.org/whats-the-matter-with-dystopia/

Hicks, Heather, *The Post-Apocalyptic Novel in the Twenty-First Century: Modernity beyond Salvage* (New York: Palgrave Macmillan, 2016), https://doi.org/10.1057/9781137545848

Höhler, Sabine, '"Carrying Capacity": The Moral Economy of the "Coming Spaceship Earth"', *Atenea*, 26 (2006), 59–74.

Horn, Eva, *The Future as Catastrophe: Imagining Disaster in the Modern Age* (New York: Columbia University Press, 2018).

Jameson, Fredric, *Postmodernism, Or, The Cultural Logic of Late Capitalism* (Durham: Duke University Press, 1989).

Jameson, Fredric, *The Seeds of Time* (New York: Columbia University Press, 1996).

Jameson, Fredric, *Archaeologies of the Future: The Desire Called Utopia and Other Science Fictions* (London: Verso, 2007).

Kornbluh, Anna, *The Order of Forms: Realism, Formalism, and Social Space* (Chicago: University of Chicago Press, 2019).

Kornbluh, Anna, 'Climate Realism, Capitalist and Otherwise', *Mediations*, 33 (2020), 99–118.

Ma, Ling, *Severance* (London: Picador, 2019).

Mandel, Emily St. John, *Station Eleven* (Basingstoke: Picador, 2015).

McCarthy, Cormac, *The Road* (New York: Vintage, 2006).

McCauley, Alex, 'The Promise of Disaster: Specters of Malthus in Marxist Dreams', *Ecozon@*, 9 (2018), 53–65, https://doi.org/10.37536/ECOZONA.2018.9.1.1649

Murphy, Michelle, *The Economization of Life* (Durham: Duke University Press, 2017).

Pearce, Fred, *Peoplequake: Mass Migration, Ageing Nations and the Coming Population Crash* (New York: Eden Project, 2011).

'The Population Delusion', *New Scientist* (23 September 2009), https://www.newscientist.com/round-up/population/

Population Matters, *Homepage* (2021), https://populationmatters.org/.

Sargent, Lyman Tower, 'The Three Faces of Utopianism Revisited', *Utopian Studies*, 5 (1994), 1–37.

Sargisson, Lucy, 'Utopia and Intentional Communities', unpublished manuscript (Uppsala: ECPR Conference, 2004), https://perma.cc/HB77-N2QG

Shriver, Lionel, 'Population in Literature', *Population and Development Review*, 29 (2003), 153–62, https://doi.org/10.1111/j.1728-4457.2003.00153.x

Steinlight, Emily, *Populating the Novel: Literary Form and the Politics of Surplus Life* (Ithaca: Cornell University Press, 2018).

Tate, Andrew, *Apocalyptic Fiction* (London: Bloomsbury, 2017), https://doi.org/10.5040/9781474233545-004.

Uncrowded, Homepage (Fair Start Movement, 2021), https://havingkids.org/uncrowded-home/

Vollset, Stein Emil, et al., 'Fertility, Mortality, Migration, and Population Scenarios for 195 Countries and Territories from 2017 to 2100: A Forecasting Analysis for the Global Burden of Disease Study', *The Lancet*, 396 (2020), 1285–306, https://doi.org/10.1016/S0140-6736(20)30677-2

IV. ECOLOGICAL SCALES

10. The Everyday Pluriverse
Ecosystem Modelling in *Reservoir 13*

Ben De Bruyn

Introduction: The Rural Mesocosm

This chapter examines the narrative strategies and environmental imagination of a novel that rewards closer scrutiny for literary scholars interested in scale and biology, since it evokes a deliberately circumscribed rural storyworld but also expands the conventional spatial and temporal parameters of the novel in a multiscalar, more-than-human direction. As scholars such as Mark McGurl and Timothy Clark have asserted in the last decade, contemporary life forces us to grapple with questions of scale for multiple reasons, including the ecological challenges that we now routinely, if uneasily, group under the label of the Anthropocene, the geological age of the humanised planet. Readers, writers, and teachers of fiction are hence encouraged to interrogate the scalar limitations of literature and abandon its conventional anthropocentric architecture. One project that responds to such calls is David Herman's foray into bionarratology, a newly expansive form of analysis that is equipped to address 'storytelling at species scale'.[1] In doing so, Herman clarifies,

> My guiding hypothesis is that narrative, even though it is [...] optimally calibrated for human-scale phenomena, furnishes routes of access to [...] processes extending beyond the size limits of the lifeworld. To

---

[1] David Herman, *Narratology beyond the Human: Storytelling and Animal Life* (Oxford: Oxford University Press, 2018), p. 22, https://doi.org/10.1093/oso/9780190850401.001.0001.

> test this hypothesis, I [will] fles[h] out [...] the concept of multiscale narration, in which storytellers cross the boundaries of scale separating the mesophysics of everyday life from subpersonal domains situated at the level of microphysics as well as suprapersonal domains situated at the level of macrophysics.[2]

As this project indicates, we can respond to the current ecological crisis and mitigate the complicity of traditional literary formats by retooling our narrative strategies and their implicit 'cultural ontologies'[3] in a more biocentric, multiscalar direction. In a related vein, Amitav Ghosh has asserted that we should abandon the myopic gaze of the novel, the dominant literary form, by subverting its debilitating preoccupation with the disenchanted 'rhetoric of the everyday'[4] and its concomitant focus on human individuals at the expense of 'the nonhuman' and the human 'collective' alike.[5] Several commentators maintain, in other words, that the climate emergency and the biodiversity crisis force us to rethink our way of life from the ground up and this necessitates a critical reflection on the habitual scales of our literary and cultural narratives.

These reflections on the limitations of everyday realism and the need for new forms of multiscale fiction can be enriched and recast by a close analysis of Jon McGregor's Costa Award-winning novel *Reservoir 13* (2017). McGregor's novel initially appears to centre on the disappearance of a thirteen-year-old girl in the English countryside, but the text gradually relegates this dramatic starting point to the sidelines and redirects the reader's attention to the fictional village impacted by the disappearance. In line with the insights mentioned earlier, *Reservoir 13* purposely expands the novel's traditional canvas by toggling between the intersecting subplots of dozens of villagers and by alerting readers to weather phenomena and the other life forms that inhabit the valley, tracking the lives of numerous species of plants, birds, insects, and other animals. In an interview, McGregor accentuates the text's flat ontology by asserting that 'The key thing, for me, was realizing that I wanted to tell not just the stories of the human characters but of all the wildlife and *everything*. People's working routines, the water levels in the reservoirs,

---

2 Ibid., pp. 252–3.
3 Ibid., p. 20.
4 Amitav Ghosh, *The Great Derangement: Climate Change and the Unthinkable* (Chicago: Chicago University Press, 2016), p. 19.
5 Ibid., p. 80.

the weather. I wanted all of those elements to have the same prominence'.[6] In addition, the narrative speeds up the conventional rhythm of the novel form by summarising an entire year in each of its chapters, which do not just register the experiences of its human and nonhuman characters, but also cycle through the months of a thirteen-year period in a timelapse-like pattern that is both repetitive and lyrical and refuses to prioritise dramatic occurrences at the expense of unremarkable events.[7] As a result, the sensational storyline of the girl's vanishing is integrated into a more spacious narrative and ecological tapestry, which relates how different generations of humans and other animals fall in and out of love, give birth and pass away, move to and from the village in question. In an astute summary of *Reservoir 13* that praises its 'visionary power', James Wood concludes that, while 'nothing much happens' in this text, its 'roving omniscience' and 'passive construction[s]' patiently construct a collage of 'the little ecosystem of the village', where life 'blindly goes on' and finally appears to be 'as animal as it is human'.[8] In a more comparative fashion, we might say that the novel mixes the experimental sensibility of constrained writing, the emotionally charged nature imagery of haiku, and the 'portrait-of-a-community' approach of TV series like *Broadchurch* and *The Wire*, examples of what Matt Seitz has labelled the 'slow crime procedural'.[9]

As it juxtaposes the vanishing of a young girl with a fine-grained description of the countryside, *Reservoir 13* exhibits traits of crime fiction alongside nature writing. Various critics have recently begun to unearth the ecological subtext of crime narratives and vice versa, developing Timothy Morton's brief remarks on 'Anthropocene noir' and Rob Nixon's influential analysis of 'slow violence'. If we read such environmental

---

6   Jon McGregor, 'Time Marches On: An Interview with Jon McGregor', *The Paris Review* (18 October 2017), [n.p.], www.theparisreview.org/blog/2017/10/18/time-marches-interview-jon-mcgregor/.

7   Editors' note: Different nonhuman actants shape different non-anthropocentric temporalities. While *Reservoir 13*, with its vegetal and animal actants, foregrounds seasonal temporalities, the geological or mineral actants of Gillian Clarke's poetry (see chapter 3) or of the installations and performances studied by Beaufils (chapter 13) connect humans to longer temporalities, for example that of fossilization and sedimentation.

8   James Wood, 'The Visionary Power of the Novelist Jon McGregor', *The New Yorker* (27 November 2017), np, www.newyorker.com/magazine/2017/11/27/the-visionary-power-of-the-novelist-jon-mcgregor.

9   Matt Zoller Seitz, 'The Rise of the Slow Crime Procedural', *Vulture* (6 March 2015), [n.p.], https://www.vulture.com/2015/03/american-crime-rise-of-slow-crime-procedural.html.

crime stories carefully, the argument goes, we will find traces of structural violence and its long-term effects on both humans and nonhumans and realise that everyone is implicated in systemic injustice, a tragic conclusion that is one of the hallmarks of noir fiction.[10] McGregor's text is related to pastoral literature too, for it lavishes attention on landscape descriptions and agricultural routines and refuses to develop its basic crime narrative. Building on the work of Terry Gifford, John Armstrong has pinpointed pastoral, anti-pastoral, and post-pastoral tendencies in *Reservoir 13*, which mentions sheep, shepherds, and seasonal rhythms, avoids idealisation by pinpointing the brutal and commercial aspects of farming, and gestures at the intimate ties between humans and their surroundings. Yet, like other recent rural novels, McGregor's story also departs from the pastoral tradition by infusing British landscapes with 'the violence of our global times',[11] which takes the form of physical assaults, against women and animals especially, alongside indirect modes of slow, capitalist, and racist violence. Such texts constitute a new, 'globalized pastoral',[12] Armstrong concludes, in which small villages appear as 'nodal spaces reflecting the structural and physical violence of Britain's rural development [and] its globalized present'.[13] Similarly, Robert Macfarlane has identified a new rural imaginary in English culture, in which the iconic green and pleasant countryside appears as a haunted, eerie landscape that contains traces of colonial injustice and hints at recent 'environmental damage', which '[i]n England [...] has not taken the form of sudden catastrophe, but rather a slow grinding away of species and of subtlety'.[14] These arguments imply that *Reservoir 13* splices crime and pastoral fiction to reveal the subterranean ties between multiple forms of violence and to contemplate topical anxieties about biodiversity and national identity.

My summary also intimates that McGregor's novel performs valuable cultural work along the lines described by recent research on scale. Illustrating Herman's claims about multiscale narration, the text

---

10 Lucas Hollister, 'The Green and the Black: Ecological Awareness and the Darkness of Noir', *PMLA*, 134.5 (2019), 1012–27, https://doi.org/10.1632/pmla.2019.134.5.1012.
11 John Armstrong, 'Pastoral Place and Violence in Contemporary British Fiction', *Concentric: Literary and Cultural Studies*, 44.2 (2018), 51–79 (p. 53).
12 Ibid., p. 74.
13 Ibid., p. 76.
14 Robert Macfarlane, 'The Eeriness of the English Countryside', *The Guardian* (10 April 2015), p. 6. https://www.theguardian.com/books/2015/apr/10/eeriness-english-countryside-robert-macfarlane

does not just feature diverse species and plots, but regularly zooms in as well as out, abruptly switching between close-ups of minute biological events—'[i]n the dead grass around the cricket field the eggs of the skippers turned from white to yellow, and the larvae span themselves into cocoons'[15]—and wide-angle views of geological and biological processes, including the transnational migrations of bird species—'The fieldfares were away in Scandinavia, building nests and laying eggs' (41); 'The swallows which had left a few days earlier were most of the way to South Africa by now' (302). A more extended example is the novel's intermittent description of small, flea-like springtails, which results in what we might call a nonhuman subplot, a miniature narrative that seems a far cry from domestic realism, as it integrates snapshots of individual moments into a repetitive account of lifecycles and scales up from the level of a single organism to that of regional populations that reuse and convert organic waste:

> There were springtails under the beech trees behind the Close, feeding on fragments of fallen leaves. (10)
>
> There were springtails in the old hay at the back of the lambing shed, feeding and laying eggs […] and at the end of a long stem a single male sat poised […], ready to spring into the air for the first time in his life. There was a moment's hesitation. (147)
>
> There were springtails in the compost heap in Mr Wilson's garden, and in the morning Nelson [the dog] sat watching while they leapt and popped from the surface. (206–07)
>
> There were springtails in the crumbling wood of the fallen ash by the river, moulting and feeding and getting ready to lay more eggs. (234)
>
> There were springtails in the rotting sheets of plywood stacked against the wall in Fletcher's orchard, and the juveniles among them were shedding the first of their many shell-like skins. (258)
>
> There were springtails in the soil of the cricket ground, a million or more, […] and amongst them a female springtail laid the last eggs of her life. (306)

Not unlike the other examples of multispecies fiction inspected by Herman, McGregor's text experiments with scale—and with

---

15 Jon McGregor, *Reservoir 13* (London: 4th Estate, 2018), p. 252. Further quotations from the novel will be cited internally.

dispassionate declarative statements—to underline how human lives are embedded in a patchwork of ecological relations, regularly highlighting biological communities at the expense of isolated creatures. This narrative therefore illustrates the overlooked fact that novels do not only deal with individuals but also with populations, as Emily Steinlight has demonstrated—though *Reservoir 13* refuses to limit itself to human aggregrates.[16] Such projects merit closer attention, as other life forms have received limited attention in earlier publications on scale and the Anthropocene, which have all-too-quickly abandoned biological questions for geological concerns, according to Cary Wolfe.[17] As stories like *Reservoir 13* impress on their readers, we should monitor and nourish biodiversity even as we are worried about the planet's climate—and experiments in scalar poetics can help us to address both of these concerns.

Though McGregor's narrative is an example of multiscale narration, we should not overlook the fact that it simultaneously limits its focus to a small village and refuses to abandon the novel's traditional mission of capturing the human everyday. Ultimately, it uses a synoptic narrator to capture the cyclical rhythms and cross-species interactions of community life rather than the outer limits of reality's scalar continuum. While its gaze is not limited to the human, and even notices animal and human diseases (79 and 213), the narrative does not systematically travel down to a microbial or up to a cosmic perspective. Further accentuating the text's human dimension, several characters function as avatars of *Reservoir 13*'s omniscient narrator, from the journalist responsible for the *Valley Echo* newsletter and the ceramicist discreetly working with local materials to the intrusive judgments of the parish council, and the struggling vicar who carries the weight of people's 'confidences' around (128). This rural novel exhibits deliberate scalar modesty alongside scalar flexibility, in short. We might describe its ambivalent position by saying that it evokes a so-called 'mesocosm'. As Alenda Chang explains, '[i]n ecology, mesocosms are experimental enclosures intermediate in size and complexity between small, highly

---

16 Emily Steinlight, *Populating the Novel: Literary Form and the Politics of Surplus Life* (Ithaca, NY: Cornell University Press, 2018), https://doi.org/10.7591/9781501710728.
17 Cary Wolfe, 'What "the Animal" Can Teach "the Anthropocene"', *Angelaki*, 25.3 (2020), 131–45, https://doi.org/10.1080/0969725X.2020.1754033.

controlled lab experiments and large, often unpredictable real-world environments'.[18] This notion illuminates the ecological imagination of video games, Chang believes, because 'games, like mesocosms, are "mini-ecosystems"—functional arenas of a size usefully intermediate between field experiments and laboratory conditions, which replicate select aspects of the surrounding world'.[19] This notion is relevant here too, seeing that *Reservoir 13* runs a narrative simulation, as I explain further below, that pinpoints and replicates the social and ecological interactions taking place in its fictional world—though I should immediately add that the valley evoked in McGregor's text is not a segregated, enclosed retreat, as it regularly highlights ties between the country and the city, this place and the planet. Yet I want to draw attention to the novel's scalar modesty here because it discloses the peculiar affordances of mesocosmic storyworlds for ecological storytelling.

The novel's rural focus is not the only reason we should treat the question of scale with caution in this context. As Neal Alexander has elucidated,[20] McGregor's writing typically traces the unremarkable routines of ordinary lives. My observations about nonhuman scale appear to imply that *Reservoir 13* relinquishes this signature interest in everyday life for a sweeping view of our more-than-human reality. But it would be more accurate to say that the novel portrays ecological processes in order to enlarge and enrich our conception of the everyday. In thinking about scale in culture and science, Joshua DiCaglio has warned that we should not confuse the level of 'objects' being observed with the level of 'observation'.[21] It follows that the lesson of scalar thinking for cultural critics is not simply that we should contemplate unfamiliar small- and large-scale phenomena like viruses and ecosystems, but should appreciate that human-scale phenomena have multiscalar aspects and have always been implicated in microscopic and macroscopic processes we failed to acknowledge because our level of observation remained at

---

18 Alenda Y. Chang, *Playing Nature: Ecology in Video Games* (Minneapolis: University of Minnesota Press, 2019), p. 17, https://doi.org/10.5749/j.ctvthhd94.
19 Ibid., pp. 19–20.
20 Neal Alexander, 'Profoundly Ordinary: Jon McGregor and Everyday Life', *Contemporary Literature*, 54.4 (2013), 720–51, https://doi.org/10.1353/cli.2013.0046.
21 Joshua DiCaglio, 'Scale Tricks and God Tricks, or The Power of Scale in *Powers of Ten*', *Configurations*, 28.4 (2020), 459–90 (p. 466), https://doi.org/10.1353/con.2020.0021.

a single, narrow scale. Consequently, the point is less that we should look at other phenomena and more that we should rethink the same phenomena in a more scale-sensitive manner. What DiCaglio's warning allows us to see is that *Reservoir 13* does not renounce everyday life to foreground biological processes, but intentionally enlarges our conception of quotidian reality to arrive at a more encompassing, gently multiscalar understanding of the everyday. This dynamic portrait of a rural mesocosm provides resources to rethink what DiCaglio calls 'our this-scale lifeworld',[22] in other words, and the result recalibrates the parameters of the realist novel while reimagining McGregor's typical concerns. Indeed, the novel's scalar experiments stage a new version of the central insight and paradox at the heart of writing about ordinary life: the fact that mundane reality is profoundly weird and that, as Andrew Epstein clarifies, 'the ordinary becomes extraordinary the moment one begins to talk about it'.[23] In doing so, *Reservoir 13* invites us to reflect on the ties between biology and contemporary fiction and to ponder urgent debates about the intersecting climate and biodiversity crises. The rest of this chapter expands on these provisional remarks about mesocosmic narratives and ordinary lives by analysing the environmental imagination of McGregor's text further; as I will show, this post-pastoral post-crime novel incorporates nonhuman subplots to alert us to the other worlds that surround us and offers a narrative form of ecosystem modelling that captures the rhythms of this rural mesocosm and its dynamic assemblage of species. The result, I explain at the end, offers a particular, multispecies and ecostoicist perspective on life in the Anthropocene.

Noticing Nonhuman Narratives

This first section analyses three interconnected features of *Reservoir 13*, namely its interest in the reader's attention, its juxtaposition of nonhuman subplots, and its evocation of a multiperspectival reality made up of phenomenologically distinct 'bubble worlds'. Though other

---

22 Ibid., p. 474.
23 Andrew Epstein, '"The Rhapsody of Things as They Are": Stevens, Francis Ponge, and the Impossible Everyday', *Wallace Stevens Journal*, 36.1 (2012), 47–77 (p. 67), https://doi.org/10.1353/wsj.2012.0008.

scholars have linked the novel's interest in crime and ecology via notions such as slow violence, they can also be related via the topic of attention. As I noted, *Reservoir 13* slowly but surely redirects the reader's focus from the dramatic events of the opening to the humdrum routines of its large cast of characters, finally refusing to provide definitive answers about the child's disappearance. At the end we do not even know whether it should be considered a crime at all. Though the missing girl accordingly seems to play no critical role, the unresolved event ensures that readers are eager to scrutinise the whereabouts and behaviours of characters and cross-reference disjointed passages. Because vital clues always appear to be on the horizon: 'Around the deep pond at the far end of Thompson's land a ring of willow trees were in full leaf, shielding the pond as though something shameful had once happened there' (71). We are rewarded for our efforts too; the characters may not recognise the importance of a 'body-warmer, navy blue' (141) but readers are likely to remember that Becky wore 'a navy body-warmer' (8) when she went missing. Her unresolved disappearance also has an environmental function, crucially, because it gives new meaning to the everyday landscape. We will not learn more about this vanishing by noticing the local willow trees, but our vigilance ends up disclosing that these plants have lives and narratives too: 'By the river a willow came down in a storm and carried on growing as though nothing had changed' (188). The modern detective story is characterised by the presence of decodable clues, according to Franco Moretti,[24] and I would argue that McGregor uses this expectation for ecological purposes, first mobilising the reader's suspicion via its crime plot and then rerouting this newly acute attention from the search for human-oriented clues towards a new appreciation of the nonhuman landscape. The result encourages us to adopt a position akin to that of an 'ecological detective' who parses 'the human traces on the landscape',[25] as Sara Crosby explains in her reading of Edgar Allan Poe—though this search finally leads beyond the human in *Reservoir 13*.

This variation on the crime narrative template implies that McGregor's novel raises topical questions about reading and attention. At the start

---

24 Franco Moretti, *Distant Reading* (London: Verso Books, 2013), p. 74.
25 Sara L. Crosby, 'Beyond Ecophilia: Edgar Allan Poe and the American Tradition of Ecohorror', *ISLE*, 21.3 (2014), 513–25 (p. 515), https://doi.org/10.1093/isle/isu080.

of the twenty-first century, as Caleb Smith has observed, literary reading and analysis are typically promoted less as a form of interpretation and more as 'a kind of therapy for distraction' in the 'new attention economy'.[26] This view of reading as a special form of '[m]indfulness training' (884) has its roots in older debates, Smith shows, including nineteenth-century polemics on education and Thoreau's famous experiments in deliberate living. McGregor's novel is related to these older debates on reading and watchfulness, for it constitutes one elaborate exercise in attention management, as I have implied, and reflects explicitly on the art of noticing. Consider the following scene from *Reservoir 13*, where a character visits the village allotments and attempts to exercise his mindfulness skills by meditating about his multispecies surroundings:

> At the allotments Martin sat on the bench at the top end of their plot. [...] [His ex-wife Ruth] was making a better job of the place on her own. [...] Or perhaps she was getting help. From someone he didn't know about. [...] *This wasn't a line of thinking that helped, of course. He'd been advised. There were steps he could take, to steer around this line of thinking.* [...] *He looked outside himself and took other sensory information on board.* He listed the plants he could see. Gooseberries and strawberries and currants; sweetcorn and courgettes and beans [...] Nettles, cow parsley, thistles, bindweed. Plenty of bloody bindweed. Whoever the bugger was he wasn't much of a gardener after all, leaving all that weeding to be done. [...] *He had another go at being mindful but mostly he minded a drink*. (96–97, emphasis added)

Martin Fowler's bitter observations humorously undercut the fantasy of mindful attention he has been advised to cultivate. Yet it is clear that his failed attempt to absorb sensory information and identify local plant species runs parallel to McGregor's overall project in *Reservoir 13*. As we have seen, the text systematically interrupts the thoughts of its human characters with descriptions of local fauna and flora, integrating the valley's assorted inhabitants into a mode of writing that celebrates small acts of noticing and might aptly be characterised as mindful. In line with the debates sketched by Smith, this novel responds to our current attention deficit and its commercial solutions with its own alertness exercises. It is worth underlining that these exercises are

---

26 Caleb Smith, 'Disciplines of Attention in a Secular Age', *Critical Inquiry*, 45.4 (2019), 884–909 (p. 885), https://doi.org/10.1086/703963.

integrated into an unusual crime narrative, furthermore, seeing that popular 'novels of crime' are traditionally considered to be prompts for diametrically opposed, distracted modes of reading, as Smith reminds us (896). Initially, *Reservoir 13* seems to demand the anthropocentric and adrenaline-fuelled reading protocols of crime fiction, but it ends up training the reader's attention skills by rerouting them in a slower and more-than-human direction.

To be more precise, McGregor encourages us to notice other forms of life by integrating various nonhuman subplots into his narrative. Elsewhere, I have argued that the notion of the subplot might be productively applied to stories like *Ducks, Newburyport* (2019), which systematically alert us to minor animal characters.[27] A similar lesson applies here, seeing that *Reservoir 13* offers a form of multiplot as well as multiscale narration, and replaces the idea of a centripetal main narrative with a centrifugal collection of human and nonhuman subplots. Underlining the importance of these storylines, McGregor has claimed that he wrote the novel by stitching together autonomous mini-narratives: 'I wrote everything out of sequence [...] and then put it together across the thirteen years of the novel. So for each of the human characters, I wrote these episodic storylines. For each of the animals, as well'.[28] We have seen that readers are invited to follow a subplot involving springtails and we are likewise tasked with tracking the routines of local blackbirds—training our easily distracted faculty of attention by concentrating on uneventful, everyday stories:

> A blackbird dipped across Mr Wilson's garden with a beakful of dead grass for a nest. (10)
>
> In the hedge outside Mr Wilson's window a blackbird waited on its grassy bowl of blue-green eggs as the chicks chipped away at the shells. (37)
>
> In the churchyard a pair of blackbirds courted, fanning their tails [...] and watching each other bright-eyed. (62)

27　Ben De Bruyn, 'The Mom and the Many: Animal Subplots and Vulnerable Characters in *Ducks, Newburyport*', *Genre*, 54.2 (2021), 265–92, https://doi.org/10.1215/00166928-9263104.

28　McGregor, 'Time Marches On', [n.p.].

> There were blackbirds going in and out of the hedge in Jones's garden, yanking up earthworms [...] and fetching them back. Jones's sister sat at the window [...] and watched them. (70)
>
> In [Mr Wilson's] garden a pair of blackbirds were feeding together on the hawthorn, their young long gone. (101)
>
> The young blackbirds had put on their adult feathers. (121)
>
> In Fletcher's orchard the blackbirds were fattening on the early windfalls, lazy about territory and forgetting to sing. Sally watched them from the kitchen window while she made an omelette for dinner. (147)
>
> Three young blackbirds appeared on Mr Wilson's lawn, plump and bristle-feathered, and were taken by crows. (171)
>
> A blackbird's nest was blown from the elder tree at the entrance to the Hunter place, the mud mortar crumbled and the grasses scattered as chaff. (175)
>
> In the beer garden a blackbird poked at the crumbs beneath their table. (321)

Though these disconnected scenes do not necessarily feature the same character(s), they fit into a plot of sorts that travels across the village and along human houses while describing the day-to-day habits of a population of blackbirds, birds who are born and pass away, make nests and look for food, and occasionally forget to mark their territory. The novel features several similar subplots that reveal the behaviours and movements of creatures living alongside the human villagers, unnoticed, yielding a rich multiplot picture of this rural mesocosm. Gradually discontinuing its crime narrative for unconnected human and nonhuman subplots, *Reservoir 13* functions as an attention exercise for its readers because it confronts us with multifarious storylines about unexceptional routines.

The attempt to map an entire mesocosm might yield anthropocentric results if it approaches these nonhuman populations in a distanced, top-down manner or, conversely, if it renders such lives in an overly intimate, carelessly anthropomorphic fashion. But McGregor's novel largely refrains from speculating about the inner lives of plants and animals and indicates that these separate subplots do not add up to a single, implicitly anthropocentric reality. As the narrative stresses throughout, this storyworld is a *multiperspectival* environment. Cary Wolfe has tackled a similar topic in *Ecological Poetics* (2020), an analysis of Wallace Stevens' poetry that explains how these poems force us to jettison reductive conceptions of nature and environmental literature. Crucial in this respect is a remark by Jacques Derrida that Wolfe quotes at length, twice. This remark about 'worlds' consists of three parts:

> 1. Incontestably, animals and humans inhabit [...] the same objective world even if they do not have the same experience of the objectivity of the object. 2. Incontestably, animals and humans do not inhabit the same world, for the human world will never be [...] simply identical to the world of animals. 3. In spite of this identity and this difference, neither animals of different species, nor humans of different cultures, nor any animal or human individual inhabit the same world [...] because [it] is always constructed [...] by [...] language in the broad sense [...] among all living beings, [and the resulting unity] is always deconstructible [...] There is no world, there are only islands.[29]

Or, reformulating these claims, humans and animals inhabit the same reality but experience it differently, not just because they belong to distinct cultures and species but because their individual experience of the world is shaped by the logic of language 'in the broad sense', which unifies impressions in a species- and individual-specific way that remains provisional by definition. In the terminology of systems theory that Wolfe draws on too, we inevitably find ourselves in an environment shaped by observations that make certain aspects of reality available yet render others unavailable, and that unending process of 'worlding' is not restricted to the human. These lessons are at the heart of Stevens' project, Wolfe claims in his reading of the classic poem 'Thirteen Ways of Looking at a Blackbird' (1917) and its 'cubist audacity': 'Stevens' poem [reveals] that the problem of observation (and the "worlds" that emerge therefrom) is not exclusively a *human* problem'.[30] Wolfe concludes that Stevens alerts us not to a separate realm of 'nature' but to the fact that individual organisms inhabit shared yet different worlds that remain in flux as reality is observed and reobserved from distinct vantage points.

This argument sheds additional light on *Reservoir 13*, for the novel represents its storyworld through a collage of more-than-human

---

29   Cary Wolfe, *Ecological Poetics; or, Wallace Stevens's Birds* (Chicago: Chicago University Press, 2020), pp. 107–8, see also p. 64), https://doi.org/10.7208/9780226688022.
30   Ibid., p. 124.

viewpoints. McGregor's text invites us to contemplate a multifaceted ecological web, not by observing it from a single, top-down perspective, but by jumping from one perceptual bubble to another, travelling back and forth between close-ups of various critters that disclose their particular, nonhuman perception of space: 'From the eaves of the church the first bats were seen leaving at dusk, hungry from a long winter's sleep and listening for food' (112); 'In the beech wood the foxes were ready to mate. There had been scent-marking and fighting and now the pairs were established' (136). Like the Stevens described by Wolfe, McGregor is fascinated by the ways in which assorted life forms—springtails, willows, bats, foxes, etcetera—make their worlds in a shared landscape. It is hence no coincidence that one of the subplots in *Reservoir 13*, as we have seen, involves *blackbirds*. As a matter of fact, the novel accentuates its more-than-human cubism by using two lines from Stevens' classic poem as a motto: 'The river is moving./The blackbird must be flying' (viii). In offering its own observations and subplot about blackbirds, McGregor's narrative demonstrates that human and animal worlds overlap and that birds behave in everyday ways while being exposed to finitude, especially at a young age—much as the missing girl. It further emphasises that ordinary reality is composed of divergent viewpoints; as my earlier blackbird quotation shows, these birds look for something to eat, they court and watch other 'bright-eyed', they are fatally seen by crows, and are briefly observed by Jones' sister and Sally Fletcher (and remember, in this context, the dog who watches the springtails). The bird's-eye view is not a panoramic, top-down perspective, moreover, but a view that shows glimpses of reality seen from the tree, the hedge, the garden, and beneath the table. While looking for clues about a human disappearance, the novel's reader progressively becomes aware of other plots and nonhuman perspectives—the worlds of the everyday pluriverse.

Visualising Coexistence, Part I

Before returning to McGregor's mesocosm, I will further contextualise *Reservoir 13* with the help of two contemporary projects that paint a similarly wide-lens view of our everyday reality with the help of images made available by new tracking technologies. The maps and graphics collected by James Cheshire and Oliver Uberti in *Where the Animals Go* (2016) draw on various 'bio-logging' techniques to map the journeys and behaviours of creatures such as bees, cougars, and sharks.[31] The result is a textbook case of the biodiversity narratives analysed by Ursula Heise, in which 'artists and writers [...] resort to lists or catalogs [and employ the] structure of the travelogue, [which] gives such enumerations a narrative progression by juxtaposing different species and locations'.[32] *Where the Animals Go* explores multispecies coexistence in ways that differ from *Reservoir 13*, to be clear. For the illustrations in this atlas are made with the help of large-scale data sets collected by high-tech sensors, and construct a beautiful top-down view of charismatic forms of wildlife and their extraordinary mobility in locations from across the entire planet—not to mention that it delivers an explicit environmental message about biodiversity. McGregor's text evinces no such emphasis on digital tools, scientific protocols, and planetary oversight. But the novel exhibits a related environmental sensibility in the sense that its high-resolution account of animal routines and pathways implicitly mines our growing knowledge of population and movement ecology for its narrative potential. Reading this narrative is hence akin to viewing images like the ones in *Where the Animals Go*, especially those that prioritise everyday creatures, like the visualisation that maps the trips of individual bumblebees (see Figure 34).

The second project might be an even better visual analogy. As the celebrated novelist Karl Ove Knausgård asserts in the two essays that accompany *Night Procession* (2017) and *The Pillar* (2019), the photographs of Stephen Gill manage to reimagine the quiet village they both call home. For these grainy images of nocturnal wildlife and of various birds on the same wooden pillar were made automatically with motion-activated cameras and seem to present us with 'the animals and the world that is theirs when they are on their own'.[33] These pictures are examples of 'nonhuman photography', to use Joanna Zylinska's

---

31 James Cheshire and Oliver Uberti, *Where the Animals Go: Tracking Wildlife with Technology in 50 Maps and Graphics* (London: Penguin Books, 2018), p. 22.
32 Ursula Heise, *Imagining Extinction: The Cultural Meanings of Endangered Species* (Chicago: Chicago University Press, 2016), p. 55, https://doi.org/10.7208/chicago/9780226358338.001.0001.
33 Karl Ove Knausgård, 'Birdland', in *The Pillar*, Stephen Gill (Nobody Books, 2019), p. 10.

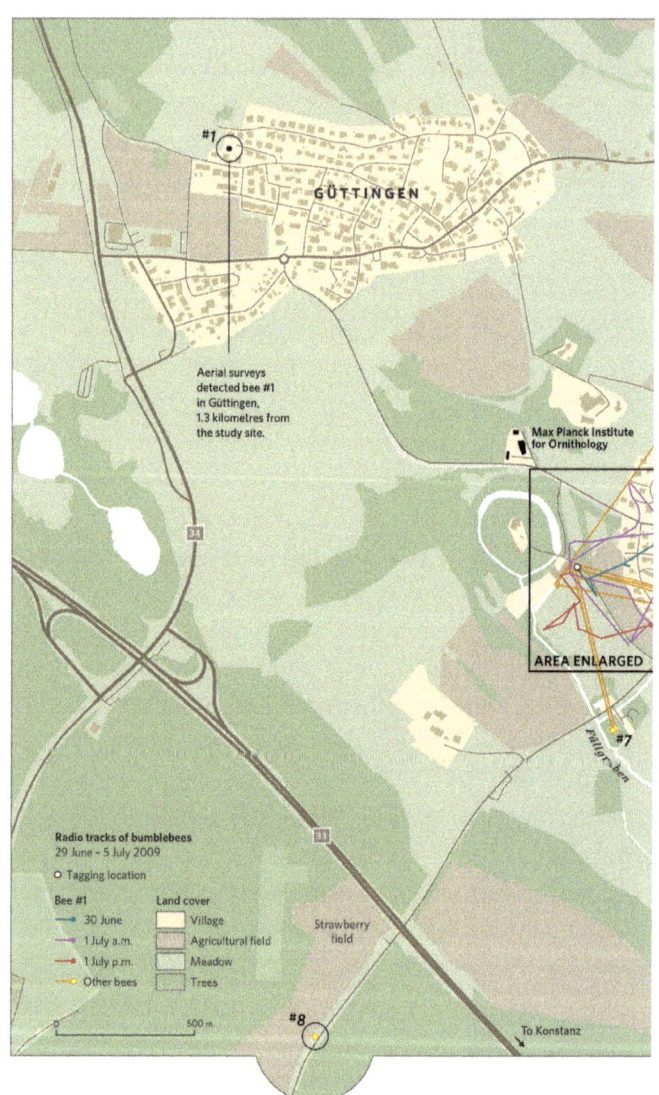

FLIGHTS OF A BUMBLEBEE

More than half of the point locations in this study came from a single queen bee (#1, above). You can follow three of her flights in the box to the right.

30 June, 1–3:45 p.m.
After receiving her tag, Bee #1 spends an hour on a tree, cleaning herself. The first day ends with a trip into town to visit some lavender, linden and a blossoming tree.

1 July, 8:45 a.m.–2:55 p.m.
She begins the day on a pear tree. After a loop around the cornfield, she returns to the same tree to rest for 80 minutes. Then she's off to a walnut tree in west Möggingen for another long break.

1 July, 4:25–7:50 p.m.
The queen's long day ends more leisurely with a buzz through the clover in Bee Marie meadow. At 7:50 p.m., she lands on a flower stalk in a garden and stays there overnight.

SOURCES: MELANIE HAGEN, UNIVERSITY OF BIELEFELD; DANIEL KISSLING, AARHUS UNIVERSITY; MARTIN WIKELSKI, MAX PLANCK INSTITUTE FOR ORNITHOLOGY; OSM

Fig. 34 'Flights of a Bumblebee', from *Where The Animals Go* by James Cheshire and Oliver Uberti. Illustration Copyright © James Cheshire and Oliver Uberti, 2016, published by Particular Books, 2016, in North America by W.W. Norton & Co Inc, 2016, Penguin Press, 2018. Reprinted by permission of Penguin Books Limited. All rights reserved.

terminology; the human appears to be absent 'as [their] subject, agent, or addressee', seeing that these images do not feature humans, were not taken by humans, and do not seem to be directed at human spectators.[34] Yet these photos still perform valuable cultural work, Knausgård holds, as they expose a world that is both nearby and unfamiliar—expanding or re-scaling, again, the everyday landscape. The images of *Night Procession* were 'mostly captured within a radius of perhaps ten kilometres, [yet] seeing them was like having lived in a house for many years and then suddenly becoming aware of a [...] hitherto concealed room'.[35] The photographs of *The Pillar* likewise foreground overlooked animals and invite us to hone our attention skills: 'despite our living so close together [...] we barely notice [birds]'.[36] These photographs reimagine our experience of time and space too: 'We see the same landscape in spring and summer, in autumn and winter [y]et there is not the slightest monotony about these pictures, for [...] each of these birds opens up a unique moment in time'[37] (see Figure 35). These points can, again, be aligned with *Reservoir 13*, for the novel likewise concentrates on a small village, reveals the adjacent but overlooked worlds of animals, and features repetitive, seasonal descriptions that nevertheless feel remarkably alive.

Gill's images also accentuate the condition of finitude that is central to McGregor's narrative about vanishing children and vulnerable animals. While reflecting on these bird images, Knausgård shares an anecdote related by the photographer which illustrates that animals live in the now and thereby lay bare our shared condition as mortal beings with remarkable intensity. As Gill narrates, a pair of woodpigeons had been nesting on his property for several years, 'building their nest, laying eggs, hatching them, feeding their young' and '[e]very year, just before the young reached the fledgling age, a goshawk had come and taken them' but 'this didn't stop the pair from returning' and starting anew every time (8). This anecdote underscores the repetitive improvisations of vulnerable creatures and exhibits the same aesthetic as Gill's bird portraits. Woodpigeons appear in a similar light in McGregor's novel:

---

34   Joanna Zylinska, *Nonhuman Photography* (Cambridge, MA: MIT Press, 2017), p. 5.
35   Karl Ove Knausgård, 'The World Inside the World', in *Night Procession*, Stephen Gill (Nobody Books, 2017), p. 8.
36   Knausgård, 'Birdland', pp. 4–5.
37   Ibid., p. 3.

Fig. 35 'Untitled', p. 53, from *The Pillar*, a series of photographs by Stephen Gill, with words by Karl Ove Knausgård, published by Nobody Books (nobodybooks. com). *The Pillar*, 2015–2019 © Stephen Gill. All rights reserved. Any further use or distribution of the photographs is strictly prohibited. Please contact The Wylie Agency (UK) Ltd to request permission for any further use or distribution of the photographs.

The woodpigeons built their nests in the trees by the river. The thin frame of sticks seemed barely enough to take the weight of one fat bird. But it was assumed they knew what they were doing. (88)

In the horse chestnut tree by the cricket ground the woodpigeons were fighting, rearing up at each other with rattling wings. It wasn't always clear what kept them from falling out of the tree. (166–67)

The woodpigeons laid eggs in their nests in the beech wood and in the horse chestnut by the cricket ground. They took turns sitting on the eggs, but there were still plenty stolen by magpies and crows. (248)

Like Gill's anecdote and photographs, McGregor's pigeon subplot illustrates that we share a condition of finitude with other animals and inhabit an everyday pluriverse that is profoundly strange even as its residents execute repetitive routines. The work of Gill and McGregor seems to differ from the high-tech panorama painted in *Where the Animals Go*. Yet we should not overemphasise this difference, as the atlas does not only use bio-logging projects to communicate its conservation message, but also to develop a new mode of storytelling, arguably, that reconstructs the everyday lifeworlds of other animals in unprecedented detail. One radiotagged bee, for example, 'spends an hour on a tree, cleaning herself' and her day 'ends with a trip into town to visit some lavender, linden and a blossoming tree' (154). Tracking technology likewise allows us to witness how a songbird moves between different bird feeders: 'At 7:20 a.m., it arrived at feeder 4 and joined the [other birds] feeding there for over an hour. It then flew to feeder 1, via feeder 3, for a mid-morning snack. After ping-ponging between feeders 1 and 2, it took wing to the woods until its last feed of the day at feeder 4 at around 2:15 p.m'. (152) Even though these scientific images and projects often draw attention to the remarkable feats of exotic species, such diary-like entries imply that they record the animal everyday with remarkable granularity too.

Modelling Interspecies Assemblages

The previous sections have argued that *Reservoir 13* displaces the reader's attention from a human vanishing towards a more-than-human mesocosm by integrating descriptions of small scenes and bubble worlds that develop into uneventful subplots about human characters and animal communities. While other novels encompass nonhuman subplots and perspectives too, McGregor's text stands out because it stubbornly downplays the narratable quality of animal and plant lives and foregrounds their repetitive quality via its looping, longitudinal approach. Keeping track of its characters across the seasons and a thirteen-year period, this novel about everyday routines ultimately starts to function less like a traditional narrative and more like a textual algorithm—one that uses the resources of literature to perform a particular version of ecosystem modelling. Explaining this point will enrich my analysis of the novel's subplots and worlds by taking into account its seasonal form and cross-species interactions.

McGregor's interest in intricate ecosystems rather than single species or plots is underscored by the fact that several characters again function like avatars of the narrator because they are engaged in a form of landscape monitoring that is akin to what we find in *Reservoir 13*. The clearest example is the National Park ranger who keeps track of local butterfly populations: 'The first small tortoiseshells began mating, flying after each other above the nettle beds [...] [He] spent an enjoyable hour watching them, and making a record, and when he got back to the office he filed it carefully away' (38). Graham's butterfly safaris even manage to convert another character into a citizen scientist, as we later learn, in a scene that accentuates the complexity of these lives: 'They found half a dozen species but he seemed to be talking about two dozen more, describing their lifecycles, migrations, feeding habits, mating styles. [...] She'd had no idea there was so much to it' (45). The river keeper in turn is tasked with checking the river's level and water quality; twice we hear how he 'dropped the [cage of] sample bottles into the water', '[a]lways the same spot' and 'same time of day' (90, 247). Winnie, similarly, knows when the time is ripe to harvest wild plants for jams, soups, and syrups, picking wild fennel (15), wild apples (75), hazel nuts (203), nettles (242), and elderflowers (249). These storylines are small examples of a logic adopted by the novel as a whole, as it regularly informs us about the weather, the water level, and the condition of the local fauna and flora—even when human concerns seem less central, if not simply absent altogether. Without further explanation, we hear that 'The nettles grew up around the dead oak in Thompson's yard' (312), for instance, and that 'the young foxes lit out for new territory and were killed on the roads in great number' (302). To return to the butterflies, Graham's counts prompt us to notice other references to these creatures, which can be read as standalone lyrical moments, but also as traces of yet another subplot devoted to local critters, in line with my earlier remarks about springtails, willows, blackbirds, and woodpigeons: 'In the quarry by the main road the small coppers were mating again' (92); 'In the old quarry by the main road the larvae of small coppers were feeding on the sorrel plants where they'd hatched' (145); 'In the evenings through the beech wood the last small coppers were seen, roosting head down on the grasses beside the track' (321). Like certain characters, in short, the narrator monitors multiple aspects of this regional ecosystem, including animal lifecycles, water levels, and flowering times.

These observations imply that *Reservoir 13* is attuned to so-called phenological events, periodic changes linked to the seasons such as changing weather patterns, blooming and hibernation cycles, the departure and arrival of migratory birds. In her account of 'seasonal form' in literature, Sarah Dimick explains that such phenological events functioned as signs of nature's regularity in classic nonfiction texts by Thoreau and Leopold, and that these seasonal patterns provided a model for literary form too, since they produce 'momentum without generating suspense'.[38] Yet because global heating is changing these patterns, Dimick reflects, the seasonal rhythms of older texts are starting to appear as historical artefacts and to point towards 'environmental disorder' instead of 'environmental perpetuity'.[39] The reassuring and non-narrative quality of phenological records will evaporate, moreover, as biological cycles become dramatic and narratable because of 'environmental arrhythmias'.[40] Like the writings of Thoreau and Leopold, *Reservoir 13* frequently alerts readers to phenological events. In several chapters, the January section notes the emergence of the first snowdrops (5, 136, 159), for instance, and the swallows are described as leaving in September (101, 253, 277) and returning in April (11, 139, 313). While these events appear reliable across the period covered in *Reservoir 13*, certain passages destabilise the appearance of seasonal regularity and ecological balance. Alongside news bulletins about 'forests burning in Malaysia' (69), and vague subplots about ecological protest (97, 230) and the installation of wind turbines (92, 272), we are told at one point that 'the reservoirs were as low as they'd been in forty years [...] [t]here were hosepipe bans over four counties; the hills were drying out' (221). Like the novel's animal and vegetal subplots, seasonal patterns accordingly hover just above or below the threshold of narrativity proper; they do not appear dramatic and denaturalised, but neither do they really function as confident signs of natural perpetuity.

---

38 Sarah Dimick, 'Disordered Environmental Time: Phenology, Climate Change, and Seasonal Form in the Work of Henry David Thoreau and Aldo Leopold', *ISLE*, 25.4 (2018), 700–21 (p. 710), https://doi.org/10.1093/isle/isy053.
39 Ibid., p. 714.
40 Ibid., p. 716.

The narrative may not represent forms of environmental arrhythmia, yet it hints at this looming threat by suggesting that seasonal events—like the nonhuman lifeforms identified earlier—merit sustained attention and monitoring.

Because *Reservoir 13* exhibits seasonal form and tracks ongoing biological processes, it offers a dynamic simulation rather than a static portrait of the village's environment—a simulation that performs acts of 'ecosystem modelling'. In contrast to the literal meaning of this phrase, McGregor's narrative obviously does not offer a quantitative representation of ecological interactions aimed at comparing the outcomes of particular environmental management strategies.[41] I have already noted that the novel does not describe its landscape in the distanced manner of experts or privilege human goals. Vinciane Despret has warned us, moreover, that the notion of the ecosystem is linked to the antiquated 'machine analogy' of 'the balance of nature'[42] and Elizabeth DeLoughrey has pinpointed disquieting ties between this notion and devastating nuclear tests: '[e]cosystem ecology, with its emphasis on closed systems, management, control, and equilibrium, drew tremendous support in part because it was appealing to the military'.[43] Nevertheless, the term captures the fact that McGregor's nonhuman (and human) characters do not simply coexist but interact. What is more, these interactions signify that it might be misleading to describe this storyworld in terms of separate subplots and worlds. As Anna Tsing remarks in *The Mushroom at the End of the World* (2015), her influential account of economic and ecological mushroom networks, the notion of 'world'—derived from Uexküll but also used by Derrida and Wolfe, as we have seen—is less helpful if we want to describe interspecies ties. If we apply what Tsing calls 'arts of noticing'—and train our attention again—we can develop alternative stories in which 'landscape' becomes 'the protagonist of an adventure in which humans

---

41 See: William L. Geary et al., 'A guide to ecosystem models and their environmental applications', *Nature Ecology & Evolution*, 4 (2020), 1459–71, https://doi.org/10.1038/s41559-020-01298-8.

42 Vinciane Despret and Michel Meuret, 'Cosmoecological Sheep and the Arts of Living on a Damaged Planet', *Environmental Humanities*, 8.1 (2016), 24–36 (p. 26), https://doi.org/10.1215/22011919-3527704.

43 Elizabeth DeLoughrey, 'The Myth of Isolates: Ecosystem Ecologies in the Nuclear Pacific', *Cultural Geographies*, 20.2 (2013), 167–84 (p. 178), https://doi.org/10.1177/1474474012463664.

are only one kind of participant'.⁴⁴ Yet that involves abandoning the focus on individual worlds, Tsing stresses, because this fails to present organisms as 'participant[s] in the wider rhythms and histories of the landscape' (156). To tell landscape stories, we need to trace fragile cross-species 'assemblages', a strategy that draws us 'beyond bubble worlds into shifting cascades of collaboration and complexity' (157). While the notion of the ecosystem might be faulted for its outdated connotations and troubling roots, I use it here to refer to similar landscape assemblages and their contingent histories. Because both Tsing's account of interspecies networks and McGregor's ecosystem model involve 'watching the interplay of temporal rhythms and scales in the divergent lifeways that gather' (23). In the storyworld of *Reservoir 13*, springtails eat fallen leaves, coppers eat sorrel plants, goldcrests live in yew trees, wood pigeons glean kale and grain produced by humans, and blackbirds eat earthworms even as their young are eaten by crows that feed on young woodpigeons too. The real story does not centre on individual worlds and plots but on the landscape, the ecosystem.

Because I describe this aspect of the text in terms of ecosystem modelling, we should here also review existing literary research on models. Developing her earlier reflections on the political affordances of sociocultural forms, Caroline Levine has recommended what she calls 'model thinking', 'a reading practice that involves designing generalizable political models for the common good',⁴⁵ in an explicit attempt to avoid the monoscalar approach of the humanities, which typically privileges 'the emancipatory exception'⁴⁶ at the expense of broader social norms that help to address the thorny challenges of collective life. Levine defines models as follows (and recall Chang's definition of the mesocosm in this context):

> Models are made to repeat [b]ut [they] do not have to be static: [...] the task is to test out multiple scenarios. [...] All models are aesthetic in that they are imagined and constructed, and all are material and social in that they are put to use to design and redesign our world. For my own

---

44 Anna Tsing, *The Mushroom at the End of the World: On the Possibility of Life in Capitalist Ruins* (Princeton: Princeton University Press, 2015), p. 155.
45 Caroline Levine, 'Model Thinking: Generalization, Political Form, and the Common Good', *New Literary History*, 48.4 (2017), 633–53 (p. 636), https://doi.org/10.1353/nlh.2017.0033.
46 Ibid., p. 641.

purposes [...] models are [...] useful because they deliberately abstract relationships so that we can grasp those relationships apart from their details.⁴⁷

We should treat literary texts as attempts to model particular forms of social organisation, Levine proposes, even if novels especially have not always been successful in representing labour, repetition, and the 'maintenance of daily life',⁴⁸ she believes—with the exception of examples that 'push at the limits of their affordances'.⁴⁹ Paying more attention to ecology, Heidi C. M. Scott has asserted that the literary modelling of ecological processes 'maintains complexity, holism, and in situ context, and [...] tends toward enhancing the vitality of the system it describes', 'choreograph[ing] the discrete components of a natural system [while] leav[ing] space for the little wildernesses that lurk within the model'.⁵⁰ Scott's argument is akin to Tsing's in the sense that both ask us to ponder narratives about cross-species assemblages that are attuned to specific historical contexts and are therefore more appropriate tools for ecosystem modelling, arguably, than abstract mathematical accounts. As researchers in population ecology gradually discovered, Sharon Kingsland explains, 'similar [...] local ecological conditions [...] did not support floras and faunas of similar diversity [...] forcing ecologists to recognise that [such] differences were the result of the peculiar historical circumstances that had operated in these locations'.⁵¹ Decontextualised mathematical laws and models fail to capture this contribution of the contingent histories at play in particular settings and their ecological effects. That does not mean that general reflections are unhelpful; indeed, and returning to Levine, I maintain that *Reservoir 13* hints at abstraction as well as specificity, seeing that it points both towards local, contingent interactions and a more general, abstract view of interspecies exchanges. Braiding its subplots into a dynamic model of cross-species

---

47   Ibid., p. 643.
48   Ibid., p. 645.
49   Ibid., p. 646.
50   Heidi C. M. Scott, *Chaos and Cosmos: Literary Roots of Modern Ecology in the British Nineteenth Century* (University Park: Pennsylvania State University Press, 2014), p. 191.
51   Sharon Kingsland, *Modeling Nature: Episodes in the History of Population Biology*, 2nd ed. (Chicago: Chicago University Press, 1995), p. 227.

entanglements, the novel constitutes a narrative, ecological version of model thinking.

In characterising this landscape story as an ecosystem model, I am not claiming that there is a one-to-one correspondence between fictional and biological ecosystems (nor is there such a correspondence between physical environments and quantitative models, as Kingsland is at pains to emphasise). My point is rather that *Reservoir 13* performs acts of modelling that are related in a non-trivial manner to the descriptive protocols of certain forms of biological science. The dynamic, networked quality of its descriptions is more important in this respect than its fidelity to particular facts or to specific ecosystems. We have already learned that the story highlights seasonal patterns and interspecies attachments. Yet another reason we can characterise the result as a model is that it accentuates the repetitive, routine occurrences of everyday life. Consider the following set of quotations from McGregor's text, which collects some of the phrases that are repeated as the story unfolds—or rather, as this ecosystem model cycles through its iterations (repetitions are in bold):

[A] heron stood and watched the water. (9)

The cow parsley was thick along the footpaths and the shade deepened under the trees. (14)

The goldcrests fed busily deep in the branches of the churchyard yew. (29)

In the eaves of the church the bats were folded deeply into hibernation and the air around them was still. (31)

In the beech wood the fox cubs were moved away from their dens. (66)

The cow parsley was thick along the footpaths and the shade deepened under the trees. (67)

[A] heron stood and watched the water. (72)

By the packhorse bridge a heron paced through the mud at the river's edge [...] It stopped, and settled, and watched the water. (86).

In the beech wood the fox cubs were [moved] away from their dens and taught to find food for themselves. (97)

> **By the packhorse bridge a heron paced through the mud at the river's edge […] It stopped, and settled, and watched the water.** (125)
>
> **The cow parsley was thick along the footpaths and the shade deepened under the trees.** (294)
>
> **The goldcrests fed busily deep in the branches of the churchyard yew.** (306)
>
> **In the eaves of the church the bats were folded deeply** in their **hibernation and the air around them was still.** (324)

The looping, monotonous quality of these passages reconnects the novel's environmental imagination to its exploration of everyday life—and to the serial portraits of birds from *The Pillar* and the high-tech animal diaries sketched in *Where the Animals Go*. Because, like their human counterparts, the nonhuman characters of *Reservoir 13* turn out to be creatures of habit, whose existence consists of predictable routines that can be described in strikingly similar ways, despite the passage of time—resulting in haiku-like refrains that fit into distinct subplots but also into a dynamic ecosystem model, a landscape story.

Visualising Coexistence, Part II

Zooming out once more, my previous remarks can also be linked to citizen science and to a final visual analogue, diagrams of ecological food webs. My starting point here is Elizabeth Callaway's intriguing analysis of fictional ecosystems, which maps the interactions between various species in science fiction novels set on imaginary planets. As she notes, these 'puzzle planets'[52] accentuate mysterious, half-understood interspecies relations and invite a form of ecosystemic reading. For these novels 'feature ecosystems composed of a variety of unique life forms that interact in ways that force observers to consider multiple types of organisms together' and typically '[c]haracters and readers must piece together clues that […] force them to reevaluate what they think they know about the more-than-human worlds of which they are

---

52 Elizabeth Callaway, 'Islands in the Aether Ocean: Speculative Ecosystems in Science Fiction', *Contemporary Literature*, 59.2 (2018), 232–60 (p. 237).

a part'.⁵³ It goes without saying that *Reservoir 13* is not a science fiction novel, and it differs from Callaway's examples in other respects too, seeing that these scifi novels involve deliberately simplified 'speculative ecosystems' and urgent quests for knowledge about 'planetary-scale patterns'⁵⁴ and 'bizarre food chain[s]'⁵⁵ that help human characters survive in alien environments while raising moral questions about terraforming. Nevertheless, I would argue that our earth is now, and has always been, a 'puzzle planet' too, and have tried to articulate that McGregor's narrative invites a similar form of biological curiosity and ecosystemic reading.

These parallels point us in the direction of a final visual analogue for *Reservoir 13*. In the course of her analysis, Callaway proceeds to 'plo[t] out'⁵⁶ fictional ecosystems visually to compare their properties and environmental imaginations. We can apply the same strategy to *Reservoir 13* and visualise the predator-prey interactions that characterise this particular environment, in line with Callaway's example and with the 'food web[s]' and 'conceptual interaction networks' mentioned in a recent overview of ecological modelling techniques.⁵⁷ Though this visualisation can surely be fleshed out in several respects, McGregor's text evokes an ecosystem that looks more or less like Figure 36.

What makes such diagrams extra interesting here is that, as I argued above, McGregor's text cannot simply be visualised via such a network but actually *already functions* as a diagram or abstract ecosystem model of sorts. In contrast to other ecosystem narratives, its chapters do not simply describe this network but run different iterations of the associated algorithm, as it were, conjuring up the storyworld's multispecies assemblage by travelling along the paths of this diagram and repeating that badgers feed (and hibernate), that worms interact with buzzards and badgers and blackbirds, and that these creatures in their turn interact with beetles, rabbits and sheep, not to mention crows and humans and *Dichelobacter nodosus*, the bacterium that causes 'scald' disease in sheep (79). *Et cetera*. A full visualisation of the narrative's interspecies ties would be even more complicated, incidentally, not

---

53 Ibid., p. 233.
54 Ibid., p. 234.
55 Ibid., p. 248.
56 Ibid., p. 246.
57 Geary, 'Ecosystem models', p. 1460.

10. The Everyday Pluriverse: Ecosystem Modelling in Reservoir 13

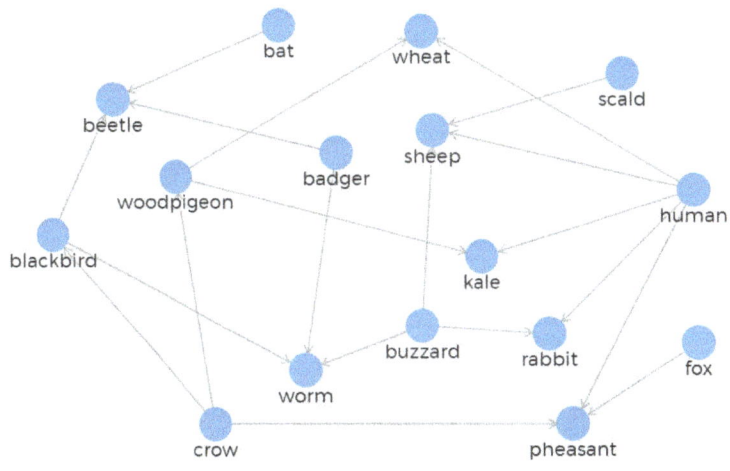

Fig. 36 Ben De Bruyn, *Conceptual interaction network of the food web in* Reservoir 13 (2021) © Ben De Bruyn. CC BY-NC.

simply because it would include more creatures (think of the dog, the heron, the swallows, the springtails) but because it would obviously need to integrate other forms of connection beyond simple food-based and predator-prey interactions.

Such diagrams enable us to map the ecosystems of stories featuring multiple life forms. They could even be used to compare the density, cohesion, and other characteristics of different fictional ecosystems. But the iterative quality of McGregor's writing implies that it approximates the texture of such networks to an unusual degree. And I believe that when novels begin to model ecosystems in such detail, we can make an even stronger case for the expansion of our methodological toolkit and the value of such visualisation techniques. When our writing techniques evolve in this direction, our reading strategies should follow suit. A detailed comparison of my foodweb and Callaway's plots falls outside the scope of this paper, but I should underline in closing that *Reservoir 13* evokes a highly everyday landscape and assemblage, in contrast to its exotic counterparts in scifi narratives (see Figure 37).

Fig. 37 'Untitled', p. 113, from *The Pillar*, a series of photographs by Stephen Gill, with words by Karl Ove Knausgård, published by Nobody Books (nobodybooks.com). *The Pillar*, 2015–2019 © Stephen Gill. All rights reserved. Any further use or distribution of the photographs is strictly prohibited. Please contact The Wylie Agency (UK) Ltd to request permission for any further use or distribution of the photographs.

In reading this novel, furthermore, we are not asked to adopt the position of a cutting-edge scientist at the outskirts of the universe, but to execute the unspectacular routines of ordinary, amateur observers. So it rather recalls Akiko Busch's *The Incidental Steward* (2013), a nonfiction text that sketches Busch's participation in various citizen science projects that monitor animal and plant populations along the Hudson River valley. As in *Reservoir 13*, the main lesson of her text involves paying attention to phenological rhythms and quotidian creatures that resist human noticing: 'A swallow skimmed along the surface of the creek, a great blue heron took flight from the opposite bank, the level of the water dropped imperceptibly as the tide receded. At the edge of water, [...] you would have to be a fool not to want to be the one who is doing the noticing' (120). The environmental imagination of McGregor's rural novel is related, in short, to bio-logging and citizen science projects, to artistic photos of everyday birds and abstract diagrams of interspecies assemblages.

Conclusion: Scale and Stoicism in the Everyday Anthropocene

Building on existing research about McGregor's everyday and post-pastoral aesthetics, this chapter has analysed *Reservoir 13* in four interconnected steps. First, I have explained that this novel evokes a reality that is larger and more densely inhabited than that of traditional forms of realism, even if it disregards truly micro- and macroscale phenomena to contemplate a rural mesocosm. Second, we have seen that the novel hones the reader's noticing skills by reorienting our gaze from linear plots to cyclical subplots and from a singular human perspective to a more-than-human pluriverse composed of overlapping worlds. Third, I have elucidated how the novel's references to phenology and predation fit into its peculiar, seasonal form and dynamic portrait of a cross-species assemblage. Ultimately, the text conjures up this everyday ecology via a repetitive representational strategy that we can call ecosystem modelling, a phrase that is not simply metaphorical even if McGregor uses narrative rather than mathematical techniques to do so. Alongside these observations, finally, I have identified parallels between the novel and similar attempts to render the adjacent lives of nonhuman creatures in verbal and visual form, pointing towards artistic photographs, scientific maps, citizen science projects, and diagrams of ecological interactions. These claims imply that *Reservoir 13* helps us to rethink the everyday in a multispecies direction and that debates on nonhuman scale should not underestimate mesocosmic and polyrhythmic modes of storytelling. By way of conclusion, I will draw together these points about scale, ecology, and everyday life by briefly pinpointing the position of McGregor's work in debates on biodiversity, climate change, and stoicism. As we have learned, *Reservoir 13* encourages us to modify the conversation on our environmental crisis by making more room for nonhuman worlds and assemblages, in line with recent claims by Cary Wolfe and Ursula Heise on the undiminished importance of animal studies and the need for new forms of multispecies justice at the start of the twenty-first century. The novel also underlines the role of everyday life in a way that resonates with the approach Stephanie LeMenager has labelled the 'everyday Anthropocene'.[58] Like the texts she mentions, *Reservoir 13* intimates that a narrow focus on spectacular events and responses fails to do justice to the present situation. At the same time, McGregor's novel subtly differs from these examples, I believe, because it deliberately backgrounds planetary concerns and accentuates the continuing claims of everyday life on humans and nonhumans alike, finding solace in mundane routines even as seasons and lifeworlds come under pressure. While the narrative alludes to ecological disturbance, its use of subplots and seasonal form is designed to reorient our gaze towards larger patterns and the uncaring cycles of nature. Even as young girls disappear, people are exposed to racism, women are stalked by ex-husbands, people lose their jobs and lives, and new quarries open despite environmental protest. And in the background, life goes on, snowdrops emerge, and swallows return. Even Becky's disappearance turns out to be part of a pattern; when another girl goes missing and turns up dead, the text bleakly notes that 'These things just kept happening, it seemed' (161).

It follows from these observations that we can interpret *Reservoir 13* as an example of what Caren Irr calls 'ecostoicism'. As she notes, ideas from the Stoicist tradition are compatible with an environmental mindset, and that explains why writers are returning to these ideas at a time of planetary crisis. Famous thinkers like Marcus Aurelius elaborate that Stoicism involves 'distantiation tactics' designed to foster 'an observant, meditative mind embedded in the long and impersonal cycles of nature'.[59] Hinting at the importance of scale in this context, Aurelius recommends 'the habit of taking the long view—i.e., asking how significant particular concerns might appear after a century or from afar'.[60] Irr finds traces of this mindset in the writings of Richard Powers and Jonathan Franzen, but there are multiple strands of ecostoicism in

---

58 Stephanie LeMenager, 'Climate Change and the Struggle for Genre', in *Anthropocene Reading: Literary History in Geologic Times*, ed. by Tobias Menely and Jesse Oak Taylor (University Park: Penn State University Press, 2017), pp. 220–38 (p. 223), https://doi.org/10.1515/9780271080390-013.

59 Caren Irr, 'Ecostoicism, or Notes on Franzen', *Post 45* (30 May 2018), p. 12, https://post45.org/2018/05/ecostoicism-or-notes-on-franzen/.

60 Ibid., p. 11.

contemporary fiction, she believes, from explicitly scientific visions to 'more meditative' ones.[61] *Reservoir 13* exemplifies the latter approach via its repetitive mode of modelling and its mesocosmic perspective. When we take a longer view and adjust the scale of our narratives, it reveals that losses are to be expected:

> When he got back to the yard his brothers were all inside the shed. They'd lost a ewe while he'd been gone. (10)
>
> Late in the month there was snow and the Jacksons went out on the hills looking for ewes. [...] There were no losses yet but if this weather kept up it was likely. (106)
>
> A storm came and blew snow sideways across the valley, and when it had passed the trees were edged with white. The Jacksons had losses in the hills. (156)
>
> In February it snowed solidly for a week [...] Jackson's boys had to go up on foot to pull out as many sheep as they could find. Most of them were easy enough to find, [...] but the losses were high. (187)
>
> Will Jackson kept Tom out of school and took him up to look for lost ewes. [...] It was likely some would be dead by now, and Will thought that Tom was old enough to see. (241)

If we zoom out, such patterns and repetitions yield reassurance as well as grief. A scene after a funeral suggests, in fact, that the vicar is comforted by phenological patterns rather than religious promises: 'The snowdrops were up and the crows flew overhead and the wind moved through the trees. Jane had to keep herself from smiling' (136). As such passages spell out, the ecosystem interactions modelled in *Reservoir 13* encourage a meditative, mindful version of ecostoicism that is inextricably linked to its mesocosmic approach. The novel distances us from individual human characters and their concerns, not to dismiss such suffering, but to integrate it into a panorama of creaturely finitude— and to reformulate the detective novel in doing so.

In closing, I should note that this stoicist version of the everyday Anthropocene returns in McGregor's most recent novel *Lean Fall Stand* (2021). Whereas *Reservoir 13* is mainly preoccupied with biodiversity, this text explicitly refers to the climate crisis. One of the main characters

---

61   Ibid., p. 20.

is a female climate scientist who is committed to her research but has become sceptical about its impact:

> The students were so committed to their studies, and so impatient to move into research. She wondered what they might achieve. They could only establish the same things all over again, with ever-increasing certainty and detail. Yes, there is a clear link between CO2 emissions and temperature rise. No, there is no historical precedent. Yes, immediate action is required.[62]

As if to underline this dispiriting lesson, the scientist is forced to abandon her professional commitments because her husband has a dangerous accident while working on Antarctica—a setting that again hints at climate destabilisation—which compels her to become a full-time caregiver. The implied message seems to be that we should renounce spectacular narratives for accounts that integrate climate change into our day-to-day routines, along the lines proposed by LeMenager. Yet the text also differs subtly from that proposition because it implies that everyday struggles demand our attention in ways that end up sidelining planetary concerns, however real and urgent they might be. And in the process, snowdrops and seasonal rhythms continue to produce a measure of comfort, much as in *Reservoir 13*:

> Anna spent the afternoon barrowing thick layers of mulch to the vegetable beds and fruit cages. She had been turning the pile all through the autumn, and the decaying leaves were full of worms and beetles and all manner of microbia. The just-right smell of it was an immense pleasure. The bright green nubs of snowdrops and crocuses were nudging through the soil.[63]

These brief notes prove that seasonal form and stoicist endurance reappear in *Lean Fall Stand*—though this novel does not flesh out such scenes into extended subplots or systematically perform acts of ecosystem modelling. It also confirms that McGregor's writing constitutes a subtle variation on the 'everyday Anthropocene' via its meditative form of ecostoicism. Even at a time of planetary crisis, it asks us to consider the ongoing rhythms of ordinary ecologies and quotidian responsibilities, and to inhabit a multispecies and *even more everyday* Anthropocene.

---

62 Jon McGregor, *Lean Fall Stand* (London: 4th Estate, 2021), p. 107.
63 Ibid., p. 260.

Works Cited

Alexander, Neal, 'Profoundly Ordinary: Jon McGregor and Everyday Life', *Contemporary Literature*, 54.4 (2013), 720–51, https://doi.org/10.1353/cli.2013.0046

Armstrong, John, 'Pastoral Place and Violence in Contemporary British Fiction', *Concentric: Literary and Cultural Studies*, 44.2 (2018), 51–79.

Busch, Akiko, *The Incidental Steward: Reflections on Citizen Science* (New Haven: Yale University Press, 2013).

Callaway, Elizabeth, 'Islands in the Aether Ocean: Speculative Ecosystems in Science Fiction', *Contemporary Literature*, 59.2 (2018), 232–60.

Chang, Alenda Y., *Playing Nature: Ecology in Video Games* (Minneapolis: University of Minnesota Press, 2019), https://doi.org/10.5749/j.ctvthhd94

Cheshire, James and Oliver Uberti, *Where the Animals Go: Tracking Wildlife with Technology in 50 Maps and Graphics* (London: Penguin Books, 2018 [2016]).

Crosby, Sara L., 'Beyond Ecophilia: Edgar Allan Poe and the American Tradition of Ecohorror', *ISLE*, 21.3 (2014), 513–25, https://doi.org/10.1093/isle/isu080

De Bruyn, Ben, 'The Mom and the Many: Animal Subplots and Vulnerable Characters in *Ducks, Newburyport*', *Genre*, 54.2 (2021), 265–92, https://doi.org/10.1215/00166928-9263104

DeLoughrey, Elizabeth, 'The Myth of Isolates: Ecosystem Ecologies in the Nuclear Pacific', *Cultural Geographies*, 20.2 (2013), 167–84, https://doi.org/10.1177/1474474012463664

Despret, Vinciane & Michel Meuret, 'Cosmoecological Sheep and the Arts of Living on a Damaged Planet', *Environmental Humanities*, 8.1 (2016), 24–36, https://doi.org/10.1215/22011919-3527704

DiCaglio, Joshua, 'Scale Tricks and God Tricks, or The Power of Scale in *Powers of Ten*', *Configurations*, 28.4 (2020), pp. 459–90, https://doi.org/10.1353/con.2020.0021

Dimick, Sarah. 'Disordered Environmental Time: Phenology, Climate Change, and Seasonal Form in the Work of Henry David Thoreau and Aldo Leopold', *ISLE*, 25.4 (2018), 700–21, https://doi.org/10.1093/isle/isy053

Epstein, Andrew, '"The Rhapsody of Things as They Are": Stevens, Francis Ponge, and the Impossible Everyday', *Wallace Stevens Journal*, 36.1 (2012), 47–77, https://doi.org/10.1353/wsj.2012.0008

Geary, William L. et al., 'A guide to ecosystem models and their environmental applications', *Nature Ecology & Evolution*, 4 (2020), 1459–71, https://doi.org/10.1038/s41559-020-01298-8

Ghosh, Amitav, *The Great Derangement: Climate Change and the Unthinkable* (Chicago: Chicago University Press, 2016).

Heise, Ursula, *Imagining Extinction: The Cultural Meanings of Endangered Species* (Chicago: Chicago University Press, 2016), https://doi.org/10.7208/chicago/9780226358338.001.0001

Herman, David, *Narratology beyond the Human: Storytelling and Animal Life* (Oxford: Oxford University Press, 2018), https://doi.org/10.1093/oso/9780190850401.001.0001

Hollister, Lucas, 'The Green and the Black: Ecological Awareness and the Darkness of Noir', *PMLA*, 134.5 (2019), 1012–27, https://doi.org/10.1632/pmla.2019.134.5.1012

Irr, Caren, 'Ecostoicism, or Notes on Franzen', *Post 45* (30 May 2018), https://post45.org/2018/05/ecostoicism-or-notes-on-franzen/

Kingsland, Sharon, *Modelling Nature: Episodes in the History of Population Biology*, 2nd ed. (Chicago: Chicago University Press, 1995).

Knausgård, Karl Ove, 'The World Inside the World', in *Night Procession*, Stephen Gill (Nobody Books, 2017).

Knausgård, Karl Ove, 'Birdland', in *The Pillar*, Stephen Gill (Nobody Books, 2019).

LeMenager, Stephanie, 'Climate Change and the Struggle for Genre', in *Anthropocene Reading: Literary History in Geologic Times*, ed. by Tobias Menely and Jesse Oak Taylor (University Park: Penn State University Press, 2017), 220–38, https://doi.org/10.1515/9780271080390-013

Levine, Caroline, 'Model Thinking: Generalization, Political Form, and the Common Good', *New Literary History*, 48.4 (2017), 633–53, https://doi.org/10.1353/nlh.2017.0033

Macfarlane, Robert, 'The Eeriness of the English Countryside', *The Guardian* (10 April 2015), https://www.theguardian.com/books/2015/apr/10/eeriness-english-countryside-robert-macfarlane

McGregor, Jon, 'Time Marches On: An Interview with Jon McGregor', *The Paris Review* (18 October 2017), https://www.theparisreview.org/blog/2017/10/18/time-marches-interview-jon-mcgregor/

McGregor, Jon, *Reservoir 13* (London: 4th Estate, 2018 [2017]).

McGregor, Jon, *Lean Fall Stand* (London: 4th Estate, 2021).

Moretti, Franco, *Distant Reading* (London: Verso Books, 2013).

Scott, Heidi C.M., *Chaos and Cosmos: Literary Roots of Modern Ecology in the British Nineteenth Century* (University Park: Pennsylvania State University Press, 2014).

Seitz, Matt Zoller, 'The Rise of the Slow Crime Procedural', *Vulture* (6 March 2015), https://www.vulture.com/2015/03/american-crime-rise-of-slow-crime-procedural.html

Smith, Caleb, 'Disciplines of Attention in a Secular Age', *Critical Inquiry*, 45.4 (2019), 884–909, https://doi.org/10.1086/703963

Steinlight, Emily, *Populating the Novel: Literary Form and the Politics of Surplus Life* (Ithaca, NY: Cornell University Press, 2018), https://doi.org/10.7591/9781501710728

Tsing, Anna, *The Mushroom at the End of the World: On the Possibility of Life in Capitalist Ruins* (Princeton: Princeton University Press, 2015).

Wolfe, Cary, *Ecological Poetics; or, Wallace Stevens's Birds* (Chicago: Chicago University Press, 2020), https://doi.org/10.7208/9780226688022

Wolfe, Cary, 'What "the Animal" Can Teach "the Anthropocene"', *Angelaki*, 25.3 (2020), 131–45, https://doi.org/10.1080/0969725X.2020.1754033

Wood, James, 'The Visionary Power of the Novelist Jon McGregor', *The New Yorker* (27 November 2017).

Zylinska, Joanna, *Nonhuman Photography* (Cambridge: MIT Press, 2017).

11. The Narrative and Aesthetic Strategies of Climate Change Comics

Susan M. Squier

Global climate change is an inescapable reality. While this phenomenon may seem the purview of physicists, atmospheric scientists, geologists, and geophysicists, the effects of climate change touch every living being. Yet many people find climate change too big to grasp and too traumatic to envision. Timothy Morton's concept of hyperobjects, 'things… massively distributed in time and space relative to humans', provides one useful framing for this wicked problem.[1] In a critical review of Morton's work, Elizabeth Boulton argues that his hyperobject frame can be helpful because it draws together all of the disparate details about climate change (which can otherwise be overwhelming) into one all-encompassing notion.[2] Yet she contends that art and literature can also play a crucial role in helping us face climate change, because 'humans

---

1   Timothy Morton, *Hyperobjects: Philosophy and Ecology after the End of the World* (Minneapolis: The University of Minnesota Press, 2013). Horst Rittel and Melvin Webber have coined the term 'wicked problems' to describe issues with an indeterminate scope and scale that are too complex, multi-factorial, and shifting to admit of simple solutions. Horst W. J. Rittel and Melvin M. Webber, 'Dilemmas in a General Theory of Planning', *Policy Sciences*, 4.2 (June 1973), 155–69.

2   Editors' note: This view resonates with Kristin M. Ferebee's characterization of the hyperobject as marked by 'partedness' (chapter 8)—the irreconcilable heterogeneity of the parts forming a whole. Such characterisation is coherent with the hypothesis, formulated in this chapter, of an affinity between climate change—as an hyperobject—and the aesthetic heterogeneity through which artists choose to represent it.

[also] learn from stories, fable, and myths, which often describe dangerous or unwanted scenarios'.³ A survey of some recent graphic narratives about climate change reveals that they address the challenges it poses not with abstraction but with scaled specificity, using narrative and aesthetic strategies afforded by the medium of comics. These climate change comics feature a narrative structure which see-saws between large scale non-fiction reporting and small-scale fictional interludes and an aesthetic strategy which toggles between a focus on the individual life course and attention to expansive complexities. As they move beyond generalities these comics can challenge our climate denial by showing us very specific, detailed, and affectively charged narratives that incorporate the biological discourses of ecology, environmentalism, epigenetics, and extinction.

Making the Global Threat Personal

'The Story of Kram and Ailat', an eight-part comic interspersed between the eleven chapters of Mark Kurlansky's young adult book on the problem of fish extinction, *World Without Fish*, presents the effects of climate change through several perspectives.⁴ In Part I, Kram, an ocean biologist, takes his six-year-old daughter, Ailat, out fishing. The scene seems idyllic: they hear a humpback whale singing, and then it breaches right in front of them. Circling birds signal that fish are below and Ailat even catches a fish. Yet when she wants to take it back to her mother, her father, the ocean biologist, answers with the metaphor at the center of this comic: 'Sorry, my little sardine. There aren't enough of them left! We'll let this one go back to his family'. That evening, they dine out at Captain Leo's fish restaurant. 'Will you cut this halibut steak for me, daddy?' Ailat asks.

Thus begins a sequence of fishing trips in which Kram and Ailat join their friend Serafino, a commercial fisherman, and his sons, Frank and Salvy. Serafino has switched from line-fishing to netting, despite Kram's

---

3   Elizabeth Boulton, 'Climate change as a "hyperobject": a critical review of Timothy Morton's reframing narrative', *WIREs Climate Change*, 7 (2016), 772–85, https://doi.org/10.1002/wcc.410.

4   Mark Kurlansky and Frank Stockton, 'The Story of Kram and Ailat', in *World Without Fish* (New York: Workman publishing, 2011).

disapproval. 'Kram, I have to make a living', Sarafino explains. 'You won't be able to when all the fish are gone'. Kram counters. As Serafino and his sons adjust to the environmental collapse of the oceans, the comic juxtaposes their resourceful practices to Kram's increasing despair that his conservationist message is not being heeded. As Serafino's sons Frank and Salvy throw back the dead flounders that are bycatch, Kram broods that 'The whole system is out of balance... If we reorder the food-chain the whole thing could collapse!'

As the years pass, Serafino and his sons must alter their fishing methods to accommodate the shifting fish populations. The comic documents how their catch changes from halibut to Parrotfish, then herring, then krill, and finally to jellyfish and sea turtles. Kram and his daughter also alter their consuming practices, from their initial meal of highly valued halibut at Captain Leo's restaurant to the dystopian dinner that Kram and his daughter share near the comic's end. Their meal is krill, the food of baleen whales that is also used by industry to produce dog food and fish oil. With this final meal, the comic asks us to make connections along the food chain, from charismatic megafauna like whales, who are often over-represented in the imagination of extinction, to less charismatic species like sardines, and ultimately to human children, including Kram's 'little sardine', his daughter Ailat. In this narrative, sardines are the middle species, the mediators that make visible for both characters and readers the ecological relations that entangle and endanger all species, including humans.

'The Story of Kram and Ailat' is about more than the loss of fish populations. Through the eyes of Kram and his daughter we see not only the steady decline to a world without fish, but also the distal results of climate change: vanishing birds and insects, the ocean now orange and slimy from plankton, and fishing people struggling to maintain an endangered livelihood. Within each panel, we watch as age interacts with ethnicity and education, revealing economic, environmental, and social tensions, rendering some confrontations impossible and some questions unasked.

Kurlansky's comic addresses the tension between the fishing rights of indigenous peoples and the goal of scientific fish management. Yet it also demonstrates the limits of both forms of knowledge. From Kram the ocean scientist, to Serafino the fisherman and his sons, to the newscaster and Dr. Kessel, who works for government fishery management, none of the adult men is able to fully formulate the social and environmental implications of fish decline. Instead, it takes the two young girls in the comic to articulate the situation: the declining ocean, with its diminishing tourism, is intimately linked to the social injustices resulting from climate change in the context of a global economy that requires fishermen to overfish. This augurs badly for human beings, because what happens to fish, birds, and insects can also happen—indeed, is already happening—to people. Ailat's daughter looks directly at the reader to ask a question that carries the punch of extinction: 'What is a fish?' [Figure 38] Addressing the biological theme of ecology, particularly the inescapable impact of the dwindling food chain, this comic encourages us to understand the affective and social impact of climate-change-based extinction.

Fig. 38 Frank Stockton, 'What's a fish?', from Mark Kurlansky, *World Without Fish* illustrated by Frank Stockton, p. 142 © Frank Stockton. All rights reserved.

'The Story of Kram and Ailat' uses a fictional narrative to encourage identification across species, making the threat of extinction personal. Josh Neufeld's *A.D. New Orleans After the Deluge* (2009) and Meredith Li-Vollmer and Mita Mahato's *Climate Changes Health: How Your Health Is at Stake and What You Can Do* (2016) rely on non-fiction narratives to press home the point that climate change adversely affects the health of the most vulnerable populations. The main actor in *A.D. New Orleans*

11. The Narrative and Aesthetic Strategies of Climate Change Comics 303

after the Deluge is clearly 'The Storm' as the first chapter reveals, in a twenty-one-page sequence of nearly wordless panels. Scene-to-scene panel transitions providing aerial images of New Orleans, Louisiana and Biloxi, Mississippi give way to satellite images of Hurricane Katrina itself as it spirals towards the coast in a series of moment-to-moment splash pages labeled August 23 through August 29. Finally, a two-page spread reveals the hyperobject itself, as the hurricane towers above New Orleans and (at the far right, in the most significant place) the Superdome. In the pages that follow, we see the devastation the storm causes, in panels offering sequential glimpses of a street sign being ripped away by the wind, buildings and houses battered by the cresting waters, and entire neighborhoods (and their inhabitants) under water.

While the first chapter follows a nonhuman actor, the next chapter reveals the many human agents who also figure in this story. In 'The City', Neufeld introduces us to the seven individuals whose interviews form the basis of the narrative: a doctor, a publisher and comic book fan, a waitress, a counselor, a high-school-aged Pastor's son, a Ukrainian-born convenience store owner and his fishing friend; they are all individuals who experience Hurricane Katrina, though in very different ways. To take one example, the story of Abbas the Ukranian store owner and his African-American friend Darnell demonstrates how cross-caste empathy and connection give rise to social support in a shared disaster. The men try to ride out the hurricane in New Orleans, but a two-page spread reveals the outcome:[5] the men stand waist high together in polluted water preparing to do battle against the expected rioters and looters. Their battle never comes; instead, the concentric circles of water around them draw the reader's attention to the expanding circles of people affected negatively by climate change. These concentric circles, in their scalar nature, are a visual strategy particularly well suited for the representation of the hyperobject Katrina. The men move beyond belligerence to access their own bravery, finding a way not only to survive but to help others, bringing bottled water to the other abandoned flood victims of the Third Ward. Neufeld's aesthetic decision to employ a

---

5   This image appears on pp. 100–01 of Josh Neufeld, *A.D. NEW ORLEANS: AFTER THE DELUGE* (New York: Pantheon Books, 2009) and can also be found in the original web comic: http://www.smithmag.net/afterthedeluge/2007/11/14/chapter-8/14/

progressive series of color washes as the comic progresses, representing all the people and places of a chapter in the same pastel color, challenges a conventional, racially binarized view of Katrina to illustrate the ever-expanding impact of this environmental disaster.

A similar expansive attention to the impact of climate change appears in Meredith Li-Vollmer and Mita Mahato's beautiful cut paper comic, *Climate Changes Health: How Your Health Is at Stake and What You Can Do*. Created under the auspices of Seattle and King County Public Health, by a public health educator and a cut paper artist, this comic links climate (often erroneously conceptualized as a non-biological entity) to health, and thus to both human and animal biology. The images of Seattle, Washington in the panels reveal the impact of climate change close to home. An asthma puffer juxtaposed to the familiar Seattle Sky Needle registers the respiratory impact of the changed Western Washington climate. Four panels invoke the most urgent public health concerns linked to climate change: 'mosquitoes, ticks, and other disease vectors' result in more cases of West Nile Virus and lyme disease, there is an increased incidence of toxic algae and polluted shellfish while 'wildfires, heavy rainfall, flooding, and windstorms' resulting from climate change trigger stress and anxiety; specific populations are at greater risk for health disparities, including low-income workers and pregnant women whose fetuses may incur the epigenetic impact of those climate-related stressors [Figure 39]. Yet, four final panels illustrate that human beings *can* make a difference, at the scale of the individual and the population. People can choose to use a drying rack for laundry, eat less meat and dairy and more vegetarian meals, repair and recycle clothes and electronic devices, and carry reusable water bottles. Together, they can ride buses or join van pools, plant trees, create community gardens, and engage in 'smart community planning and clean energy policies'.

Anthropomorphic Figures

Many climate change comics adopt a scientific perspective while also incorporating strategies to render the information more accessible and palatable to their readers. Yet even here the strategies differ widely. Darryl Cunningham's 'Climate Change', in his volume *Science Tales: Lies, Hoaxes, and Scams* (2011), reframes the scientific question and its human

11. The Narrative and Aesthetic Strategies of Climate Change Comics 305

Fig. 39 Li-Vollmer and Mahato, from *Climate Changes Health*, written by Meredith Li-Vollmer and artwork by Mita Mahato, published by Public Health—Seattle & King County (2019) © Li-Vollmer and Mahato. All rights reserved.

reception by bringing its male protagonist into conversation with a polar penguin who explains 'the argument for human-driven climate change' in a punchy sequence of panels. The penguin distinguishes between the two sorts of people who resist the theory of human-created climate change: those who 'don't know all the information and are therefore doubtful of the theory. But... tend to be aware of the limits of their knowledge and so remain open to the experience that climate change might be real', and 'those who... reject climate change, not on scientific grounds, but on the grounds of ideology and dogma'.[6] Despite the 'funny animals' tone, there is a historical heft to this penguin preaching. The comic draws on an article in the 2010 *Proceedings of the National Academy of Sciences* surveying the climate change denial groups funded by the Heartland Institute, the American Enterprise Institute, and Koch Industries, the latter being the major donor to climate denialist organizations 'motivated by an ideological commitment to minimal government and free markets'.[7] We move then to the Climategate e-mail controversy of November 2009, in which climate scientists were charged with manipulating research by major media outlets. Although the scientists were ultimately exonerated, the damage was already done, since 'media outlets had devoted five to eleven times more stories to the accusations against the scientists than they had to the resulting exonerations'.[8] The comic ends with a bleak page of penguin exhortation that mingles photographs (close up and views from space) with lyrical cartooning to conclude: 'The science predicts that these events could happen. Let's not leave it to the super-rich to decide who lives and who dies'.[9]

The Great Transformation—Can We Beat the Heat? engages its readers by returning to one of the most venerable comics characters, Little Nemo, protagonist of Winsor McKay's *Little Nemo in Slumberland*, which ran in the *New York Herald* from 1905 to 1911. The plot framework is always the same: Little Nemo eats a dinner of Welsh rarebit and it disagrees with him, giving him a wild array of disastrous dreams. Struggling in 'slumberland' to save himself, he thrashes around, falls

---

6 Darryl Cunningham, 'Climate Change', *Science Tales: Lies, Hoaxes, and Scams* (Brighton, UK: Myriad Press, 2011), pp. 143–44.
7 Cunningham, 'Climate Change', pp. 147–48.
8 Ibid., p. 151.
9 Ibid., p. 153.

Fig. 40 Image reproduced from the chapter 'Climate Change' from *Science Tales: Lies, Hoaxes and Scams* by Darryl Cunningham © Myriad Editions, UK, 2019. www.myriadeditions.com. All rights reserved.

out of bed, and finally wakes up. Feverish Little Nemo reappears in *The Great Transformation* as the overheating earth. Its dreams are populated by climate scientists and politicians who compare the impact of two degrees of earth warming to the organ breakdown a fever causes in the human body. Unlike Darryl Cunningham's preaching penguin who makes the grim climate change story more palatable, the image of the earth as Little Nemo elevates the climate scientists' predictions into our collective fever dream.[10]

Biography and Autobiography

Biographical and autobiographical comics enable a reader to draw near the overwhelming topic of climate change by taking a route that is measured, specific, and local. In Mary M. Talbot and Brian Talbot's *Rain* (2019), we open with images of the devastating 2015 Boxing Day flood in Thrushcross, Yorkshire: flooded streets, canal boats beached on cars half-submerged in mud; houses flooded and their drenched contents out on the street for pickup. Then, moving back in time, we follow Catherine, an English teacher from London, as she is drawn into environmental activism over a series of visits to the North of England to see her environmentalist lover, Mitch. We see things through Catherine's eyes, as she slowly comes to understand how traditional practices of burning the heath heather, digging drainage ditches, intensive grouse farming, and even grouse hunting create environmental destruction in the once idyllic Yorkshire moorlands. As Cath comes to understand the relationship between the local environmental issues and the flood risk all over the United Kingdom linked to climate change, she abandons her initial impatience with activism and joins a massive protest demanding 'Climate Action Now!'. *Rain*'s epigraph comes from Alexander von Humboldt: 'Our imagination is struck only by what is great; but the lover of natural philosophy should reflect equally on little things'.[11] Moving from the local story of one person's environmental activism to the more 'universal' narrative of natural philosophy animating von Humboldt's travel journal (1799–1804), the comic moves across cultural and temporal scales to portray the ecological scales of hyperobjects.

---

10   Alexandra Hamann, Claudia Zea-Schmidt, Reinhold Leinfelder, Jörg Hartmann, Jörg Hülsmann, Robert Nippoldt, Studio Nippoldt, and Iris Urgurel, *The Great Transformation: Climate—Can We Beat the Heat?* (Berlin: WGBU, 2014). I discuss this comic at greater length in 'Scaling Graphic Medicine: The Porous Pathography, a New Kind of Illness Narrative', in *PathoGraphics: Narrative, Aesthetics, Contention, Community*, ed. by Susan M. Squier and Irmela Marei Krüger-Fürhoff (University Park: Penn State University Press, 2020), pp. 205–25, https://doi.org/10.1515/9780271087337-014.

11   Mary M. Talbot and Brian Talbot, *Rain* (London: Jonathan Cape, 2019), p. 2.

In *Climate Changed: A Personal Journey through the Science* (2014), Philippe Squarzoni also combines a macro-scale perspective on climate change with a focus on micro-scale, or individual, experience.[12] The comic incorporates research and discussions with members of the IPCC (Intergovernmental Panel on Climate Change), as well as consultations with six other experts in economics, environmental management, sustainable development, nuclear physics, and development. Their accumulated knowledge is dense and frightening. But it is the personal narrative that brings it into focus.

As the book begins, Squarzoni ponders the aesthetic and narrative challenges of beginning 'a book, or a film, or a graphic novel', noting 'there are beginnings we never forget... openings that set the tone... the colors that impregnate the rest of the work... and the memories we keep'.[13] He settles on the memory of a country house he visited as a child, and acknowledges that 'for this book, it's not the beginning that's the most difficult. The hardest thing is... how to end it'.[14] On a return trip with his wife in 2006 he muses over his childhood memories: 'As if all those events happened simultaneously. As if they merged over time. But there must have been a sequence. A beginning, a middle... an end'. The questions of origins, causes, and outcomes stay with Squarzoni as he begins research for a new book that will help him understand not only what 'climate change', 'greenhouse gases', and 'reducing emissions' mean, but more importantly 'if it was global warming, do we have any leeway? How much longer can we just let ourselves go on doing nothing?'.[15] As he continues his research in that bucolic setting, sensual images of walks in the Swiss countryside clash with abstract images of charts and graphs and brassy visuals from commercial advertising in a gut-wrenching collision of different visual modes. 'Talking heads' interviews with climate change experts are interspersed with the author's personal diary, in which he struggles to overcome his own climate denial and acknowledge the trauma he feels in the face of impending disaster caused by our out-of-control consumerist lifestyle.

---

12   Philippe Squarzoni, *Climate Changed: A Personal Journey Through the Science* (New York: Abrams Comic Arts, 2014), translated by Ivanka Hahnenberger. Originally published in 2012 as *Saison brune* (Paris: Editions Delcourt).
13   Squarzoni, *Climate Changed*, p. 9.
14   Ibid., p. 14.
15   Ibid., p. 32.

The densely argued pages of interviews give a visual and verbal structure to Squarzoni's growing understanding of the ways the interlocked environmental and fuel crises put the entire globe at risk. Yet the emotional heft of the book is felt elsewhere. Squarzoni's most eloquent rhetorical strategy lies in the wordless panels in which his own personal response to the scientific research is open to the reader, so he (and we) can sit gazing at the mountain landscape and pondering the personal impact of climate change. And pondering. And pondering. The seemingly abstract challenge posed by his research is made fully specific when, midway through the volume, we find ourselves with Squarzoni still frozen at the beginning, trying to imagine an adequate response. Lyrical panels show the summer home, and the grassy lanes beyond it, followed by images of the road diving steeply downhill into a tunnel. Each quiet panel features a line or two of text whose poetic tone captures the metaphoric implications of the landscape:

> How to begin? / The clock is ticking. And time is running out./ It's already too late to go back. / The change has taken place. / A new story is beginning. / A story that we cannot avoid. / A time when we've already run out of time. / A time that's rushing away from us, expired. We're committed in spite of ourselves. / If we want to avoid the worst of the consequences of climate change, we have only a few decades to reduce our greenhouse gases significantly. / The coming years will be crucial. / When and where shall we begin?[16]

The sequential deferrals ('when and where shall we begin?') in this volume create a mood of hopelessness that lingers in the concluding sequence, a walk in the stark and snowy woods whose sequence of etched panels features thin lines of text that read like a haiku.

> I understand the desire to point out the answer. To finish this on a positive note. / But if I'm being honest with myself, I believe three things. / One, there's a doorway we need to pass through. / Technically, it's still possible to avoid the worst consequences of climate change and to take the necessary measure to manage the upheavals that are already inevitable. / Two: the doorway is not very wide. It closes a little more each day. And we have only a little time to pass through it./ Three: I don't think we'll pick that door.[17]

---

16 Squarzoni *Climate Changed*, pp. 148–51.
17 Ibid., p. 453.

11. The Narrative and Aesthetic Strategies of Climate Change Comics   311

At the end of the book, the attempt to conclude is repeated, as was the attempt to begin in earlier episodes. Close to the end, we find three tiered panels of stark black branches against the sky. The text returns us to where we began: 'So, how to end this book? / Just because the answer is filled with gloom doesn't mean the question was pointless. / To care how these questions are being asked shows that we care about the future'. The following page, two spare line drawings of branches against the sky and a snowy precipice, still seems to offer hope: 'And who knows? I could be wrong. / The story isn't over' (see Figure 41). The comic ends with an ambiguous image of sun breaking through clouds over the mountain range.[18]

Fig. 41 From *Climate Changed* by Philippe Squarzoni. *Saison brune* by Philippe Squarzoni © Éditions Delcourt. 2012. Text and illustrations by Philippe Squarzoni. Translation by Ivanka Hahnenberger. English translation copyright © 2014 Harry N. Abrams, Inc. All rights reserved.

18   Ibid., pp. 456–58.

Scientific Distance Versus Intimate Experience

Two final climate change graphic novels take a more indirect route to the topic, but their impact is just as intense. Lauren Redniss' exquisite *Thunder & Lightning: Weather Past, Present, Future* introduces climate change in the eighth chapter, 'Dominion', following chapters addressing Chaos, Cold, Rain, Fog, Wind, Heat, and Sky, and followed by chapters addressing War, Profit, Pleasure, and Forecasting.[19] Although '[f]or millennia, people have found meaning, and divinity, in weather', Redniss first casts climate change as a problem approached from the technical side. For the IPCC, 'Warming of the climate system is unequivocal'. For Harold Brooks of the National Oceanic and Atmospheric Administration's National Severe Storms Lab, it's simple: 'The climate is warming, and the planet will continue to warm. That's almost an uninteresting statement, it's so obvious'.[20] The Military Board of CNA Corporation publishes a report stating that 'climate change will "place key elements of our national power at risk and threaten our homeland security"'. That report counsels its readers, 'When it comes to thinking about how the world will respond to projected changes in the climate, … it is important to guard against a failure of imagination'. Or, to put it directly, 'Can mankind and technology replace God and magic to claim dominion over the weather?'.[21]

Redniss introduces one of the prime movers in the argument for geoengineering to combat climate change: the mathematical economist and theoretical physicist Nathan Myhrvold. In a densely written text, she sets out his argument for what he calls 'the Stratoshield', a method for 'bounc[ing] some of the sunlight back out into space'. 'The simplest approach', which he admits 'sounds really dumb', involves 'a series of balloons that hold up a pipe' that pumps sulfate (or sulfur dioxide) into the sky. The Stratoshield would make it possible, Myhrvold argues, 'to

---

19   While this volume stretches the definition of comics by alternating between heavy pages of text and full-page and two-page images, it also incorporates many of the strategies of comics, from speech balloons to iconic animals. I also include it here because of the power and argumentative force that its images hold. They are far from merely illustrative.
20   Lauren Redniss, *Thunder and Lightning: Weather Past, Present, Future* (New York: Random House, 2015), p 142.
21   Ibid.

stabilize the temperature, or stabilize the climate, at any temperature you wanted. So you could say, "Let's stabilize it at today's temperature." That's probably the best thing you could do. But you could also say, "Let's take it back to the preindustrial climate. Let's just negate all global warming"'.[22] Despite fear that this invention will turn our blue skies cloudy and grey, Redniss reveals that what was once an idea too absurd to countenance publicly is now being taken seriously.

In the following double page spread, Myhrvold describes the way that hardcore environmentalists, or those he terms 'John Muirs', respond to his geoengineering proposals. Myhrvold, journalists Elizabeth Kolbert and Emma Maris, climatologist Alan Robock and environmental scientist David Keith sit around a grey conference table debating the ethics of intervening in the climate. Threadlike speech balloons form a climate of their own, a dark brown atmosphere under which the group considers possibilities: what the economic impact would be were a multinational corporation to develop the technology and weaponize it; what the geopolitical impact would be should the technique be used by the Maldives or Brazil to produce a climate that suits their national needs; what the impact is of 'consciously admitting that we are living on a managed planet'; and finally whether, as distasteful as we find it to interfere with nature, a time will come when 'we are morally obligated to do it'. In his closing statement, Myhrvold brings the debate down to brass tacks: 'Look, geoengineering should only be deployed if we think there's really going to be a problem. If a disaster scenario doesn't occur, there's no reason to do this shit'.[23] Such directness is surprising in a book whose aesthetic strategy seems more aimed at soothing and lulling. Redniss shows that not only are both strategies compatible, but they are also surprisingly effective in presenting the complexity of this wicked problem. The contrasting styles, tones, and genres appear to be a feature shared by many climate change comics. Perhaps the 'hyperobject' that is climate change demands such formal and aesthetic heterogeneity.

---

22 Ibid., p. 145.
23 Ibid., p. 149.

Nathan Myhrvold: "When our scheme came out, I got all kinds of hate mail, one of them from someone who said, 'You are worse than baby killers.'"

Nathan Myhrvold described the reactions he hears to geoengineering proposals. "I've discussed this with a lot of people who are hardcore environmentalists. One kind I call the 'John Muirs.'"

Muir, the celebrated naturalist and activist, co-founded the Sierra Club in 1892 and is regarded as the "Father of National Parks."

Nathan Myhrvold: "John Muir loved mountains. Some of my environmentalist friends, they love mountains. They love wilderness. And so they say, 'This is great! You've got a way to prevent this thing I love from being destroyed.' They look at geoengineering, and they may have some reservations, which is all fine, but basically, they don't want the planet fucked up."

Fig. 42 Lauren Redniss, *Thunder & Lightning: Weather Past, Present, Future* (Random House, 2015), pp. 146–47 © Lauren Redniss. All rights reserved.

Conveying the impact of climate change also requires careful contextualizing. Images of mountains and rivers, a swan, a shorebird, a vulture, an antelope, a moth, and a snail precede the double-page spread that pictures the round-table discussion between environmentalists, journalists, and engineers. These natural images dramatize the tension between Myhrvold's aspirations to dominate nature and environmentalists' commitment to protect the wilderness and its creatures. The comment of Richard Pearson, a scientist at the American Museum of Natural History's Center for Biodiversity and Conservation, takes on additional force when considered in relation to the round-table discussion to follow: 'Flora and fauna, skies and sea, are all affected. Today the question of whether to play God and intervene in earth's systems has moved from the margins and become a mainstream debate'[24] [Figure 42]. Perhaps these iconic images of mountains and rivers, birds, and mammals are a way of convoking a 'Parliament of Things' in which 'all affected' can participate in the debate. Humans now occupy the margins while the mainstream is now the whole society of nonhumans, who will debate how best to address climate change.[25] A reviewer captures the impact of the deliberate contrast between visual form and conceptual content:

> On one question, [Redniss] is unequivocal: 'Scientists agree that we are living in an age of global climate change,' Redniss writes in 'Dominion.' 'Human activities are transforming the planet. The consequences, scientists contend, include warmer temperatures, extreme events, wildfires, floods and droughts, rising sea levels and species extinction.'
>
> One can imagine, at this point, certain dissenting readers sitting up in alarm, having been beguiled, perhaps, by the many personal anecdotes, the colors, the seemingly artless font.... Such a reader might find an approach like this particularly irresponsible, or even dangerous. Certainly, anyone will recognize its power.[26]

While interviews are used as a documentary method in the works by Li-Vollmer and Mahato and Squarzoni, and as a rhetorical strategy in

---

24 Ibid., p. 147.
25 My thanks to Liliane Campos and Pierre-Louis Patoine for this, and other, insightful suggestions. For 'Parliament of Things', see Bruno Latour, *We Have Never Been Modern* (Cambridge, MA: Harvard University Press, 1993), pp. 142–45.
26 Sadie Stein, '"Stormy Weather", review of *Thunder and Lightning* by Lauren Redniss', *New York Times* (11 October 2015).

the comics by Redniss and Hamman et al., interviews function as an authenticating strategy in Brian Fies' *A Fire Story*. This volume first began as a blog entry, a twenty-page online comic Fies created days after he was forced to flee his family home in 2017 to escape the Northern California wildfire that killed forty-four people and destroyed more than 6,000 homes, including Fies' own.[27] The wide readership and public acclaim this comic received led Fies to expand it, within the year, into 'a full-length graphic novel, including environmental insight and the stories of others affected by the disaster'.[28] Fies is one of the founders of the field of Graphic Medicine, as well as the author of the Eisner-winning comic *Mom's Cancer*, and this venture into documentary cartooning arguably stretches the field of graphic medicine to include the medical impact of a devastating fire. I am pushing the field still farther when I nominate *A Fire Story* as a climate change comic, so let me set out the reasons for making that leap.

One way to claim something as a climate change comic would be to identify explicit mentions of climate change in its argument or narrative. Most of the comics I have dealt with already make it explicit that their mission is to transmit information about climate change in such a way that the reader realizes the need to respond to its challenge, even if they differ in the responses they imagine. Fies' memoir is different: the challenges this comic documents are the forced evacuation from his home, its destruction by fire, and the fire's impact on his broader community, which ranges from forced displacement and economic devastation to social fragmentation. Only once Fies and his family have evacuated and are beginning the process of rebuilding the records of their lives does the concept of global climate change emerge.

A Fire Story avoids almost completely the graphs, talking heads, and over-arching explanations of climate change that characterize many of the comics I have discussed. Rather than abstract discussions of its ecological, environmental, and emotional impacts—the ways that fossil fuel use has led to rising greenhouse gases, and thus to droughts, floods, and fire, and finally to helpless despair—Fies' comic focuses on the impact of a climate-change related fire *in medias res*. In its attention to the

---

27 The Fies family fled on Monday 9 October 2017. The house was destroyed by the fire.
28 Brian Fies, flyleaf copy, *A Fire Story* (New York: Abrams Comic Arts, 2019).

objects, practices, and moments associated with one family's experience of the 2017 California fires, it represents human beings coping by telling themselves the oldest story of all: we will endure.

An early image in the comic links the fire to climate change, picturing the multiple 'weather and climate disasters' of 2017: fires in the West and hurricanes in the South and East.[29] But the story moves away from that macro-scale concept to concentrate on the local and specific practices essential to rebuilding. Like Squarzoni, Fies acknowledges the reader's need for hope and completion, yet admits he can't deliver. 'People seem to want a story with uplift and closure, but I have no uplift to give, and anyone who says "closure" around me gets a punch in the nose'. Like Squarzoni's comic, the conclusion of *A Fire Story* is ambiguous: 'But even if you lose the place and the stuff, home can still be the memory and hope and promise of those things. Sometimes home is nothing but a bare patch of dirt. This is mine'.[30]

So, what does this mean about the role of art and literature in communicating about climate change, or more specifically about the narrative and aesthetic strategies adopted by comics dealing with climate change? Perhaps that it is no longer possible to refuse the abstract perspective, to refuse to countenance the hyperobject and all its implications in a deliberate decision to focus on the smaller scale human story. That is, if we want to change our own narrative, and build a different future. When I reread Fies' 2019 comic in 2020, during the escalation of lightning-strike fueled fires in California, I found myself doubting Elizabeth Boulton's reassuring observation that, 'humans *learn* from stories, fable, and myths, which often describe dangerous or unwanted scenarios' (emphasis added).[31] I wondered whether there is a limit to our capacity to learn from those stories we tell. As Timothy Morton tells us, the hyperobject of climate change is not accessed 'across a distance', but rather must be faced 'here, right here in my social and experiential space ... [it] becomes clearer with every passing day that

---

29 Ibid., p. 48.
30 Ibid., p. 142.
31 As Farhad Manjoo observed in the *New York Times* on the day I am concluding this essay, 'climate change has ushered in a new era of "megafires" that includes some of the largest blazes the state has ever faced'. 'California, We Can't Go On Like This', *New York Times* Opinion section (26 August 2020).

'distance' is only a psychic and ideological construct to protect me from the nearness of things'.[32]

Perhaps that is why Fies' graphic narrative refrains from situating his family's story of endurance in a form that presents the full impact of climate change. That would be discouraging at best, and downright despair-inducing at worst. A temporally and spatially scaled perspective on the California fires, one presenting them as a widespread and escalating manifestation of climate change, might have undercut a commitment to rebuild one's home and one's community. At least it might generate a different definition of rebuilding, one modeled by the work of Jem Bendell and the Deep Adaptation Forum.[33]

Instead, the Fies family functions in synecdoche, as stand-ins for the human family more broadly. We follow their story as they respond not to climate change, which seems a distant problem to be managed by the National Oceanographic and Atmospheric Administration (NOAA), but rather to the specific, multiple Northern California wildfires that drive them out of their home. As we can see, if we contrast Fies' use of a full splash page of highly detailed satellite photography to show the fires to his choice of an abstract half-page schematic map claiming to present 'a historic year of weather and climate disasters', the fires seem frighteningly close, while climate change still seems a distant abstraction.[34] (see figures 43 and 44). In its material specificity, pragmatic detail, human warmth, and social embeddedness, as well as its portrait of a family not yet able to grasp the full implications of this hyperobject, *A Fire Story* may well be the indexical climate change comic.

---

[32] Timothy Morton, *Hyperobjects: Philosophy and Ecology after the End of the World* (Minneapolis: The University of Minnesota Press, 2013), loc. 526 of 4946.

[33] See Jem Bendell's website (https://jembendell.com) and Jem Bendell, 'Deep Adaptation: A Map for Navigating Climate Tragedy' (2018, 2020), http://lifeworth.com/deepadaptation.pdf.

[34] Fies, *A Fire Story*, p. 40 and p. 48.

Fig. 43 From *A Fire Story* by Brian Fies. Text copyright © 2019. 2020 Brian Fies. Used with permission of Abrams ComicArts®, an imprint of ABRAMS, New York. All rights reserved.

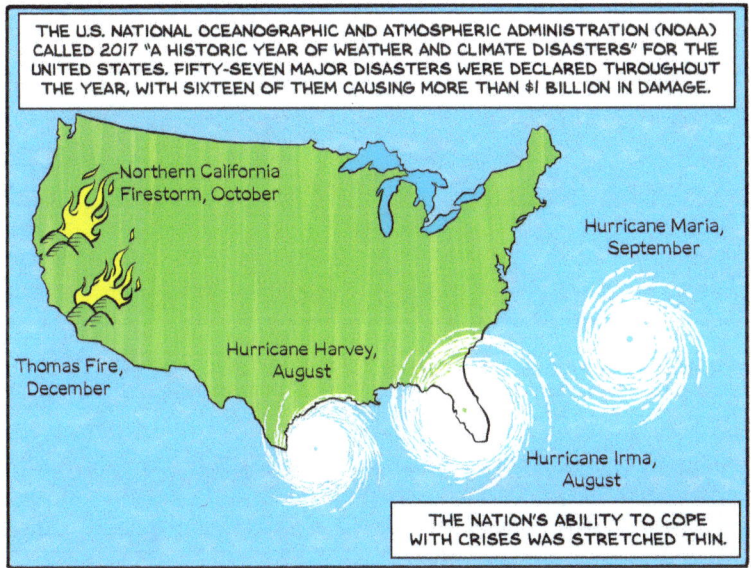

Fig. 44 From *A Fire Story* by Brian Fies. Text copyright © 2019. 2020 Brian Fies. Used by permission of Abrams ComicArts®, an imprint of ABRAMS, New York. All rights reserved.

Works Cited

Primary Sources

Cunningham, Darryl, 'Climate change', in *Science Tales: Lies, Hoaxes and Scams* (Brighton: Myriad Press, 2011), 135–54.

Fies, Brian, *A Fire Story* (New York: Abrams Comic Arts, 2019).

Hamman, Alexandra et al., *The Great Transformation: Climate—Can We Beat the Heat?* (Berlin: WBGU, 2013).

Kurlansky, Mark and Frank Stockton, 'The Story of Kram and Ailat', in *World Without Fish* by Mark Kurlansky (New York: Workman publishing, 2011).

Li-Vollmer, Meredith and Mita Mahato, *Climate Changes Health: How Your Health Is at Stake and What You Can Do* (Seattle: Seattle Public Health & King County, [n.d.]).

Neufeld, Josh, *A.D. New Orleans After the Deluge* (New York: Pantheon Books, 2009).

Redniss, Lauren, 'Dominion', *Thunder and Lightning: Weather Past, Present, Future* (New York: Random House, 2015).

Squarzoni, Philippe, *Climate Changed: A Personal Journey Through the Science*, translated by Ivanka Hahnenberger (New York: Abrams Comic Arts, 2014).

Squarzoni, Philippe, *Saison brune* (Paris: Editions Delcourt, 2012).

Talbot, Mary M. and Brian Talbot, *Rain* (London: Jonathan Cape, 2019).

Secondary Sources

Bendell, Jem, 'Deep Adaptation: A Map for Navigating Climate Tragedy' (2018, 2020), http://lifeworth.com/deepadaptation.pdf

Boulton, Elizabeth, 'Climate Change as a "hyperobject": a critical review of Timothy Morton's reframing narrative', *WIREs Clim Change*, 7 (2016), 772–85, https://doi.org/10.1002/wcc.410

Krüger-Fürhoff, Irmela Marei and Susan M. Squier, eds, *PathoGraphics: Narrative, Aesthetics, Contention, Community* (University Park: Penn State University Press, 2020), https://doi.org/10.1515/9780271087337

Latour, Bruno, *We Have Never Been Modern* (Cambridge, MA: Harvard University Press, 1993).

Manjoo, Farhad, 'California, We Can't Go On Like This', *New York Times* Opinion section (26 August 2020)

Morton, Timothy, *Hyperobjects: Philosophy and Ecology after the End of the World* (Minneapolis: The University of Minnesota Press, 2013).

Morton, Timothy, 'This is not my beautiful biosphere', in *A Cultural History of Climate Change*, ed. by Tom Bristow and Thomas H. Ford (New York and London: Routledge, 2016), pp. 229–38.

Rittel, Horst W. J. and Melvin M. Webber, 'Dilemmas in a General Theory of Planning', *Policy Sciences*, 4.2 (June 1973), 155–69.

Stein, Sadie, '"Stormy Weather", review of *Thunder and Lightning* by Lauren Redniss', *New York Times* (11 October 2015).

12. Displacing the Human
Representing Ecological Crisis on Stage

Kirsten E. Shepherd-Barr
and Hannah Simpson

The visual iconography of the COVID-19 global pandemic has been striking. Every night for at least eighteen months, BBC News showed an enormous backdrop graphic of the virus, blown up to grotesque proportions that dwarfed the news presenter and made its crown-like protein structures clearly visible. The image served as an instantaneous shorthand for two otherwise 'invisible', near-incomprehensible scales of existence: the microscopic virus itself, and the unprecedented worldwide health crisis it caused.

Our iconography for climate change is rather different, typically resorting to the more familiar scale of animal and landscape. Several recognisable images recur on our pages and screens: polar bears stranded on shrinking ice; sea birds drenched in oil spills; vast swathes of the rainforest burned or cut down. But rarely do we see images of climate change that take place beyond the more easily conceived human scale, at the level of microorganisms analogous to the icon of the COVID-19 virus, or on the vastly larger scales of global CO_2 emissions or ozone layer depletion: 'the new kind of incommensurability that is being forced on us by our ecological predicament'.[1] Endearing, easily

---

1   Una Chaudhuri, 'Anthropo-Scenes: Staging Climate Chaos in the Drama of Bad Ideas', in *Twenty-First-Century Drama: What Happens Now*, ed. by Siân Adiseshiah and Louise LePage (London: Palgrave, 2016), pp. 303–21 (p. 305), https://doi.org/10.1057/978-1-137-48403-1_15.

anthropomorphised animal beings like the tiger and the panda tend to take precedence in our collective visual imagination, and yet it is the less photogenic beings, like insects and still smaller microzoa, that are the engines of our ecosystems. Tony Juniper's environmentalist study *What Has Nature Ever Done for Us?* (2013) underlines this point via the striking image of the teaspoon of soil dug up from an ordinary plot of land, home to millions of microorganisms, a tiny ecosystem in itself recalling Darwin's iconic metaphor of the 'entangled bank' teeming with life.[2] Alan Weisman's book *The World Without Us* (2007) offers a thought-experiment about how the world would cope if all human life disappeared. (Spoiler alert: it would do just fine without us.) And Rhys Blakely's 2020 *Times* article 'Why saving the panda, but not its parasites, is really a lousy idea' cautions:

> Ticks and tapeworms may not stir the emotions in the same way as large charismatic species such as tigers and elephants. But researchers are warning that millions of parasitic animals face local declines or global extinction and that their disappearance risks throwing ecosystems out of kilter.[3]

At the other end of the scale, colossal global, atmospheric, and multi-century or even multi-millennial environmental changes are difficult to conceptualise, let alone represent—a challenge neatly articulated by Timothy Morton's influential concept of the ecological 'hyperobject', so vast, hyperdimensional, and massively distributed in time and space as to baffle the scope of human comprehension.[4] Our collective consciousness lacks ready visual iconography for the 'invisible' spectra of climate crisis that extend beyond or below our own human-centred scale.

---

2   Charles Darwin, *On the Origin of Species* (London: John Murray, 1859), p. 489. See *Darwin Online*, http://darwin-online.org.uk/Variorum/1859/1859-489-dns.html.

3   Rhys Blakely, 'Why saving the panda, but not its parasites, is a really lousy idea', *The Times* (24 August 2020), https://www.thetimes.co.uk/article/why-saving-the-panda-but-not-its-parasites-is-really-a-lousy-idea-9026rlwx2. See also Ursula Heise's exploration of how conservationist activism tends to foreground 'charismatic megafauna' in public campaigning, in *Imagining Extinction: The Cultural Meanings of Endangered Species* (Chicago: University of Chicago Press, 2017), pp. 23–25, https://doi.org/10.7208/chicago/9780226358338.001.0001.

4   See Timothy Morton, *Hyperobjects: Philosophy and Ecology after the End of the Earth* (Minneapolis: University of Minnesota Press, 2013), pp. 1–3, and *The Ecological Thought* (Cambridge, MA: Harvard University Press, 2010), pp. 130–31, https://doi.org/10.2307/j.ctvjhzskj.

The problem with this anthropocentric scope of reference is the obstacle it poses to our larger understanding of (and effectual response to) the climate crisis. As Patrick Lonergan observes, 'if our models of representational realism fail to accommodate the realities of climate change, [...] those models of realism might be inhibiting our ecological awareness, and therefore our ability to produce meaningful change'.[5] And representing climate change beyond the human scale is doubly challenging when it comes to theatre. Theatre's engagement with climate change has tended to remain human-centred and thus human-scaled. The theatrical medium traditionally relies on the human as its main referent, with human actors (and audience members) at its core, and so inevitably inclines towards the anthropocentric. Theatre's engagement with climate change suggests that we may need to update the classic *theatrum mundi* trope: no longer the 'theatre as world' but the world as theatre, *mundus ut theatrum*, climate change as a drama enacted on simultaneously microscopic and enormous scales, in which the natural world sits centre-stage as protagonist and human characters are relegated from centre to wings or backdrop. The real action that climate change has produced, after all, is a paradoxically nonhuman one: it is the earth responding to the consequences of human activity. The representational challenge is not to downplay or deny human agency in ecological crisis—or indeed human responsibility in attempting to remedy the damage done—but rather to deprioritise human action and experience as always the primary concern, to avoid 'reinforcing the anthropocentrism that got us into this mess in the first place, falling back on the habitual human exceptionalism of Western dramatic tradition', as Catherine Love puts it.[6]

To conceptualise this new *mundum ut theatrum* is an imaginative leap that requires new stage images and new scales of dramatic representation. Might there be a theatrical representation of the vast macro scale of global climate change, atmospheric pollution, and mass species extinction, analogous to the iconic COVID-19 microbe: a powerful representation of

---

5 Patrick Lonergan, 'A Twisted, Looping Form: Staging Dark Ecologies in Ella Hickson's *Oil*', *Performance Research*, 25.2 (2020), 38–44 (p. 41), https://doi.org/10.1080/13528165.2020.1752575.

6 Catherine Love, 'From Facts to Feelings: The Development of Katie Mitchell's Ecodramaturgy', *Contemporary Theatre Review*, 30.2 (2020), 226–35 (p. 226), https://doi.org/10.1080/10486801.2020.1731495.

a biological phenomenon that becomes a part of our shared imaginary, crossing cultural and linguistic borders? Why does so much theatre remain narrowly anthropocentric, when there are so many other ways it might convey the problems and challenges of climate change? In this chapter, we explore some of the efforts to wrestle with innovative stage images and scales that theatre-makers have produced in their creative endeavours to capture and convey the current ecological crisis, and offer some models and theoretical approaches that might be productive in thinking about the theatre's representations of climate change, and of the science of climate change.[7] Theatre's unique functioning in a range of spatial and temporal dimensions and sensory modes, we argue, offers rich potential for representations of climate change that might move us into a more-than-human scope of thought.

'It's Actually Not About Us': The Paradox of Human-Centric Ecological Drama

It is hardly surprising that in the realms of prose fiction and non-fiction, where representation is conjured in the mind of the reader via narrative, we find some innovative imaginings of the nonhuman, the non-anthropomorphic, and of micro- and macroscopic ecological scales. Climate change fiction specialist James Berger cites Richard Powers' Pulitzer Prize-winning novel *The Overstory* (2018) as a compelling recent example of a prose fiction work that reaches beyond traditional human-centric expectations: 'It's about forests. And it really is about forests. There are interesting human characters too. But in a sense, it's actually not about us'.[8] New narrative and framing conventions are required to represent environmental crisis, which in turn require new ways of reading such works.

---

7   A comprehensive catalogue of climate change drama and performance lies beyond the scope of this chapter. Readers seeking such a list can consult Chantal Bilodeau's list of climate change plays on the *Artists and Climate Change* website (https://artistsandclimatechange.com/2014/11/01/creating-a-list-of-climate-change-plays/) and Julie Hudson's article '"If You Want to be Green Hold Your Breath": Climate Change in British Theatre', *New Theatre Quarterly*, 28.3 (2012), 260–71, https://doi.org/10.1017/s0266464x12000449.

8   James Berger and Peter Cunningham, 'Meditations in an emergency: James Berger on climate, fiction, and apocalypse', *Yale News* (27 January 2020), https://news.yale.edu/2020/01/27/james-berger-climate-fiction-and-apocalypse.

However, in dealing with ecological crisis, theatre has tended to be very much 'about us'. Realism, which typically forces a focus on the human story, may be in part to blame, and science fiction and fantasy seem the obvious route out of the realist trap. Yet how does this play out on stage, in practical terms? Even when plays have featured talking, sentient, or otherwise agentive plants—for example in Susan Glaspell's 1921 *The Verge* or Alan Menken and Howard Ashman's rock musical *Little Shop of Horrors* (1982)—these ostensibly nonhuman creatures have been played by people wearing plant suits. Similarly, George Bernard Shaw's 1932 play *Too True to Be Good* introduces the microscopic scale of existence into its condemnation of luxury consumption with the appearance of a jelly-like talking microbe—enacted by a human in a microbe suit. Fast-forward to 2018, and Punctuate! Theatre's *Bears* enacts a range of Trans Mountain Pipeline flora and fauna through the movement of the choreographed eight-person chorus and its combination of animal mimicry and interpretative dance (Fig. 45).[9]

Fig. 45 Matthew MacKenzie, *Bears* (2018) © Alexis McKeown and Punctuate! Theatre. Environmental design by T. Erin Gruber. Pictured: Sheldon Elter. Chorus: Lara Ebata, Gianna Vacirca, Skye Demas, Alida Kendell, Zoe Glassman, Kendra Shorter, Rebecca Sadowski. Photograph by Alexis McKeown. All rights reserved.

---

9   For fuller analysis of Punctuate! Theatre's *Bears*, see Gabriel Levine's 'Black-Light Ecologies: Punctuate! Theatre's *Bears* Wipes Off its Oil', *Performance Research*, 25.2 (2020), 45–52, https://doi.org/10.1080/13528165.2020.1752576.

Again and again, the reference point remains the human body, and the scale remains human. What usually gets put on stage is the results or effects of climate change felt by human inhabitants of the earth, rather than its often difficult-to-see (and indeed, difficult-to-conceive) causes or mechanisms. The drama comes from the conflict between human and nature: a new mode of revenge tragedy that sees humans battling hugely destructive natural disasters brought about by our own disregard for the planet. This dynamic was effectively portrayed in the National Theatre's 2015 version of the medieval morality play *Everyman* in a new adaptation by Carol Ann Duffy that gave it a climate change context. Reimagining older plays as relevant to the current ecological crisis has proved fertile ground for twenty-first-century theatre: Samuel Beckett's *Happy Days* (1961), for example, has been restaged as a 'climate change drama', most recently in Sarah Franckcom's 2018 production at the Royal Exchange Manchester, which staged Winnie's mound surrounded by a puddle of water and an assortment of plastic bags and bottles, and in Katie Mitchell's 2015 production at the Deutsches Schauspielhaus in Hamburg, which saw Winnie submerged in floodwater rather than sand.[10]

Beckett's blighted landscape is given new ecological impetus by these restagings—but the vividly characterised human figure of Winnie remains literally central to the stage focus, and our spectatorial access to the play's landscape itself is strictly limited to Winnie's own periods of consciousness. Much 1890s Symbolist theatre struggled with the human-inclined materiality of the theatre medium by hinting at a world devoid of all human activity, a blueprint for Samuel Beckett's later barren theatrical landscapes in *Endgame* (1957), *Act without Words I* (1957) and *Happy Days* (1961).[11] Maurice Maeterlinck's *Les Aveugles* (*The*

---

10 On Beckett's evolutionary and geological vision across his theatre work and in *Happy Days* in particular, see Kirsten E. Shepherd-Barr's 'Beckett's Old Muckball', in *Theatre and Evolution from Ibsen to Beckett* (New York: Columbia University Press, 2015), pp. 237–72, https://doi.org/10.7312/shep16470. See also Anna McMullan, 'Katie Mitchell on Staging Beckett', *Contemporary Theatre Review*, 28.1 (2018), 127–32, https://doi.org/10.1080/10486801.2018.1426822, and Joe Kelleher, 'Recycling Beckett', in *Rethinking the Theatre of the Absurd: Ecology, the Environment and the Greening of the Modern Stage*, ed. by Carl Lavery and Clare Finburgh (London: Bloomsbury, 2015), pp. 127–46, https://doi.org/10.5040/9781472511072.0009.

11 For fuller discussion of older plays reimagined as contemporary climate-change theatre, see Shepherd-Barr, *Theatre and Evolution*.

Blind, 1890), for example, shifts attention away from the recognisably human by staging depersonalised beings in place of any realistically sketched characters. In a similar gradual deprioritising of localised human action, Shaw's *Back to Methuselah: A Microbiological Pentateuch* (1922) stages a vastly protracted scope of action that extends from the Garden of Eden to 31,920 AD and a final vision of human beings transformed in immaterial luminescence. Yet even where the threat of human extinction is dramatised, or where the natural world encroaches more determinedly into the scene, it is still the human body and the facticity of human experience that structures these stage texts.

Even in much recent contemporary theatre, the human has remained the central focal point of environmental anxieties. The profusion of new British and American plays focused on climate change that premiered between the years 2008 and 2011 repeatedly turned to human relationships to focus or metaphorise their ecocritical stakes. In Duncan Macmillan's *Lungs* (2011), for example, climate-change anxiety is framed through a young liberal couple's discussion as to whether or not to have children; Macmillan has spoken of *Lungs* as the solution to his struggle 'to write about some of the bigger issues facing our species', to 'distil' large-scale environmental concern 'into a pinpoint, compelling dramatic metaphor'.[12] Richard Bean's *The Heretic* (2011) filters the threat of climate crisis through a heady rush of romantic affairs, broken marriages, mother-daughter discord, and a climate change scientist being subjected to professional pressure, death threats and stalking because of her research. *Greenland* (2011), co-authored by Moira Buffini, Penelope Skinner, Matt Charman and Jack Thorne, similarly dramatises ecological anxiety through a medley of intersecting narratives featuring multiple instances of parent-child conflict, romantic discord, the Copenhagen COP 15 climate change conference punctuated with casual sex, and an Artic explorer who communes in flashback with

---

12 Duncan Macmillan, 'Some Thoughts on Lungs', *The Old Vic* (11 October 2019), https://www.oldvictheatre.com/news/2019/10/duncan-macmillan-some-thoughts-on-lungs. Catherine Love discusses how Katie Mitchell's 2013 production of *Lungs* at Berlin's Schaubühne theatre introduced an enlarged sense of ecological scale by means of a backdrop screen that 'displayed the global population—a number that changed in real time throughout the performance', emphasising the individual and the global simultaneously. However, as Love points out, this widened lens 'still excluded the more-than-human world' (2020, p. 233).

his younger self. Wallace Shawn's *Grasses of a Thousand Colors* (2009) focalises its Anthropocene concerns through the figure of an arrogant GM food scientist, whose main concern is satisfying his erotic urges by both extra-marital and trans-species means. And Simon Stephens' *Wastwater* (2011) employs a triptych of troubled couples, their fractured relationships edged with the gradually encroaching forces of mud, water, plant life, and pollution to disquietingly suggest environmental as well as personal disaster. In each case, the human figure is the central focal point onstage: interpersonal human crisis translates larger ecological catastrophe while at the same time relegating it to the margins of the action.

A sharply different strand of ecological performance is the hybrid form of the 'lecture-performance' or 'dramatised lecture', at least in part a legacy of Al Gore's documentary-lecture *An Inconvenient Truth* (2006). Stephen Emmott's *Ten Billion* (2013) and Chris Rapley and Duncan Macmillan's *2071* (2014), both directed in the première performances by Katie Mitchell at the Royal Court, offer key examples of the form: in place of actors, professional scientists Emmott and Rapley appeared as 'themselves' on stage, delivering something more akin to a university lecture or keynote address in order to foreground a climate scientist's perspective on our environmental crisis. The dynamics of dramatic performance are not entirely absent from either endeavour: in the Royal Court's production of *Ten Billion*, for example, Emmott appeared in a replica of his University of Cambridge office reconstructed on the Jerwood Theatre Upstairs stage, and playwright Duncan Macmillan was responsible for crafting Rapley's words into a monologue with dramatic and affective impact for *2071*. But most recognisable theatrical conventions have been expunged. There is no scenic action, scripted plot or choreographed movement, no mimetic construction or enactment of character. The emotional muddle and noise of interpersonal human conflict are elided in favour of the delivery of as unmediated a form of fact-based scientific communication as possible, with all the attendant promises of authority and authenticity. Here, the idea of conventional theatrical performance becomes something suspect, unnecessary or distracting 'ornamentation', something 'extraneous—if not opposed—to fact', as Ashley Chang puts it.[13]

---

13   Ashley Chang, 'Staging Climate Science: No Drama, Just the Facts', *PAJ: A Journal of Performance and Art*, 43.1 (2021), 66–76 (p. 67), https://doi.org/10.1162/pajj_a_00547.

12. Displacing the Human: Representing Ecological Crisis on Stage

French sociologist and philosopher Bruno Latour's performance lectures such as *Inside* (2017) and *Moving Earths* (2020), developed in collaboration with scenographer Frédérique Aït-Touati, similarly placed Latour himself on stage in keynote-lecture style, accompanied by large-scale, high-quality visual and animated projections, lighting changes and occasional sound effects that complement his discussion of human interaction with, and perception of, the environment. Just as director Katie Mitchell explains the performance lecture as a response to conventional theatre forms that threaten 'to diminish and oversimplify and sensationalise the subject',[14] Latour and Aït-Touati also emphasise the form's potential to refocus attention on the scientific and more broadly intellectual dimensions of our discussions about ecological catastrophe—with a particular emphasis on the deprioritising of the human figure within the grand scheme of ecological catastrophe. Their collaborative work, Aït-Touati explains, asks 'how can we deconstruct theatre as a traditionally human-centred art form?', and Latour elaborates, 'We are seeking another relationship to scenography by—yes—decentring the human, moving him or her slightly off stage so to speak'.[15] Rather as the flurry of large-scale animated projections in *2071* drew the eye from Rapley's onstage body, similarly in *Inside*, Latour is a small figure on stage, visually dwarfed and often blurred by the large projections of the earth and its various ecosystems that overlie both the stage backdrop and his own body; he occasionally disappears from view altogether in the dark as the moving projected images shift and coalesce behind him.[16]

Nevertheless, in each of these lectures, the onstage individual still remains the centre of attention (often quite literally, in terms of stage imagery), almost inevitably coming to dominate audience focus. As recognisable individual figures of intellectual expertise—and,

---

14 Mitchell quoted in Stephanie Merritt, 'Climate change play *2071* aims to make data dramatic', *The Guardian* (5 November 2014), https://www.theguardian.com/stage/2014/nov/05/climate-change-theatre-2071-katie-mitchell-duncan-macmillan.

15 Latour quoted in 'Décor as Protagonist: Bruno Latour and Frédérique Aït-Touati on Theatre and the New Climate Regime', Sébastien Hendrickx and Kristof van Baarle, *The Theatre Times* (18 February 2019), https://thetheatretimes.com/decor-is-not-decor-anymore-bruno-latour-and-frederique-ait-touati-on-theatre-and-the-new-climate-regime/.

16 Frédérique Aït-Touati and Bruno Latour, 'Inside—a performance lecture', http://www.bruno-latour.fr/node/755.html.

particularly in Latour's case, of the rare 'star power' of celebrity public intellectual status—Emmott, Rapley and Latour sit authoritatively at the heart of their performances, the central axis around which the lecture unfolds. Even visually, the ostensible minimising of the human figures in these lecture-performances is inconsistently enacted. *Ten Billion*'s stage setting of Emmott's own office calls attention to his personality as an individual within a larger institution—his choice of pot plants, his conference lanyards displayed behind his desk—and Royal Court press images frequently focused in on Emmott's face, illuminated by or shadowing his projected data; likewise, the theatre's promotional material for *2071* typically foregrounded Rapley's seated figure.[17] Early in *Inside*, the dark backdrop curtains open slightly directly behind Latour's figure, illuminating him in a floor-to-ceiling shaft of bright white, almost angelic light. Latour's individual features are indiscernible, but his human figure is clearly, dramatically silhouetted in the centre of the stage. His audience applauds wildly. We have returned from our virtual planetary travels to the comfortable and familiar human scale.

Shifting the Boundaries: The Spatial, the Temporal, and the Sensory

However, other recent plays have shifted theatrical imagery away from the human body and towards nonhuman scales. These plays often draw successfully on theatre's near-unique functioning in a range of spatial and temporal dimensions and sensory modes, exploiting the consequent opportunity for multi-elemental, boundary-defying representation that moves us into a more-than-human scope of thought. While the human figure may not disappear from the stage altogether, slippages disrupt the borders between human and nonhuman, biotic and abiotic, organism and environment. These are plays that, in stretching the boundaries of both human action and theatrical staging conventions, offer a new perspective on other ecological forces.

---

17 See, for example, https://www.thetimes.co.uk/article/ten-billion-at-the-royal-court-sw1w-v7l5bqj0jx0, https://www.theguardian.com/stage/2012/dec/10/best-theatre-2012-ten-billion, and https://www.britishtheatreguide.info/reviews/ten-billion-royal-court-the-7701.

The Contingency Plan by Steve Waters, which premièred at the Bush Theatre, London, in April 2009, is structured in two parts—*On the Beach* followed by *Resilience*—and is to some degree rooted in similarly family-centric (that is, human-centric) structure as *Lungs*, *The Heretic*, and the other plays mentioned in the foregoing discussion. *On the Beach* focuses on the antagonistic dynamic between a father-son pair of glaciologists; *Resilience* sees the son Will ignore his father Robin's advice that he avoid mixing science with politics as Will moves into a governmental advisory role in an effort to prevent catastrophic flooding of England, caused by melting glaciers and rising sea levels. However, *The Contingency Plan* self-consciously diverges from the human-centric theatrical image and scale in several notable ways. *On the Beach* looks constantly towards the inevitable wiping of human presence from the earth by geological forces far beyond our scope of power or lifespan, by staging a miniature ecosystem, a kind of environmentalist metatheatrical device: a fish tank, '*a stage-design model box*' constructed by Robin, that demonstrates what will happen to the ecosystem of the Norfolk coastal area—where the characters live—if sea levels rise. The resultant catastrophic flooding is illustrated in miniature in the fish tank. Robin's human body becomes an onstage visual surrogate for larger geological forces as he operates the fish-tank model: 'and we have to factor in the surge— (*He is splashing and moving water about with his hands*)'.[18] In one simple but effective stage image, this fish-tank scenario shifts the focus and scale of the play's concern from the human to the planetary: the localised emotional catastrophe of the family's home being destroyed is literally minimised before the audience's eyes, reduced conceptually to 'a Petri dish' for wider, global-scale extrapolations, gesturing towards the scope of unseen forces beyond the human eye. '[B]ack comes the pristine landscape of the Holocene era', Robin predicts. '[W]e are a disturbance in the sleep of the world and we're gonna be brushed away, sweated out. The sea rises, the land goes, the cities go, the people are gone. You can't fight that'.[19] His wife Jenny's expression of fear shortly before the play's end—'It's the sea, Rob. […] I don't recognise it. Fine. It doesn't recognise me'—redefines the sea as the dominant agential subject and herself as the passive object of its gaze.[20] Fulfilling Robin's prophecy, by the end

---

18   Steve Waters, *The Contingency Plan* (London: Nick Hern Books, 2009), p. 52, p. 54.
19   Ibid., p. 55, p. 83.
20   Ibid., p. 81.

of *On the Beach* the coastal family home and both parents are poised to be swept away by the rising tide, into the vast scope of 'continental time. Geological time' that Robin invokes in his final words.²¹ Likewise, at the end of *Resilience*, the babble of human voices attempting vainly to counter the force of ecological apocalypse is replaced, abruptly, mercilessly, by *'the sound of an enormous storm. Blackout'*—represented onstage, we should note, by 'invisible' sonic rather than any human-scaled visual modality.²² The diptych drama stretches ecological scale across the two plays, in an *Alice-in-Wonderland*-style shrinking and growing cycle enabled by specifically theatrical means.

Temporal and spatial dislocation interrupt the human-centric scope of climate-change theatre in several recent plays. Andrew Bovell's *When the Rain Stops Falling* (2009) focalises ecological disaster through a father-son estrangement, mother-son conflict, and fracturing marriage, but the scope of the play extends across four generations of intercontinental family shift, flitting between 1959 and 2039, and between London and various urban and outback spaces in Australia. Thus, although much of the play's ecological concern is once again metaphorised through a human-centric model, Bovell expands the localised, individualised human scope to a vaster multi-generational and international scale. Ella Hickson's *Oil* (2016) offers a still more striking version of this expansion of theatrical time and space in grappling with the hyperobject nature of ecological shifts. Moving between 1889 Cornwall, 1908 Tehran, 1970 London, 2021 Baghdad, and back to 2051 Cornwall, it explores the global use of oil at domestic and industrial levels and associated geopolitical and ecological consequences.²³ The mother-daughter relationship between Amy and May ostensibly structures the play, but the exaggeratedly non-realist course of their lives—May is twenty and pregnant with Amy in 1889, and approaching old age in 2051, cared for by the young adult Amy—shifts the informing temporal scope of the play from the recognisably human to the larger-scale, beyond-human chronology of the slow calamity of environmental breakdown.²⁴ Patrick

---

21 Ibid., p. 84.
22 Ibid. p. 182.
23 Editors' note: See also Ferebee's discussion of oil as a hyperobject (chapter 8). Here oil might be considered as a hyperactor rather than a hyperobject (see Beaufils's distinction in chapter 13).
24 The anagrammatic overlap of mother May and daughter Amy's names further destabilises our sense of these characters as not-quite-real human figures, as in

Lonergan has connected the colliding chronologies of *Oil* to Morton's argument that ecological catastrophe demands that we think 'at the level of both individual and species simultaneously'.[25] Hickson enacts this challenge by way of the seemingly unstageable scripting of the human body in the interscenes that mark the passing of time between acts. For example:

> She walks through lands, through empires, through time.
>
> A woman walks across a desert.
> The air is hot; the night is black.
>
> One newborn baby gasps for breath.
> A million newborn babies gasp for breath.[26]

Or later:

> A child flies backwards into the future.
>
> A child drives backwards.
> A child walks backwards
>
> Retreats, returns, retracts
> Yestermorrow.
>
> A child returns, retreats, contracts
> A child sits.
>
> Home in time for bed.[27]

Representing the vertiginous passage of ecological time between acts, these interscenes unsettle any straightforward staging of the human body. 'One newborn baby gasp[ing] for breath' already poses a challenge to even the most innovative director; the sudden expansion of the direction to a 'million newborn babies' pushes into the bounds of the seemingly impossible. Hickson's interscenes chafe at the scope of what the theatre medium can stage, the very possibility of staging the human figure in

---

the similar case of the couple Gabriel and Gabrielle in Bovell's *When the Rain Stops Falling*.

25 Lonergan, 'A Twisted, Looping Form', p. 38. Lonergan quotes here from Timothy Morton's *Dark Ecology: For a Logic of Future Coexistence* (New York: Columbia University Press, 2016), p. 40, https://doi.org/10.7312/mort17752.

26 Ella Hickson, *Oil* (London: Nick Hern Books, 2016), p. 26.

27 Ibid., p. 101.

line with the vast scale of ecological change, eventually seeming to erase the human referent altogether in the subject-less direction *'Retreats, returns, contracts / Yestermorrow'*, before demanding the reappearance of the child's figure onstage again. By challenging any staged production 'to work physically with impossibility', to borrow theatre scholar Karen Quigley's evocative expression,[28] Hickson's interscenes in *Oil* offer an alternative iteration of the 'unrepresentable' scale of the climate crisis in the theatre medium.

Hickson's provocative intertwining of environmental and biological timescales might shift our attention to other recent theatrical works that have attempted to engage with ecological frameworks of thought by breaking out of the confines of the traditional proscenium stage space.[29] Where *The Contingency Plan* used *'a stage-design model box'* to miniaturise the human-scaled stage space, these plays stretch theatrical scale by literally expanding it beyond the limits of the proscenium stage itself. Mike Bartlett's carnivalesque extravaganza *Earthquakes in London* (2010), for instance, paralleled its inflated temporal scope (the play's timeframe ran from 1969 to 2525) by taking over the auditorium itself as a site of action. *'The play is about excess'*, Bartlett's opening stage directions note, and the stage *'should overflow with scenery, sound, backdrops, lighting, projection, etc'*.[30] Accordingly, Miriam Buether's award-winning set design for the play's première production at the National's Cottesloe Theatre connected two stages at the front and back of the auditorium with a bright orange, winding catwalk, with the audience seated on barstools running the length of both its sides (Fig. 46). Other productions have played on the theatre medium's functioning in multiple sensory dimensions to expand beyond the visual confines of the proscenium stage. Ian Rickson's 2018/19 production of Brian Friel's *Translations* (1980) at the National Theatre, London, for example, filled the amphitheatre-style Olivier stage with cut turf; the strong smell of the turf suffused the auditorium, blending an ecocritical dimension into the playtext's concern with land colonisation. Similarly, Rickson's première

---

28　Karen Quigley, *Performing the Unstageable: Success, Imagination, Failure* (London, Methuen, 2020), p. 5, https://doi.org/10.5040/9781350055483.

29　Editors' note: In chapter 13, Eliane Beaufils studies the strategies used by three contemporary performances to turn this traditional proscenium stage space into a bio-geological milieu, allowing the deployment of a spectatorial gaze 'from inside'.

30　Mike Bartlett, *Earthquakes in London* (London: Methuen, 2021), p. 5.

production of Jez Butterworth's *Jerusalem* (2009) at the Royal Court saw set designer Ultz bring live chickens, tortoises, freshly cut logs and real trees onto the stage; the noise and smell of the animals and plant material that permeated the auditorium added a competing ecological sensory-scape to the human-centric threads of the play (Fig. 47).[31] These plays reach beyond the proscenium stage confines—whether by extending the tangible material elements of performance into the auditorium, or by employing more subtle multi-sensory mechanisms—to re-imagine the possible scales of theatrical representation in the context of climate change.

Fig. 46 Miriam Buether's auditorium set for Mike Bartlett's *Earthquakes in London* at the Cottesloe, National Theatre, London (2010). Photograph by Manuel Harlan © Manuel Harlan. All rights reserved.

---

31   Anna Harpin notes that, in Rickson's production, 'the trees were rotten and so already earmarked for felling and thus they got a new lease of life through the production'. See Anna Harpin, 'Land of Hope and Glory: Jez Butterworth's Tragic Landscapes', *Studies in Theatre and Performance*, 31.1 (2011), 61–73 (p. 69), https://doi.org/10.1386/stap.31.1.61_1. However, beyond this immediate eco-friendly concern, the presence of literally rotting trees on *Jerusalem* stage adds a material ecological dimension to a play centrally concerned with the demise of 'traditional England' and its outdoor spaces.

Fig. 47 Mark Rylance on the Royal Court stage set of *Jerusalem* (2009). Photograph by Tristram Kenton. © Tristram Kenton. All rights reserved.

While these plays break through the proscenium stage space, other ecologically inclined performance pieces have rejected the confines of the theatre building itself, reaching out into the outdoor world that is their subject. In Carole Kim's immersive site-specific work *The Seed Will Search...*, performed in the Descanso Gardens in California, participants were led at night through the gardens to the three-hundred-year-old Heritage Oak, where Butoh dancer Oguri moved in and out of the tree alongside the projected image of Rozanne Steinberg, who was dancing in real time elsewhere in the garden and projected in virtual space onto the oak tree. Once again, the human body did not vanish entirely from the performance site, but here the dancers' bodies facilitated rather than focalised the performance; the central unfolding focus remained the oak tree itself—a subject in its own right, rather than a theatrical backdrop—and the audience's direct encounter with the exterior natural world. In a more explicitly politicised context, Earth Ensemble, the playwriting and performance-making arm of environmental protest group Extinction Rebellion, regularly stage outdoor 'guerrilla theatre' plays specially written for mobile protest performance, such as April de Angelis' *Mrs Noah*, *Plane Truth* and *We Hear Bird Song*, Bec Boey's *An*

Apology, and John Farndon's *The Silence Ends* and *Paper Cranes*. Earth Ensemble's outdoor protest plays blur a line between indoor theatrical performance and the equally theatrical protest gestures devised by Extinction Rebellion: dumping manure outside *Daily Mail* newspaper offices in London; the hundreds of litres of fake blood spilled outside Downing Street; co-ordinated 'die-ins' on city streets around the world; regular parades of sackcloth 'penitent' figures (Fig. 48) and the iconic Red Rebel Brigade protestors (Figs. 49 and 50).[32]

Fig. 48 Extinction Rebellion 'penitent' performer in Cornwall, protesting the G7 Summit (June 2021). Photograph by Greg Martin © Greg Martin / Cornwall Live. All rights reserved.

---

32 Extinction Rebellion, 'From Monday 15 April: Extinction Rebellion to block Marble Arch, Oxford Circus, Waterloo Bridge & Parliament Sq round the clock until Government acts on Climate Emergency', https://extinctionrebellion.uk/2019/04/09/from-monday-15-april-extinction-rebellion-to-block-marble-arch-oxford-circus-waterloo-bridge-parliament-sq-round-the-clock-until-government-acts-on-climate-emergency%EF%BB%BF/.

Fig. 49 Red Rebel Brigade in Cornwall, protesting the G7 Summit (June 2021). Photograph by Joao Daniel Pereira © Joao Daniel Pereira / Extinction Rebellion. All rights reserved.

Fig. 50 Red Rebel Brigade protesting in London (2019). Photograph by Connor Newson © Connor Newson/Adams Creative Media. All rights reserved (no financial gain).

Earth Ensemble's outdoor guerrilla plays, like Extinction Rebellion's protest performances, formally replicate several key properties that Morton has ascribed to ecological hyperobjects. Permeating the common spaces of daily life, they mimic the hyperobject's viscosity, simultaneously 'near' but 'uncanny'.[33] As Morton explains, 'along with this vivid intimacy goes a sense of unreality',[34] and the Earth Ensemble and Extinction Rebellion performances specialise in this heady combination of intimate unreality: a scripted speaker in character suddenly appears beside you in the crowd, transforming you from anonymous pedestrian to disoriented audience member, abruptly involved in theatrical action; or you turn a street corner or a bend in the beach and are confronted with the otherworldly sight of the Red Rebel Brigade. These are affectively sticky performances, to borrow (and slightly adapt) Morton's articulation of hyperobjects as viscous phenomena which '"stick" to beings that are involved with them'.[35] They are designed to impel immediate and enduring affective response—it is hard to forget these stark little vignettes, particularly when unexpectedly encountered in one's everyday environment—and indeed they often solicit mass and even spontaneous participation: John Farndon's verse performance *The Silence Ends* (2019), for example, is a group spoken-word chorus that stages a medley of overlapping voices, designed to accommodate as many participants as can be persuaded to join. These protest performances are also frequently *'nonlocal'*, a term Morton uses to describe the ecological hyperobject, 'massively distributed in time and space' and visible only in their localised effect or partial appearance.[36] Many of Earth Ensemble's plays are devised specifically to lend themselves to multiple concurrent performance across various locales, paralleling Extinction Rebellion's co-ordinated global protests that take place simultaneously in cities across the globe.[37] Similarly, the

---

33 Morton, *Hyperobjects*, p. 28.
34 Ibid., p. 32.
35 Ibid, p. 1.
36 Ibid., p. 1 (emphasis in the original), p. 48.
37 The #EverybodyNow International Rebellion in October 2019, for example, occurred simultaneously in more than fifty cities, including London, Berlin, Paris, New York Los Angeles, Washington DC, Santiago, Buenos Aires, and Montreal, and was estimated to have involved up to 30,000 protestors worldwide. See the XR plans for the event at https://extinctionrebellion.uk/2019/10/04/everybodynow-londons-rebellion-on-track-to-be-five-times-bigger-than-april/.

sackcloth 'penitent' (Fig. 48) and Red Rebel Bridge attire (Figs. 49 and 50) are designed not only to offer a visually striking presence in public, but also to provide an easily replicable costume, allowing protestors around the globe to join the ongoing performance.[38] With an estimated 5,000 Red Rebel Brigade members worldwide, any of us who have witnessed one of their protest parades cannot be said to have 'seen the Red Rebel Brigade' in any total sense, just as any one performance of one of Earth Ensemble's multiple simultaneous guerrilla productions is only a fraction of the full simultaneous performance, impossible for any individual to see in its total form—and just as 'any "local manifestation" of a hyperobject is not directly the hyperobject'.[39] By shifting from the enclosed space of the theatre auditorium into direct contact with exterior public spaces, and exaggerating the common theatrical phenomenon of repeated and reproduced stagings into simultaneous and/or globally ranging performances, the Earth Ensemble and Extinction Rebellion protest works push into a scale and affective range comparable to Morton's ecological 'hyperobject' dimensions.

'Fragments, Shards, Whispers': Imagining the Impossible Other

Finally, we turn to Deke Weaver's *The Unreliable Bestiary* (2009–), a collection of immersive performance pieces that work beyond the typical temporal, spatial and sensory dimensions of conventional theatre in their staging of multi-species extinction. The ongoing project plans an 'ark' of twenty-six individual site-specific and multi-media performance pieces, focusing on one endangered animal or habitat for every letter of the English alphabet. At time of writing, Weaver has produced five of the projected twenty-six pieces: MONKEY (2009), scripted for black-box theatre performance; ELEPHANT (2010, Fig. 51), produced on the dirt floor and towering walls of the University of Illinois' amphitheatre-style Stock Pavilion; WOLF (2013), in which audience members are guided by

---

38 The Red Rebel Brigade provide instructions for members on how to craft the costume and replicate the group's make-up and choreography themselves. See https://www.youtube.com/watch?v=_qIfSNP2nuo and https://www.youtube.com/watch?v=hGPv0wALfWw. Video footage of the first public appearance of the penitents and Red Rebel Brigade, at St. Ives, Cornwall, August 2019, is available online: https://www.youtube.com/watch?v=_Z7DbR9VY9E.

39 Morton, *Hyperobjects*, p. 1.

a park ranger by bus to Allerton Park in Monticello, Illinois, across the park's river and woods to a barn performance space; BEAR (2016–2017, Fig. 52), spread over three parts in a six-month period, played out in Meadowbrook Park in Urbana, Illinois, in online performance videos, and in a mass group banquet; and TIGER (2019), an intimate séance- or salon-esque piece devised for small theatre, living room and museum performance.[40] Weaver's *Bestiary* opens up the space of ecological performance at the level of both site and scale. Focusing his attention beyond 'strictly human stories', Weaver seeks to produce work 'about the big questions [...] about moments where you know there's something bigger going on—bigger than email and Facebook, coolness and hipness, bigger than shopping, bigger than politics, bigger than power struggles, bigger than your career or your family'.[41] He correspondingly rejects the limiting constraints of conventional theatre form, theatre spaces, and single-media presentation, drawing on a medley of claymation, mechanical puppetry, realistic and exaggeratedly non-realistic costume, dance, song, documentary footage, film clip, animated video projection, photographs, sketches, and often the presence of the unchoreographable natural world. Commenting on the chaotic intermedial texture of his pieces, Weaver explains:

> In putting these pieces together I wonder what it will be like when these animals are gone. We're going to be left with these fragments, shards, whispers and cartoons of what the animal must have been like. As far as 'accurate' and 'truthful' representations, it's always always always going to fall short of the real thing. So I'm more interested in the failure of representation, as a way to point out the absence of the real animal.[42]

---

40   Further details of individual performances can be found online at https://www.unreliablebestiary.org/projects/. The full text of MONKEY and excerpts of ELEPHANT have been printed in *Animal Acts: Performing Species Today*, ed. by Una Chaudhuri and Holly Hughes (Ann Arbor: University of Michigan Press, 2014), pp. 141–55 and pp. 163–81, https://doi.org/10.3998/mpub.5633302. Digital video documentaries of MONKEY, ELEPHANT, and WOLF have been archived at Princeton University, New York University, University of Michigan, University of Iowa, University of Georgia, and University of Kentucky. Una Chaudhuri and Joshua Williams offer a detailed overview of participating as an audience member in several *Unreliable Bestiary* performances in 'The Play at the End of the World: Deke Weaver's *Unreliable Bestiary* and the Theatre of Extinction', *The Cambridge Companion to Theatre and Science*, ed. by Kirsten E. Shepherd-Barr (Cambridge: Cambridge University Press, 2020), pp. 70–84, https://doi.org/10.1017/9781108676533.006.

41   Deke Weaver and Maria Lux, 'Interview: *The Unreliable Bestiary*', *Antennae*, 22 (2012), 31–40 (p. 40).

42   Ibid., p. 34.

The disorienting impact of Weaver's breakdown of the theatre's traditional formal limits is intensified by the improbable, even impossible scale of *The Unreliable Bestiary*'s projected scope. As Una Chaudhuri and Joshua Williams observe, 'With twenty-one letters left to go, and averaging one performance every two years, the fifty-something Weaver's plan betrays an aching irony: the project faces the same near-impossible timetable as global efforts to stave off catastrophic species loss'.[43] For all its formal innovation and protracted temporal scope, *The Unreliable Bestiary* can hope to represent 'only a tiny sliver of our current catastrophic loss of habitat and biodiversity'.[44] At its most basic level of existence, then, Weaver's project is bound up in an impossibility that parallels both the seeming impossibility of these species' continued survival—and the audience's own inability to conceive of the total non-existence of this range of animal species, so deeply imbricated in our own cultural mythologies.

Fig. 51 Deke Weaver, *ELEPHANT* (Stock Pavilion, University of Illinois, 2010). Photograph by Valerie Oliviero © Deke Weaver (artist) and Valerie Oliviero (photographer). All rights reserved.

---

43 Chaudhuri and Williams, 'The Play', p. 72.
44 Deke Weaver, *The Unreliable Bestiary*, https://www.unreliablebestiary.org/.

12. Displacing the Human: Representing Ecological Crisis on Stage   345

Fig. 52 Deke Weaver, *BEAR* (Meadowbrook Park, Urbana, Illinois, 2016). Photograph by Nathan Keay © Deke Weaver (artist) and Nathan Keay (photographer). All rights reserved.

Furthermore, although *The Unreliable Bestiary* often still depends on the human body and voice to stage its ecological concerns, the human figure is decentred not only by the intermixing of intermedial resources, but also by the project's refusal to prioritise (or even permit) any stable human comprehension of the world it presents. Each piece mixes factual, mythical, and outright false information about the animal in question to the point that any definitive interpretation becomes impossible, and spectators must 'learn to content themselves with a base level of uncertainty about whether or not what they are hearing is objectively true'.[45] Reviewing WOLF, for example, Elizabeth Tavares recalls how '[s]eeing a trapped wolf far ahead on the path, it was unclear whether this was actor or animal',[46] and Nigel Rothfels notes how ELEPHANT creates 'an unsettling contemplation of the elephant as an animal we both might know better and will never know at all'.[47]

---

45 Chaudhuri and Williams, 'The Play', p. 75.
46 Elizabeth Tavares, 'WOLF Politics: Performance art behind the barn', *Bite Thumbnails* (12 September 2013), http://bitethumbnails.com/archives/433.
47 Nigel Rothfels, 'Commentary: A Hero's Death', in *Animal Acts: Performing Species Today*, ed. by Una Chaudhuri and Holly Hughes (Ann Arbor: University of Michigan

Weaver's work decentres the human and disrupts the human-centric scale by interrupting our sense of our own capacity for comprehending the beyond-human world. Yet he also insists on the simultaneously 'unreal' nature of these performed animals, presented as the product of human imagination. The result is a kind of structural irony: a series of performances that decentre the human yet constantly remind us that we cannot access the animal world without the mediating filter of our imagination. The destabilising of the anthropocentric experience, then, takes place at the level of the spectator's experience rather than solely on Weaver's 'stages' or in his performance imagery. By means of its affectively and cognitively disorienting forms, *The Unreliable Bestiary* engages with uncontainable, inconceivable ecological otherness.

This de-anthropocentric re-imagining is crucial, we suggest, to contemporary and future theatrical representation of climate crisis and the ecological world more broadly. Weaver's performance pieces draw together the human and the nonhuman in non-hierarchised forms, in which the human figure is present but not central or authoritative. Comparably, recent ecocritical thinkers and theorists have returned repeatedly to imagery of pluralised material unity (the woven cloth, the mesh, the web, the compost heap) to articulate the profound interconnection and unavoidable ongoing interaction between all human and nonhuman beings in the Anthropocene era. In *The Ecological Thought*, Morton employs the image of an immeasurably vast mesh to describe both 'the interconnected-ness of all living and non-living things', and the simultaneous gaps, absences, and voids between these interdependent organisms, the 'strange, even intrinsically strange' relationships between non-hierarchised beings.[48] Elsewhere ecofeminist Donna J. Haraway's conceptualising of the 'Chthulucene' as an alternative to the human-centric Anthropocene offers a similar foregrounding of the 'compost piles' of multi-species relationships, asserting the necessity of making ontologically heterogeneous kin or 'oddkin' with 'microbes, plants, animals [...] human and nonhuman bodies, at different scales of time and place' in order to more properly

---

Press, 2014), pp. 182–88 (p. 183), https://doi.org/10.3998/mpub.5633302.
48 Morton, *The Ecological Thought*, p. 28, p. 15. See also Morton, 'Queer Ecology', *PMLA*, 125.2 (2010), 273–82 (pp. 274–78), https://doi.org/10.1632/pmla.2010.125.2.273.

comprehend and exist on our shared planet.⁴⁹ Similarly, Weaver himself argues,

> The gears of an ecosystem's clockwork include air, water, animals, money, and the human imagination. Our fantasies, assumptions, and cultural mythologies literally shape the land. Animals and their stories are embedded in our environmental, economic, political, and judicial systems. It's all part of the same cloth. You tug on one corner of the bedsheet and the whole thing moves.⁵⁰

Mesh, compost pile, bedsheet: each of these unifying models demands a re-imagining of un-hierarchised co-existence rather than human-centric mastery, or indeed total human detachment. After all, if we assume that much ecologically-oriented theatre is at least partially 'activist', interested in influencing its audience's subsequent behaviour for the greener good, then ecological performance must still imbricate—if not centre—the human figure at some level. As Haraway notes, 'human beings are not the only important actors [...], and the biotic and abiotic powers of this earth are the main story. However, the doings of situated, actual human beings matter'.⁵¹ This combination of radical intimacy yet insistent strangeness, as staged by Weaver's *The Unreliable Bestiary*, proffers a new relation of uncanny interconnection that promises 'new epistemological orientations, affective entanglements, and ethical commitments' between human and nonhuman participants.⁵² It is precisely this revolutionary shift in perspective and feeling, this generative affect of intimate strangeness, that ecologically inclined theatre performance is poised to achieve by way of its uniquely multi-spatial, multi-temporal, multi-sensorial scope of play.

Conclusion

Two productions, one recent and one upcoming at time of writing, are radically relocating the natural world in theatrical performance, and physically relegating humankind to the margins of attention. On 22

---

49 Donna J. Haraway, *Staying with the Trouble: Making Kin in the Chthulucene* (Durham: Duke University Press, 2016), p. 4, 169, 16, https://doi.org/10.2307/j.ctv11cw25q.
50 Weaver, 'The Unreliable Bestiary', https://www.unreliablebestiary.org/
51 Haraway, *Staying with the Trouble*, p. 55.
52 Chaudhuri and Williams, 'The Play', p. 72.

June 2020, Barcelona's Liceu Opera re-opened for the first time since the COVID-19 pandemic forced a lockdown closure in March that year—but without a human audience. A string quartet played Giacomo Puccini's 'Cristantemi' to an auditorium in which every one of the 2,292 seats was occupied by plants from local nurseries, in what curating artist Eugenio Ampudia called a foregrounding of 'something as essential as our relationship with nature'.[53] In February 2021, director Katie Mitchell announced her plans for a new production of Anton Chekhov's *The Cherry Orchard* (1904) from the point of view of the trees. Mitchell explained:

> We'll do the play backwards, and then we'll exit out of the play, and we will keep going backwards through lots of geological time, through until the carboniferous period, which is when the trees first started. So that's going to be using that play as a threshold through which we can walk to also look at planetary time, as well as refocus[ing] the drama so it's no longer anthropocentric. [...] You will spend more minutes in a theatre with the trees than with the human beings. It will be a very strict ratio of about 80 trees, 20 human beings. [...] I think it's just really pushing ourselves to try, you know, theatrical embodiments of the more-than-human world.[54]

In both cases, the nonhuman plant forms take precedence over the human.[55] In Mitchell's planned *Cherry Orchard* production, the trees become the central subjects, rather than the backdrop or metaphorical props to the human action; in the Liceu Opera's performance, the auditorium plants become 'subjects' in that they are positioned as expectant audience, rather than the onstage 'object' of the spectatorial gaze. These are extreme examples of the repurposing of the *theatrum ut mundi* model, but they offer timely indications of the drive towards radically repositioning the human figure as—at the very most—co-habitant, rather than master of our ecological and theatrical landscapes.

---

53 https://www.liceubarcelona.cat/en/artist-eugenio-ampudia-inaugurates-activity-liceu-concert-2292-plants. A recording of the performance is available here: https://www.youtube.com/watch?v=rgvadprJFRc&t=11s.

54 'In Conversation with Katie Mitchell, Professor Fiona Stafford and Dr Catherine Love', The Oxford Research Centre in the Humanities (4 February 2021), https://www.youtube.com/watch?v=FHa-S-5XWcg.

55 Editors' note: See also the 'longterm planttheater' developed by Tobias Rausch (2010–2015) and studied by Eliane Beaufils in chapter 13.

Decentring, re-scaling, and enmeshing the human figure within the micro- and macroscopic scales of the climate crisis is the crucial work of twenty-first-century theatre and performance practice—particularly while our collective consciousness lacks any ready imaginative sense of the 'invisible', inconceivable, seemingly unrepresentable spectra of ecological forces that extend beyond or below our human-centric scale.

Works Cited

Aït-Touati, Frédérique, and Bruno Latour, 'Inside—a performance lecture', http://www.bruno-latour.fr/node/755.html

Bartlett, Mike, *Earthquakes in London* (London: Methuen, 2021).

Bean, Richard, *The Heretic* (London: Oberon, 2011).

Beckett, Samuel, *Endgame* [1957], *Act without Words I* [1957] and *Happy Days* [1961] in *The Complete Dramatic Works* (London: Faber, 2006).

Berger, James, and Peter Cunningham, 'Meditations in an emergency: James Berger on climate, fiction, and apocalypse', *Yale News* (27 January 2020), https://news.yale.edu/2020/01/27/james-berger-climate-fiction-and-apocalypse

Bilodeau, Chantal, 'Creating a List of Climate Change Plays', *Artists and Climate Change* (1 November 2014), https://artistsandclimatechange.com/2014/11/01/creating-a-list-of-climate-change-plays/

Blakley, Rhys, 'Why saving the panda, but not its parasites, is a really lousy idea', *The Times* (24 August 2020), https://www.thetimes.co.uk/article/why-saving-the-panda-but-not-its-parasites-is-really-a-lousy-idea-9026rlwx2

Bovell, Andrew, *When the Rain Stops Falling* (London: Bloomsbury, 2009).

Buffini, Moira, Penelope Skinner, Matt Charman, and Jack Thorne, *Greenland* (London: Faber and Faber, 2011).

Butterworth, Jez, *Jerusalem* (London: Nick Hern, 2009).

Chang, Ashley, 'Staging Climate Science: No Drama, Just the Facts', *PAJ: A Journal of Performance and Art*, 43.1 (2021), 66–76, https://doi.org/10.1162/pajj_a_00547

Chaudhuri, Una, 'Anthropo-Scenes: Staging Climate Chaos in the Drama of Bad Ideas', in *Twenty-First-Century Drama: What Happens Now*, ed. by Siân Adiseshiah and Louise LePage (London: Palgrave, 2016), pp. 303–21, https://doi.org/10.1057/978-1-137-48403-1_15

Chaudhuri, Una, and Holly Hughes, eds, *Animal Acts: Performing Species Today* (Ann Arbor: University of Michigan Press, 2014), https://doi.org/10.3998/mpub.5633302

Chaudhuri, Una, and Joshua Williams, 'The Play at the End of the World: Deke Weaver's *Unreliable Bestiary* and the Theatre of Extinction', in *The Cambridge Companion to Theatre and Science*, ed. by Kirsten E. Shepherd-Barr (Cambridge: Cambridge University Press, 2020), pp. 70–84, https://doi.org/10.1017/9781108676533.006

Darwin, Charles, *On the Origin of Species* (London: John Murray, 1859). See *Darwin Online*, http://darwin-online.org.uk/Variorum/1859/1859-489-dns.html

Duffy, Carol Ann, *Everyman* (London: Bloomsbury, 2017).

Emmott, Stephen, *Ten Billion* (London: Penguin, 2013).

Glaspell, Susan, *The Verge* [1921], in *Plays by Susan Glaspell*, ed. by C. W. E. Bigsby (Cambridge: Cambridge University Press, 2012).

Harpin, Anna, 'Land of Hope and Glory: Jez Butterworth's Tragic Landscapes', *Studies in Theatre and Performance*, 31.1 (2011), 61–73, https://doi.org/10.1386/stap.31.1.61_1

Haraway, Donna J., *Staying with the Trouble: Making Kin in the Chthulucene* (Durham: Duke University Press, 2016), https://doi.org/10.2307/j.ctv11cw25q

Heise, Ursula, *Imagining Extinction: The Cultural Meanings of Endangered Species* (Chicago: The University of Chicago Press, 2017), https://doi.org/10.7208/chicago/9780226358338.001.0001

Hendrickx, Sébastien, and Kristof van Baarle, 'Décor as Protagonist: Bruno Latour and Frédérique Aït-Touati on Theatre and the New Climate Regime', *The Theatre Times* (18 February 2019), https://thetheatretimes.com/decor-is-not-decor-anymore-bruno-latour-and-frederique-ait-touati-on-theatre-and-the-new-climate-regime/

Hickson, Ella, *Oil* (London: Nick Hern Books, 2016).

Hudson, Julie, '"If You Want to be Green Hold Your Breath": Climate Change in British Theatre', *New Theatre Quarterly*, 28.3 (2012), 260–71, https://doi.org/10.1017/s0266464x12000449

'In Conversation with Katie Mitchell, Professor Fiona Stafford and Dr Catherine Love', The Oxford Research Centre in the Humanities, 4 February 2021, https://www.youtube.com/watch?v=FHa-S-5XWcg

Kelleher, Joe, 'Recycling Beckett', in *Rethinking the Theatre of the Absurd: Ecology, the Environment and the Greening of the Modern Stage*, ed. by Carl Lavery and Clare Finburgh (London: Bloomsbury, 2015), pp. 127–46, https://doi.org/10.5040/9781472511072.0009

Levine, Gabriel, 'Black-Light Ecologies: Punctuate! Theatre's *Bears* Wipes Off its Oil', *Performance Research*, 25.2 (2020), 45–52, https://doi.org/10.1080/13528165.2020.1752576

Lonergan, Patrick, 'A Twisted, Looping Form: Staging Dark Ecologies in Ella Hickson's *Oil*', *Performance Research*, 25.2 (2020), 38–44, https://doi.org/10.1080/13528165.2020.1752575

Love, Catherine, 'From Facts to Feelings: The Development of Katie Mitchell's Ecodramaturgy', *Contemporary Theatre Review*, 30.2 (2020), 226–35, https://doi.org/10.1080/10486801.2020.1731495

Macmillan, Duncan, *Lungs* (London: Oberon, 2011).

Macmillan, Duncan, 'Some Thoughts on *Lungs*', *The Old Vic* (11 October 2019), https://www.oldvictheatre.com/news/2019/10/duncan-macmillan-some-thoughts-on-lungs

Maeterlinck, Maurice, *Les Aveugles* (Brussels: Lacomblez, 1890).

McMullan, Anna, 'Katie Mitchell on Staging Beckett', *Contemporary Theatre Review*, 28.1 (2018), 127–32, https://doi.org/10.1080/10486801.2018.1426822

Merritt, Stephanie, 'Climate change play *2071* aims to make data dramatic', *The Guardian* (5 November 2014), https://www.theguardian.com/stage/2014/nov/05/climate-change-theatre-2071-katie-mitchell-duncan-macmillan

Morton, Timothy, *Hyperobjects: Philosophy and Ecology after the End of the Earth* (Minneapolis: University of Minnesota Press, 2013).

Morton, Timothy, *Dark Ecology: For a Logic of Future Coexistence* (New York: Columbia University Press, 2016), https://doi.org/10.7312/mort17752

Morton, Timothy, *The Ecological Thought* (Cambridge, MA: Harvard University Press, 2010), https://doi.org/10.2307/j.ctvjhzskj

Morton, Timothy, 'Queer Ecology', *PMLA*, 125.2 (2010), 273–82, https://doi.org/10.1632/pmla.2010.125.2.273

Quigley, Karen, *Performing the Unstageable: Success, Imagination, Failure* (London: Methuen, 2020), https://doi.org/10.5040/9781350055483

Rapley, Chris, and Duncan Macmillan, *2071: The World We'll Leave Our Grandchildren* (London: John Murray, 2015).

Rothfels, Nigel, 'Commentary: A Hero's Death', in *Animal Acts: Performing Species Today*, ed. by Una Chaudhuri and Holly Hughes (Ann Arbor: University of Michigan Press, 2014), pp. 182–88, https://doi.org/10.3998/mpub.5633302

Shaw, George Bernard, *Too True to Be Good* [1932], in *Too True to Be Good: Village Wooing, & On the Rocks: Three Plays* (London: Constable & Co., 1934).

Shaw, George Bernard, *Back to Methuselah: A Microbiological Pentateuch* [1922] (London: Oberon, 2000).

Shawn, Wallace, *Grasses of a Thousand Colors* (New York: Theatre Communication Group, 2009).

Shepherd-Barr, Kirsten E., *Theatre and Evolution from Ibsen to Beckett* (New York: Columbia University Press, 2015), https://doi.org/10.7312/shep16470

Shepherd-Barr, Kirsten E., ed., *The Cambridge Companion to Theatre and Science* (Cambridge: Cambridge University Press, 2020), https://doi.org/10.1017/9781108676533

Stephens, Simon, *Wastwater* (London: Bloomsbury, 2011).

Tavares, Elizabeth, 'WOLF Politics: Performance art behind the barn', *Bite Thumbnails* (12 September 2013), http://bitethumbnails.com/archives/433

Waters, Steve, *The Contingency Plan* (London: Nick Hern Books, 2009).

Weaver, Deke, *The Unreliable Bestiary*, https://www.unreliablebestiary.org/

Weaver, Deke, and Maria Lux, 'Interview: *The Unreliable Bestiary*', *Antennae*, 22 (2012), 31–40.

13. Staging Larger Scales and Deep Entanglements

The Choice of Immersion in Four Ecological Performances

Eliane Beaufils

The Gaia hypothesis, formulated in 1974 by James Lovelock and Lynn Margulis, revealed how different biological scales were involved in climate dynamics. Just as the human body is a holobiont with about as many bacteria as cells of its own, the atmosphere and climate are indebted to the activity of the smallest organisms: cyanobacteria, plankton, and plant respiration. Most theatre continues, however, to support the worldview that was dominant until now in the physical and natural sciences: the perspective centred on autonomous actants or objects. Theatrical reductions even emphasize the separation of elements, be they actors or props. Indeed, if the stage is seen as the world, it is only because each element represents a more general, even universal element: the actor—the prince—power. From a structuralist point of view, the grammar of theatrical representation can only be studied because each actant can be identified and isolated in its own functions.[1] What happens, however, if one wishes to show that each actor interacts with others and is inseparable from a dwelling ensemble,

---

1   I am referring to actants as studied by A. J. Greimas or Anne Ubersfeld. Greimas conceived an actantial model of six actants. It is a device that can theoretically be used to analyse any real or thematised action. See Algirdas Julien Greimas, *Sémantique structurale* (Paris: Presses universitaires de France, 1966).

an ecosystem? One would have to transport a whole environment on stage;[2] no overview and no control of the interactions would be possible.

One has to consider, furthermore, that the viewer of spectacles can be considered an analogy of the scientific observer. If the twenty-first-century audience is aware of the action of its situated thought on the scientific object, and the impossibility of adopting on Gaia an outlook from Sirius,[3] the spectator should also be understood as participating in the reciprocal play of the elements. The Gaia hypothesis, widely confirmed since the 1970s, seems, in truth, to call for a theatre that abandons the pretence of overviews and disrupts the position of the spectator.

This contribution will study experiments moving in this direction. It will look at four set-ups that renounce theatrical frontality in favour of the theatre as medium: Kris Verdonck's *Exote I*, Pierre Huyghe's *After ALife Ahead*, Tobias Rausch's *Die Welt Ohne Uns*, and EdgarundAllan's *Beaming Sahara*. It will analyse how the theatre is converted into a *milieu*, and how the spectators are invited to enter these scenes of the world. This theatre creates biological and geological situations, where the main humans on stage are the spectators. The question is whether it changes the apprehension and representation of biological processes in which we humans are involved. First, I will look at the various 'actors' on stage and study the different scales that are brought into play by their interdependencies. What do these performances gather by presenting nonhuman actants?[4] I will then explore the contribution of the theatrical gaze from inside.[5] It would seem that immersion enables

---

2   Editors' note: The necessity of updating the classic *theatrum mundi* trope is presented as a way of escaping anthropocentrism by Kirsten E. Shepherd-Barr and Hannah Simpson in chapter 12. Here it takes another aspect: the 'world as theatre' is not only a way of extending theatre beyond the human, but also of redefining actants as fundamentally relational.

3   Bruno Latour, 'L'Anthropocène et la destruction de l'image du Globe', in *De l'Univers clos au monde infini*, ed. by Emilie Hâche (Paris: Dehors, 2015), pp. 29–56 (p. 42).

4   The verb 'gather' here refers to Bruno Latour, 'Why Has Critique Run Out of Steam', *Critical Inquiry*, 30 (Winter 2004), 225–48, http://www.bruno-latour.fr/sites/default/files/89-CRITICAL-INQUIRY-GB.pdf. The French sociologist points out the contemporary pitfalls of critique, and stands for its methodological renewal. Instead of denouncing facts and arguments, he suggests we gather the different meanings and issues we associate with an object, to discuss it anew.

5   The gaze from inside would then be opposed to the point of view from Sirius. This is a way to draw consequences from the philosophical idea that the Gaia hypothesis leads to the rejection of external viewpoints. This rejection has also led

the audience to experience other perceptions and relationalities with nonhumans. Perhaps the spectators can even develop new forms of reading and communication with the other-than-humans. I will suggest the hypothesis that theatre may then become a 'diplomat',[6] transforming the spectators' inability to think geobiologically into the capacity to do so.

Intermingling Life Forms and Scales

Each of the sets I will now discuss brings together different actors, scales and perceptions, around theatrical situations that can be globally linked to the Anthropocene, the Post-Anthropocene, and the Chthulucene.[7]

In his performative installation *Exote I*, Belgian director Kris Verdonck invites the audience to come and contemplate exotic plants and animals.[8] The textual part consists in signboards presenting the various specimens that are the protagonists: one can become familiar with their characteristics and regions of origin. The performance seems, however, gratifying: it promises the discovery of exotic fauna and flora in a reduced space with, it seems, interactions. An interactive garden!

On arrival, visitors are asked to put on a suit similar to that of laboratory workers. This already induces a tension with the idea of an astonishing discovery of new species, a discovery always linked to the notions of (scientific) conquest and (geographical) freedom. The various plant and animal species include Asian hornets, green parakeets, American bullfrogs, little trees carrying blue berries, a jungle of bamboo plants, Japanese knot weeds…[9] These so-called invasive species are spreading

---

Bruno Latour to entitle one of his lecture-performances *Inside*, thus underlining the epistemological shift of the scientist's position.

6 Baptiste Morizot, *Manières d'être vivant: Enquêtes sur la vie à travers nous* (Paris: Actes Sud, 2020), pp. 245–75.

7 Donna Haraway, *Staying with the Trouble* (Durham: Duke University Press, 2016), especially chapter 2 (pp. 30–57) and chapter 4 (pp. 99–103), https://doi.org/10.2307/j.ctv11cw25q.

8 *Exote I* was conceived and directed by Kris Verdonck in 2011, with the support of dramaturgist Marianne Van Kerkhoven. It was produced by Z33 & A Two Dogs Company. The artists worked together with the University of Diepenbeek and the Z33 contemporary art centre in Hasselt (Belgium).

9 Editors' note: This rich environment can be contrasted with the ways in which nineteenth- and twentieth-century theatre struggled with 'the human-inclined materiality of the theatre medium', and the resulting aesthetics analysed by

in the Belgian flora and fauna and eventually represent a threat, not only for other species, but for the balance of the whole biotope. Beauty is here associated with morbidity, as the term 'exote' reveals its two faces: the exotic and the 'exit'.

Fig. 53 *EXOTE* © A Two Dogs Company. Photography © Kristof Vrancken. All rights reserved.

The effectiveness of the performance is due, in particular, to the way it plays with scales: the local disturbances refer to planetary interactions. By constantly teleporting themselves by airplane, humans have brought back species from distant territories. This transport has often been involuntary, humans having forgotten or neglected that they themselves were capable of transporting seeds, germs, or larvae in their bodies, clothes, or suitcases. Ecosystems are now experiencing unintended local-regional-global interdependencies. The spectators may no doubt have difficulties in imagining all the spatial-biological dynamics that converge here.

A sort of temporal vortex is also produced since the performance gathers the results of several decades of tourism and commercial

---

Shepherd-Barr and Simpson (chapter 12): while writers like Maeterlinck or successors like Beckett tended to empty the stage, *Exote I* stages a world rich with a variety of organisms.

Fig. 54 *EXOTE* © A Two Dogs Company. Photography © Kristof Vrancken. All rights reserved.

exchanges. It recalls a distant time[10] when a clear distinction could be made between endogenous and exogenous species. It also represents a possible incubator of future devastation. The spectators are thus at a crossroads of temporalities, a position typical of the Anthropocene. They are at the centre of a planetary garden that they themselves have deregulated, situated in a liminal time, a moment when the garden still looks beautiful and accessible, but will soon tip over into a hostile configuration. In this position, it appears necessary to respond to the imbalances with responsible action.

Pierre Huyghe's performative installation *After ALife Ahead* confronted an even broader intersection of scales, commensurate with the size of the installation.[11] The French artist used the site of an ice rink

---

10 Maybe this time is not very ancient. Indeed Arthur Tansley coined the term ecosystem in 1935. As this notion implies a stability, it is possible to look back to this period as one where exogenous factors did not seem to threaten the balance of the planetary ecosystems. On the other hand, South American Indigenous people were devastated by illnesses and bacteria transported by the Spanish conquerors upon arrival in the sixteenth century.

11 *After ALife Ahead* was conceived as a temporary installation for the Skulptur Projekte Münster in 2017. More information is available on the exhibition's website: https://

that was to be destroyed as part of the city of Münster's urban planning policy. The building, seen from the outside, seemed to be preserved, but, in the huge hall of the ice rink, the floors were cracked. In front of the stands, a moonscape of varying levels of soil stretched out: bits of concrete floor could be seen on the tops of mounds that rose up between the cracks. The mounds composed of several layers of stones and minerals revealed the various strata of the foundations. Here again, the geological apprehension was doubled by a temporal apprehension: the number of strata referred to the stages of past construction, as well as to the time it took to erode them. In the earthy bottom between the mounds, irregular hollows were drawn, as if dug by bad weather and landslides. At the bottom of some of the furrows, puddles could be seen.[12]

Various hatches in the ceiling allowed water and sunlight to pass through but only occasionally and, it seemed, in a random way. The more water and sunlight were let in, the more life developed in the puddles, and the more the ground shifted. Life also developed in an aquarium containing bacteria and marine animals, and in another container that held cancer cells. Although it could be observed directly, the life of these marine and aquatic organisms was also measured by sensors, designed with oncologists and biologists. As they walked through the entire hall, viewers were free to observe all the elements: in the hollows of the grounds, around the aquarium that was placed in the centre of the space, or along the windows and the old pool.

Even more than *Exote*, *After ALife Ahead* mixed together different temporalities. The past time of the construction was concretely present in the mounds; an even more ancient time emerged in the disturbed soil, reminiscent of that of archaeological sites. The excavated ground also referred concretely to the possible future of the city. In the future, Münster could be gripped by great heat, forcing the population to abandon the buildings and lifestyles inherited from a temperate Europe.[13] The installation thus extended more concretely than *Exote* towards the

---

www.skulptur-projekte-archiv.de/en-us/2017/projects/186/.

12 A short video shows the different parts of the installation: https://www.dailymotion.com/video/x5r7i6s.

13 Editors' note: This engagement with temporalities beyond the usual span of drama is comparable to the performances analysed by Shepherd-Barr and Simpson in chapter 12, which use extended temporalities to shift the focus of ecological theatre away from the human.

future and towards the past. It might even refer to a post-Anthropocene, a period when humans will no longer be *anthropoï*, or even when they will have disappeared. The ice rink without its heart appeared as an archive of the future, as ruins to come. The installation also extended on larger scales because it made us perceive micro-organisms that we can hardly see ordinarily. Finally, it allowed the audience to experience the interdependence between geological and biological evolutions.

Die Welt ohne Uns, a 'longterm planttheater' developed by Tobias Rausch, also constitutes a post-Anthropocene performance concept which wishes to apprehend the very long term. This expression refers to a cycle of performances and installations, conceived and entitled after Alan Weisman's 2007 book, *The World Without Us*.[14] Although the book was somewhat controversial,[15] the German director's work is based on a cooperation with numerous scientists on the basis of the climate situation in Germany in 2010.[16] Originally, this cooperation was to develop theatrical meetings every three months, with the aim of following the evolution of nature left to itself after the disappearance of humans. The theatre was geographically immersed in the botanical garden of the city and wanted to follow its vegetal rhythm. Only the first five episodes, however, conformed to this initial idea, over a year and a half.

---

14 The company Lunatiks Production conceived *Die Welt ohne Uns* with the support of the Hannover town theatre. It was originally planned to take place from 2010 to 2015. The project is presented on the website of the company: https://lunatiks.de/produktion/die-welt-ohne-uns/.

15 Alan Weisman's book is a very well researched speculation, grounded in numerous interviews with researchers. But the author does not give the readers any really plausible reasons as to why a planetary situation without humans should occur. Such a state would be an ecologically desirable one and suggests somehow that humans would do better by disappearing from Earth.

16 Rausch collaborated with the following scientists from the Leibniz-Universität Hannover: Johannes Böttger, Institut für Landschaftsarchitektur; Hansjörg Küster, Institute for Geobotanical Studies; Henning von Alten, Institut für Gartenbauliche Produktionsysteme; Jutta Papenbrock, Institut für Botanik; Rüdiger Prasse, Institut für Umweltplanun; Michael Rode, Institut für Umweltplanung; Norbert Schittek, Faktultät für Architektur und Landschaft; Wolfgang Spethmann, Institut für Gartenbauliche Produktionsysteme. He also collaborated with Angela Kallhof, a specialist in plant ethics from the Institute for Philosophy, Vienna; Andreas Ebhardt, from the Schulbiologiezentrum Hannover; Ralf Köneke, from Fachbereich Umwelt und Stadtgrün, Landeshauptstadt Hannover; and Kaspar Klaffke, working for the Deutsche Gesellschaft für Gartenkunst und Landschaftskultur.

Fig. 55 The garden party before the withdrawal of humans. *Die Welt ohne Uns*. Directed by Tobias Rausch. © Katrin Ribbe. All rights reserved.

At the first meeting, the spectators met various plants that were staged and assigned a written text. After this sort of festive garden party, the audience retired to a container, filled with seats and equipped with a glass side opening onto the garden. The container was set up as a spectator room detached from the stage, that is, the garden. The audience saw the garden from which human life had withdrawn, with the remains of its activities, and entered into a speculation that was not devoid of mourning. This meditation of places left to themselves after the departure of humans was orchestrated through a voice-over that formed a radio landscape piece (a *Hörspiel*) of sorts.

The fourth and fifth episodes took up this audiophonic formula by matching it with a visual installation embedded in the garden: 'fifteen years after the abandonment by humans', a text related the co-evolution of a corpse and its humus companions inhabiting the installation, and '80 years later', another text told the story of the buildings disembowelled by the vegetation, frequented by some wolves and numerous birds.[17]

The two intermediate episodes, corresponding to one and five years after the end of human life, were animated by actors in the garden.

---

17   The first installation was created by Katrin Riddle, the second one by Mirko Bortsch.

Fig. 56 The view from the container: a garden without humans. *Die Welt ohne Uns*. Directed by Tobias Rausch © Katrin Ribbe. All rights reserved.

Fig. 57 The view from the container: the garden abandoned. *Die Welt ohne Uns*. Directed by Tobias Rausch © Katrin Ribbe. All rights reserved.

The actors commented on the reproductive action of the plants, their sexuality, their needs for nutrition and desire for colonization. The growth of the plants and the evolution of the garden was also simulated, artificially produced in collaboration with scientists and gardeners. Compared to *Exote* and *After ALife Ahead*, *Die Welt Ohne Uns* is much more focused on biological rhythms, and therefore on the temporal scales and mechanisms of plant development. It also takes shape in an environment that is changing, and whose changes are a sign of global upheaval. Ultimately, it preserves more familiar theatrical components, as it is based on texts of various kinds, and even on acting.

The last project I will consider takes place inside a theatre. In *Beaming Sahara*,[18] the spectators are invited to come up on the stage plunged in darkness. There they remain for a long time, so as to change their state: indeed, a voice-over invites them to take into account their breathing, position, and imagination. Most of the show will leave the spectators in the dark, except for the moments when landscapes are projected on a screen situated in the middle of the room. The performance is then articulated into four parts, producing a succession of bio-geological encounters that stretch over time, with a certain slowness. First the spectators hear the sound of branches, then they see a lush forest on the screen. Plunged back into darkness, they see a block of soil slowly appear, very dimly lit. This block appears as the body of the forest, the material of which the humans only see the efflorescence and ramifications. The voice-over is both scientific and poetic, letting the forest organism speak in the first person. At the end of its presentation, it mentions the state of crisis in which it finds itself today, before the earth fades from view. We hear sand trickling in the darkness, before seeing on the screen an expanse of soft, hot-coloured desert. The spectators are then shown, in a very weak light, a block of sand, of similar size and shape as the block of earth. The sand's monologue unfolds, succeeded by the stone's and then by that of the ice.

The piece shows the crisis of the Anthropocene through the founding elements of life on Earth. Rather than the Anthropocene, the performance

---

18 *Beaming Sahara* was first presented by the collective EdgarundAllan on 21 June 2019 in Erlangen (Germany). It is the English version of *Milo, ich hab mich in die Sahara gebeamt*, which premiered in May 2018 in Hildesheim (Germany). See https://www.edgarundallan.com/beamingsahara.

is reminiscent of the Chthulucene as defined by Donna Haraway. In fact, it plunges us into a crisis without actors, without direction or defined end, a crisis in which one is immersed. The spectators may feel trapped in a becoming, 'a thick ongoing' present with entangled actants.[19] Forests, minerals, and ice are not perceived as normal actors, and constitute, in truth, hyperactors. Crucially, they are clearly presented as actants, and not objects, not even hyperobjects.[20] The spectators are thus no longer immersed in the crisis from the anthropocentric point of view, but from that of these intermingled actants. They are not only immersed in the interdependencies of the Anthropocene but placed in another posture, both literally and figuratively. They are at the same time inside, as in the other theatrical designs presented above, and outside, as in the oblivion of humanist man, listening to 'chthonic' beings.

In this perspective, of the four set-ups I have presented, this one effects the most radical decentering of the human.[21] The decentering it visibly induces is, at the same time, performative and ideal. The spatial immersion is, however, less intense, as the frame remains theatrical, making the geobiological scales perhaps less perceptible. I will now measure the potentialities of these different forms by analysing more precisely their immersive effects.

Forms of Displacement by Immersion

The most obvious dimension of these performances is immersion: an invitation to apprehend unknown or 'unseen'[22] forms of life, extended by performatively divesting the spectators of their intellectual mastery. More than imagination, what is at stake is thus an ideal sensual

---

19 Haraway, *Staying with the Trouble*, p. 19.
20 This notion has been coined by Timothy Morton and makes clear that humans cannot apprehend ecological issues like other objects because of the multiple elements and dynamics involved in a single ecological phenomenon. See Timothy Morton, *Hyperobjects: Philosophy and Ecology After the End of the World* (Minneapolis: University of Minnesota Press, 2013).
21 Editors' note: Beaufils's analysis of this challenge to anthropocentrism echoes Shepherd-Barr & Simpson assertion, in chapter 12, that '[d]ecentring, re-scaling, and enmeshing the human figure from and within the micro- and macroscopic scales of the climate crisis is the crucial work of twenty-first-century theatre and performance practice'.
22 Jean-Luc Marion, *The Crossing of the Visible* (Stanford: Stanford University Press, 2004).

apprehension. By this means, these set-ups seem to answer the wish of many thinkers of the Anthropocene who consider that we humans are going through a crisis of sensibility.[23] This crisis is coupled with physiocide,[24] resulting from a profound lack of knowledge, i.e. a lack of knowledge about the physical world and, in particular, the plant world. Both the crisis of sensibility and the resulting physiocide prevent us from responding adequately to the ecological catastrophe. Promoting a renewed sensitivity with some information would be one of the first steps towards developing an ability-to-respond.[25]

Exote and *After ALife Ahead* immerse the viewers entirely in environments that have their own mode of functioning. *After ALife Ahead* was also evolving beyond the control of humans, which increased the impression of entering an ecosystem. In both cases, moreover, the sensitivity towards specific entities (plants, animals, bacteria, puddles, cracks, light, or rain traps) goes hand-in-hand with an atmospheric co-experiencing: the spectators share a space, share the air with nonhumans. Such a sensation is fundamental for philosopher Emanuele Coccia, who defines the atmosphere as a sharing of breath amongst the living. Through the atmosphere, he writes, the living are linked together like organs of the same organism, linked by flows of matter and by their breathing. The atmospheric impression accentuates immersion and stimulates the amalgam of 'matter and sensibility', of sensation with the whole being.[26]

The perception of an atmosphere is also at the foundation of any artistic experience according to Gernot Böhme, because it permeates all the perceptions during this experience.[27] It is constitutive of aesthetic

---

23 Baptiste Morizot and Estelle Zhong Mengual, 'L'illisibilité du paysage. Enquête sur la crise écologique comme crise de la sensibilité', *Nouvelle revue d'esthétique*, 22.2 (2019), 87–96.

24 'Physiocide' means the murder of the vegetal world. See Iain Hamilton Grant, 'Everything Is Primal Germ or Nothing Is: The Deep Field Logic of Nature', *Symposium: Canadian Journal of Continental Philosophy*, 19.1 (2015), 106–24.

25 By 'ability-to-respond' I would like to underline the need of a response in *'response-ability'*, the term coined by Haraway in her book *When Species Meet* (2007), which she later deepened and enlarged in *Staying with the Trouble* (2016).

26 Emanuele Coccia, 'In Open Air: Ontology of the Atmosphere' (chapter 7), *The Life of Plants: A Metaphysics of Mixture* (Hoboken: Wiley, 2018), pp. 35–53.

27 Gernot Böhme, 'The atmosphere as the fundamental concept of a new aesthetic', *Journal of Aesthetics*, ed. by Griffero-Somaini, 33.3 (2006), Year XLVI, pp. 5–24, https://doi.org/10.1177/072551369303600107.

apprehension because our sensitive being-in-the-world is based on it. The foregrounding of atmosphere thus accentuates the spectators' impression of being exposed to the ecoworld, their affects and their comprehension of their permeability to ecosystems. Finally, an 'ambiance', an affective tone, infuses the atmosphere of the installations: that of threat, linked to the fragility of the living and especially of the vulnerable balance. It can go as far as mourning within the ruined buildings and lands of *After ALife Ahead* or the becoming rock or sand in *Beaming Sahara*. All these affects contribute to the impression that we humans are arriving at a threshold: we are living in a liminal situation. We are potentially at the beginning of something. The sensitive and performative emergence of the artistic situation holds a metaphorical significance. In the four performances, the liminality is combined with anxiety or mourning, so that the performances could even give rise to solastalgia, this specific feeling anticipating the death of life on our planet and the mourning of the world as we know it.

The sensory immersion is less likely to overwhelm the spectators in Tobias Rausch and EdgarundAllan's performative spaces, but it is no less permanent. The two immersive forms are each of different interest. In *Beaming Sahara*, concentrating on small pieces of matter in the middle of the darkness focuses the attention towards earth, sand, or ice. The sounds of branches, the wind, the trickling of sand occur in the darkness, between the spectators, without them being able to determine their origin or location. At the same time, smells of dead branches, wet ground, dusty stones, or snow are propagated. This show thus calls upon all kinds of stimuli, so that the spectators are truly immersed in a situation that they cannot anticipate. The director aims at producing 'an atmospheric synaesthesia' similar to what she herself experienced at the Klimahaus 8° Ost in Bremerhaven.[28]

Here, again, the performative dimension converges with an idea, namely the Latourian leitmotiv mentioned in the introduction, according to which humans must realize that their panoramic perspectives or views from above (above or outside nature-culture) are not tenable.

---

28 She wrote the story of the project in her master thesis, which she kindly allowed me to read. I thank her for sending me her thesis. The 'Klimahaus 8° Ost in Bremerhaven' is an experimental ecological museum, that enables visitors to plunge into ecosystems and to apprehend the significance of their changes.

On the contrary, every human is deeply included in the interplay of biological materials and dynamics. As the interplay far exceeds individual mastery, it may generate an impression of crisis. On stage, darkness can even produce the effect of a performative encirclement and accentuate this impression. But as Butler states, one is always already in crisis when one realizes it.[29]

Rausch's conceptualized performances, on the contrary, could lead one to believe that there is no immersion that endures, since the spectators of the 'longterm planttheater' are reduced to staying in a container from the end of the first episode. This container position in the middle of a botanical garden nevertheless promotes a paradoxical impression. On the one hand, nostalgia of contact with the living nonhuman beings arises, and feeds an anticipated mourning that is the other side of desire. On the other hand, the theatre appears as a constructed situation, highly artificial and inverted. It is a very improbable thought experiment, which eventually echoes science fiction scenarios where humans are unwillingly exiled from Earth. In a certain way, through the accentuated separation from the garden, the effective immersion with the nonhuman is more keenly felt: the spectators become aware of how neglected, or even forgotten, this immersion is in everyday life. This awareness draws attention towards the living beings we do not know, we do not even feel, despite the fact that they live with us. One could even point out a metaphorical significance: does the incapacity to remain in the garden not echo the Biblical experience of Adam and Eve expelled from Eden? Does the garden not appear as a sort of Paradise, at least in the first episodes?

In the four performances, the spectators are finally immersed in a setting: an artistic setting with a metaphorical dimension, which feeds the impression of topicality, even of urgency. The spectators are placed in a position to share the evolution of living or non-living others, to test their co-existence. If the impression is particularly accentuated in the installations where there is no human presence to disturb the sensitive sharing, the four performances nurture, through immersion, the sensation of a 'becoming-with' (Haraway). The becoming-with is even palpable in *After ALife Ahead* and in *Beaming Sahara*, because the

---

29   Judith Butler, 'What is Critique? An Essay on Foucault's Virtue', *Transversal* (May 2001), https://transversal.at/transversal/0806/butler/en.

spectators can touch the ground, the soil, the puddles; in *Beaming Sahara* they are invited to do so: to knead some soil, to stroke sand, to put their hand in the water of the melted ice.

Fig. 58 A piece of forest. *Beaming Sahara*. Performed and directed by EdgarundAllan collective. ©edgar&allan. All rights reserved.

Fig. 59 Watering a piece of forest. *Beaming Sahara*. Performed and directed by EdgarundAllan collective. ©Julia von der Maur. All rights reserved.

Fig. 60 Crossing deserts, past and future: spectators caressing sand in *Beaming Sahara*. Performed and directed by EdgarundAllan collective. ©edgar&allan. All rights reserved.

The impression of becoming-with is perhaps accompanied by a certain powerlessness. Spectators become aware of the vulnerability of the living or mineral elements. They may experience inside the theatre a critical zone, that is a 'double' (Artaud) of the critical zone defined by biologists, this thin ribbon of life surrounding Earth which is only a few kilometres thick. The critical zone in which humans are immersed reveals itself to be a doubly, triply critical one. It is strategic, threatened, and threatening. Could such a revelation be incapacitating?

The fact is that increasing our sensitive capacities towards living beings and our interdependencies is probably not enough to develop *response-ability* towards the ecological crisis. A deep and entangled ecology can indeed be dark in more than one sense, as Timothy Morton writes in his eponymous book: deep ecology is difficult to grasp and potentially threatening. The call to mourn in Huyghe's and Rausch's work may not be empowering either. Certainly, many thinkers believe that it is necessary to mourn, but the grieving process can scarcely be accomplished in the course of a performance. The paralysis of mourning is *a fortiori* unlikely to be overcome in such a limited time span.

Reading Signs

But it seems to me that here the very obscurity of the nonhuman living, inherent to the performative immersive situations, is able to stir up the ability to respond. By relying on sensitive and atmospheric affection, the performances call for an attention that is also a *vital curiosity*. Unless the spectators of the four performances want to stay out of the game of the elements, they feel the materials and the influences exerted on each other or on themselves. They wonder where life is in the installations or on the stages, where the movements of life are. If the earth is dimly lit in *Beaming Sahara*, it is also because it is difficult to illuminate with the light of understanding; it is teeming with a thousand lives and possibilities. Viewers may realize that they have to concentrate to see it. Such illegibility arouses the curiosity to know *how it works*, *how I can get to it*, and *what is my relation to it*. It is impossible to remain dazzled by the actions of nature, as we face them with wonder, as Martyn Evans defines it:

> The attitude of wonder is thus one of altered, compellingly-intensified attention to something that we immediately acknowledge as somehow important—something that might be unexpected, that in its fullest sense we certainly do not yet understand, and towards which we will likely want to turn our faculty of understanding; something whose initial appearance to us engages our imagination before our understanding; something at that moment larger and more significant than ourselves; something in the face of which we momentarily set aside our own concerns (and even our self-conscious awareness, in the most powerful instances).[30]

If those four performances trigger a 'sense of wonder', then it is one that stirs up an imagination oriented towards something important, and that impels a search for what matters. The spectators are on the lookout for a bubble of air, for a movement of life. A puddle seems to contain larvae and small algae. One floor is cracked, another is teeming with mini-organisms. These various signs are not linguistic, they constitute indices in the Piercian terminology.[31] These bio-theatres invite us to read signs,

---

30 Martyn Evans, 'Wonder and the Clinical Encounter', *Theoretical Medicine and Bioethics*, 33.2 (February 2012), 123–36 (p. 123), https://doi.org/10.1007/s11017-012-9214-4.
31 Charles Pierce distinguishes three kinds of signs: icons, which establish a relation of resemblance with the objects they replace; indices, which have a metonymic

calling the spectators to be researchers, photographers, or hunters. Although they are not completely intelligible, these signs are eloquent, they translate a situation, the presence of a being, or relations. They are both performative and meaningful. The reading of signs confirms Donna Haraway's hypothesis according to which there is a 'sensual communication'[32] with the nonhuman, as well as a 'material semiotics'.[33]

According to philosopher Jean-Luc Marion, who studies the dialectics of the unseen in art and daily life, 'each mode of phenomenality is [furthermore] constitutive of a world of meaning and, therefore, calls for certain intersubjective parallels'. As in paintings, 'the visible [in the performances] is liberated from vision at the moment when it seizes its own invisibility'.[34] The spectators may be the ones who 'liberate visibility' and who develop the power to communicate. They may let the desire for the objects arise from the objects themselves, especially as they are revealed not to be objects, but rather subjects.

Therefore, the encounters and movements within these performances seem to feed complex experiences and thoughts, which are simultaneously sensitive, intuitive, imaginative, and reflexive. The human who leads this sensitive, intuitive, and reflexive research shows response-ability, but a paradoxical ability. He or she views the living without purpose or assurance. His or her plunge into the darkness of the living is marked by a kind of active passivity, going hand in hand with forgetfulness of former worlds of meaning and perceiving. This plunge paves the way towards another thinking, a slight transformation that can be called an involution. It allows, perhaps, to veer even further away from human sense and thought habits insofar as, strictly speaking, it is not undertaken in the search for agency and knowledge. The spectators scrutinize the nonhuman actants, their interrelatedness, and then become aware of their interdependence with themselves, a generic or

---

relation with the replaced objects; and symbols, which are linked by convention or in an arbitrary way (linguistic, gestural, or visual) with the signified objects. See Charles Pierce, *Semiotics and Significs* (Bloomington, IN: Indiana University Press, 1977).

32 Donna Haraway, *Conference at the Evergreen College* (Olympia, WA: Evergreen State College Productions, 2016), https://www.youtube.com/watch?v=fWQ2JYFwJWU.
33 Donna Haraway, *Staying with the Trouble*, p. 21.
34 Jean-Luc Marion, *The Crossing of the Visible* (Stanford: Stanford University Press, 2004), p. 19.

particular human self. This is spontaneous and/or interested thinking, which is not narcissistic nor egotistical.

This attitude of thoughtful perception is promoted by the regular movement of the spectators. For it is not possible to stay in place in the three most immersive situations. In both installations *After ALife Ahead* and *Die Welt Ohne Uns* and during *Beaming Sahara*, one has to move in any case. Thought cannot remain in the assurance of what it thinks, because perception cannot.

Such a constant displacement of thinking corresponds to what philosophers Judith Butler and Jean Luc Nancy, albeit in very different frameworks, both call critical thinking: a thinking 'put in crisis',[35] resulting from a 'state of ontological suspension'.[36] This may promote a slight powerlessness in the spectators. My hypothesis is that this powerlessness overlaps with a form of power.

The Place of the Spectator

Immersion is paradoxically accompanied by the feeling of not quite knowing anymore *where one is*—a question which also inspires the latest work of Bruno Latour.[37] Asking *where life is, how it is going*, certainly echoes the need to develop one's own vital competence and to know where one is. But the performances do not really answer the question, which is left to the audience to resolve. The metaphorical place of humans in the designed environments would rather be disappointing if one looks for an answer: dead humans or non-anthropoï in Huyghe's and Rausch's work; scientists dressed in laboratory clothes in Verdonck's work, or travellers fond of discoveries and responsible for the dissemination of larvae and seeds; spectators plunged into the dark and without any hold on the performed activity in EdgarundAllan's work. Needless to say, these positions do not appear to be very satisfying if one asks how to be on Earth.

---

35 Jean-Luc Nancy, 'Critique, Crise, Cri', *Diakritik* (2016), https://www.fabula.org/actualites/critique-crise-cri-par-j-l-nancy_75144.php.
36 Butler, 'What is Critique'.
37 Bruno Latour, *Où suis-je ? Leçons du confinement à l'usage des terrestres* (Paris: La Découverte, 2021).

The four set-ups thus question the place of the spectators/humans. The need to move performatively, to read, or to understand combines with the need to speculatively question one's position as a spectator and as a human. As the spectator cannot completely 'access' the surrounding biological or abiotic entities, rifts remain, all the stronger as proximity develops. We can make the hypothesis, nevertheless, that these divisions favour speculation. Indeed, it is because the spectators cannot project themselves directly into the behaviour of the plants in *Die Welt Ohne Uns* that they feel the inadequacy of the anthropocentric terms used by the actors (sexuality, attraction, colonization). They then weave together another representation of plant behaviours—without words or 'between words'. They feel empathy for the basilisk placed in the microwave (Rausch), and possibly for the invasive species involuntarily responsible for the imbalances (Verdonck), but it is an empathy that is at the same time sensitive, distanced, and failing, promoted by the representations they have at their disposal.

In the same way, the multiplication of perspectives by the means of texts or of the actors and images, stirs up reflection. This is notably the case in the installations accompanied by acousmatic voices,[38] in some episodes of *Die Welt Ohne Uns* and *Beaming Sahara*. Moreover, the text of *Beaming Sahara* is not limpid. It is not micro-organisms that speak, which would associate them with subjects and thus make the theatre a well-known ventriloquist practice. Through a small block of soil, we discover that it is the forest that speaks. But the forest is neither visible nor completely imaginable as such:

Moisture and warmth, where life is baked

What appears to you as the accumulation of thousands of interlocked beings, falling, crawling and creeping, is in reality one organism [...]

the power that flows through me is viscous and dark [...]

it pushes itself reliably into each of my tips, I am full of energy up to the top, it drips from the tips of the leaves, lines up ring around ring and piles up into trunks and columns of warmth and resistance. Further up, it networks, branches out into ever denser structures that pulsate,

---

38 By acousmatic voices I mean voices without bodies and without any possible localisation.

continue to grow and emerge. Every centimetre is filled with incessant movement. [...]

I can endure many things, I always start to sprout anew;

lately however it becomes lonely in me.

My feelings contract, everything that is dear to me turns inward,

what is happening on my surface? The lines of communication are broken, I no longer function reliably.

My juices flow viscously.[39]

The acousmatic texts of these installations are both informative ('What appears to you is [...] one organism') and poetic ('I start to sprout anew', 'it becomes lonely in me'). They give signs and make the other-than-human speak. But the very language of the 'forest' is displayed as artifice, replacing a nonhuman non-language. It raises the question of how to hear the forest, and how to read it: plunging us into the necessity and the difficulty of what Haraway calls a 'sensual communication' and 'material semiotics'. Many texts in Rausch's and EdgarundAllan's performances thus challenge the spectators by delivering to them the missing signs and languages, while asking the question of how to make 'nature' speak. How can we hear and interact with it? This *mise en abyme* mixes the power of poetic language and the impotence of the spectators and humans. The very notion of a 'subject' becomes again a complex question.

---

39   'Feuchtigkeit und Wärme, darin wird Leben gebacken, was dir wie die Ansammlung tausender ineinander verschachtelter Wesen, neben und übereinander fallender wuchender krabbelnder und kriechender wesen erscheint ist in wirklichkeit ein organismus./die kraft die durch mich fliesst ist zähflüssig und dunkel [...]/ zuverlässig schiebt sie sich in jede meiner spitzen, ich bin bis oben voll von energie, sie träufelt von den Blätterspitzen, reiht sich ring um ring umeinander und türmt sich zu Stämmen und Säulen aus Wärme und Widerstandskraft. Weiter oben vernetzt, verästelt sie sich zu immer dichteren Strukturen, die pulsieren, weiter wachsen und entstehen./Jeder Zentimeter ist aufgefüllt mir unaufhörlicher Bewegung.[...]/Ich kann vieles über mich ergehen lassen, immer fange ich von neuem zu spriessen an/ In letzter Zeit jedoch wird es einsam in mir/Meine Fühler ziehen sich zusammen, alles was mir teuer ist wendet sich nach innen,/was geschieht an meiner Oberfläche? die Kommunikationswege sind zerrissen, ich funktioniere nicht mehr zuverlässig/ zähflüssig fliessen meine Säfte' (minutes 15'41 to 19'56). The text was translated for the English presentation in Hildesheim (Germany) in May 2018.

The hyperactors who speak are, in truth, ecological subjects, as Stéphanie Posthumus defines them: 'the ecological subject [...] is constructed as a set of relationships and interactions rather than as an individual and isolated entity'.[40] They speak as ecological hyper-characters and address the audience as ecological subjects. The lack of power of the spectators is thus coupled with a lack of adequate subjectification, and a lack of language. The foreign languages of the earth or the stones are only sketched out.

A Diplomatic Theatre

The theatre could thus be what Baptiste Morizot calls 'diplomatic'. According to Morizot, the relationship of humans with animals and plants would rapidly improve through increased, desired, and fertile interdependencies, if there were more diplomatic humans. These diplomats would be Janus-faced people, one face turned toward nonhumans, one face turned toward humans and their institutions.[41] The theatre presented here often has such a double face, on the one hand, a staging turned towards the nonhuman, on the other, an address to the human. Its poetic language, shot through with unseen connections and *rapprochements*, is also endowed with a two-faced dimension: human imagination and signs meet nonhuman representations. Like the diplomat, these bio-theatres furthermore immerse us in interdependencies, they 'activate the creation of a new configuration of desire' and create 'communities of importance': communities that matter.[42] Through sensitive encounters, reading signs, and listening to ecopoetic texts, they increase our understanding of nonhumans in ways that are not solely cognitive. The spectators must abandon the point of view of the observer and the experimenter in the laboratory. Moreover, the photographs taken and projected by the members of the EdgarundAllan collective are often the result of trips to neighbouring

---

40   Stéphanie Posthumus, 'Écocritique et *ecocriticism*. Repenser le personnage écologique', in *La pensée écologique et l'espace imaginaire*, ed. by Sylvain David and Mirella Vadean (Montreal: Université du Québec à Montréal, 2014) pp. 15–33 (p. 15).

41   Baptiste Morizot, *Manières d'être vivant: Enquêtes sur la vie à travers nous* (Paris: Actes Sud, 2020), pp. 254–56.

42   Ibid., p. 256.

territories (the mountains of the Harz, for instance) and illustrate the encounter that is the basis of interdependence.

Finally, the poetic monologues by Rausch and EdgarundAllan open up even more widely the spectrum of possibilities. Indeed, in their works, each entity or actant is the fruit of interrelations, it is in inter- and intra-action with an infinity of others, when it is not itself, like the forest, a gigantic compound of interactions and organisms. The focus on interrelations contributes to making the world, it has a cosmogonic action.[43] Therefore, it seems that each poetic monologue could be continued and complexified by multiplying the relational perspectives that form the basis of any identity. Is it necessary to use language to invent other relations and to continue to change the world in a more conscious way? In any case, the poem makes it possible, it is a creative and stimulating diplomat.

This does not prevent the participants, human and nonhuman, from being anchored in the crisis. From this point of view, the works of Rausch and of EdgarundAllan are emblematic of the Chthulucene. They do not only show the Anthropocene, but constitute the first responses with knowledge of the crisis: this is precisely how Donna Haraway conceives the Chthulucene. It is the era that reacts to the Anthropocene, a difficult era, based on the interweaving of actions and lives that must be taken into account and developed in order to survive: for example, to take action for soil management, to keep it humid and teeming with life, paves the way for forests and prevents the expansion of sand or rock deserts. From this point of view, it is not surprising that the two performances that take the form of installations, accompanied by acousmatic voices, focus on soils and forests. The poetic monologues thus promote an empathy with 'chthonic' organisms, as essential as they are obscure. Plunged into the dark or into the container, humans are performatively associated with a dark, almost buried state, away from the light and glitter of epic human heroes. They are invited to act as chthonic creatures. This goes beyond the promotion of biological imagination, biological narrative, or cognitive explanation. Such forms of thought would contribute to a distant attitude and would run the risk of reinforcing human assurance of representation. The spectators here cannot stage internally what they

---

[43] Coccia, *The Life of Plants*, notably in chapter 1, 'On Plants, or the Origin of Our World'.

think, they are plunged into and jostled by the other-than-human, they must test it, cross it, and untangle the threads.

The four performances discussed here enable a new biological, atmospheric, and relational apprehension of the living. They not only immerse the spectators in complex ecosystems where many actants intermingle, they behave similarly to the Anthropocene by 'compressing space' and 'accelerating time'.[44] Moreover, they invite us to read unseen signs and to communicate with matter, sometimes by touching it. They even lead to a kind of thinking intertwined with its objects: a metaphysical interdependence, which accepts the darkness as a condition of thought. It is thus in multiple ways that bio-theatres can not only be responses to catastrophe, but also stimulate responses to it. The spectators' thoughts can then extend to an ontological, cosmogonic, and perhaps diplomatic pragmatics.

Works Cited

Böhme, Gernot, 'The atmosphere as the fundamental concept of a new aesthetic', in *Thesis Eleven*, 36 (January, 1993), 113–26, https://doi.org/10.1177/0725513693036000107

Butler, Judith, 'What is Critique? An Essay on Foucault's Virtue', *Transversal* (May 2001), https://transversal.at/transversal/0806/butler/en

Coccia, Emanuele, *The Life of Plants: A Metaphysics of Mixture* (Hoboken: Wiley, 2018).

Dahl, Darren E., 'Review of *The Crossing of The Visible*', *Journal of French and Francophone Philosophy*, 14.1 (January 2011), 110–15.

Evans, Martyn, 'Wonder and the Clinical Encounter', *Theoretical Medicine and Bioethics*, 33.2 (February 2012), 123–36, https://doi.org/10.1007/s11017-012-9214-4

Haraway, Donna, *Staying with the Trouble: Making Kin in the Chthulucene* (Durham: Duke University Press, 2016), https://doi.org/10.2307/j.ctv11cw25q

Haraway, Donna, *Conference at the Evergreen College* (Olympia, WA: Evergreen State College Productions, 2016), https://www.youtube.com/watch?v=fWQ2JYFwJWU

---

44   Eduardo Viveiros de Castro and Deborah Danowski, 'L'Arrêt de monde', in *De l'Univers clos au monde infini*, ed. by Emilie Hâche (Paris: Dehors, 2015), pp. 221–339 (p. 228). This text has been modified and published in English under the title *The Ends of the World* (Hoboken: Wiley, 2016).

Grant, Iain Hamilton, 'Everything is Primal Germ or Nothing Is: The Deep Field Logic of Nature', *Symposium: Canadian Journal of Continental Philosophy*, 19.1 (2015), 106–24

Latour, Bruno, 'Why Has Critique Run Out of Steam', *Critical Inquiry*, 30 (Winter 2004), 225–48, http://www.bruno-latour.fr/sites/default/files/89-CRITICAL-INQUIRY- GB.pdf.

Latour, Bruno, 'L'Anthropocène et la destruction de l'image du Globe', in *De l'Univers clos au monde infini*, ed. by Emilie Hâche (Paris: Dehors, 2015), pp. 29–56.

Lovelock, James and Lynn Margulis, 'Biological Modulation of the Earth's Atmosphere', *Icarus*, 21.4 (April 1974), 471–89.

Marion, Jean-Luc, *The Crossing of the Visible* (Stanford: Stanford University Press, 2004).

Morizot, Baptiste and Estelle Zhong Mengual, 'L'illisibilité du paysage. Enquête sur la crise écologique comme crise de la sensibilité', *Nouvelle revue d'esthétique*, 22.2 (2019), 87–96.

Morizot, Baptiste, *Manières d'être vivant: Enquêtes sur la vie à travers nous* (Paris: Actes Sud, 2020).

Morton, Timothy, *Hyperobjects: Philosophy and Ecology After the End of the World* (Minneapolis: University of Minnesota Press, 2013).

Nancy, Jean-Luc, 'Critique, Crise, Cri', *Diakritik* (2016), https://www.fabula.org/actualites/critique-crise-cri-par-j-l-nancy_75144.php

Posthumus, Stéphanie, 'Écocritique et ecocriticism. Repenser le personnage écologique', in *La pensée écologique et l'espace imaginaire*, ed. by Sylvain David et Mirella Vadean (Montreal: Université du Québec à Montréal, 2014), pp. 15–33.

Viveiros de Castro, Eduardo and Deborah Danowski, 'L'Arrêt de monde', in *De l'Univers clos au monde infini*, ed. by Emilie Hâche (Paris: Dehors, 2015), pp. 221–339.

List of Illustrations

Fig. 1	David B., *L'Ascension du Haut Mal* (1999) © David B. and L'Association. All rights reserved.	148
Fig. 2	Matteo Farinella and Hana Roš, *Neurocomic* (2014) © Matteo Farinella and Hana Roš. CC BY-NC-ND 4.0	150
Fig. 3	Matteo Farinella and Hana Roš, *Neurocomic* (2014) © Matteo Farinella and Hana Roš. CC BY-NC-ND 4.0	151
Fig. 4	David B., *L'Ascension du Haut Mal* (1999) © David B. and L'Association. All rights reserved.	155
Fig. 5	Matteo Farinella and Hana Roš, *Neurocomic* (2014) © Matteo Farinella and Hana Roš. CC BY-NC-ND 4.0	157
Fig. 6	Matteo Farinella and Hana Roš, *Neurocomic* (2014) © Matteo Farinella and Hana Roš. CC BY-NC-ND 4.0	165
Fig. 7	David B., *L'Ascension du Haut Mal* (1999) © David B. and L'Association. All rights reserved.	167
Fig. 8	David B., *L'Ascension du Haut Mal* (1999) © David B. and L'Association. All rights reserved.	167
Fig. 9	David B., *L'Ascension du Haut Mal* (1999) © David B. and L'Association. All rights reserved.	168
Fig. 10	David B., *L'Ascension du Haut Mal* (1999) © David B. and L'Association. All rights reserved.	168
Fig. 11	David B., *L'Ascension du Haut Mal* (1999) © David B. and L'Association. All rights reserved.	170
Fig. 12	David B., *L'Ascension du Haut Mal* (1999) © David B. and L'Association. All rights reserved.	171
Fig. 13	David B., *L'Ascension du Haut Mal* (1999) © David B. and L'Association. All rights reserved.	171
Fig. 14	David B., *L'Ascension du Haut Mal* (1999) © David B. and L'Association. All rights reserved. .	173
Fig. 15	Matteo Farinella and Hana Roš, *Neurocomic* (2014) © Matteo Farinella and Hana Roš. CC BY-NC-ND 4.0	176
Fig. 16	Sourav Chatterjee, *49th St. between 7th and 8th Ave* (2020) © Sourav Chatterjee. Editorial use.	184

Fig. 17	Reena Rupani, *Times Square* (2020) © Reena Rupani. CC BY-NC.	184
Fig. 18	NYC Health Department, *Coronavirus death rate by zip code* (2020) © Courtesy of NYC Health Department.	186
Fig. 19	George Hirose, *Protesters against Anti-Asian Hate*, Union Square Park, March 21 (2021) © George Hirose. All rights reserved.	194
Fig. 20	Philip Timms Studio, *Boarded-up businesses after race riots in Chinatown at northwest corner of Carrall Street at Pender* (1907) © Courtesy of Philip Timms Studio/Vancouver Public Library/ VPL 940.	195
Fig. 21	Jeff Lemire, *Sweet Tooth* #26 (2011) © Jeff Lemire and Vertigo Comics. All rights reserved.	214
Fig. 22	Jeff Lemire, *Sweet Tooth* #36 (2012) © Jeff Lemire and Vertigo Comics. All rights reserved.	218
Fig. 23	Jeff Lemire, *Sweet Tooth* #1 (2009) © Jeff Lemire and Vertigo Comics. All rights reserved.	223
Fig. 24	Jeff Lemire, *Sweet Tooth* #20 (2011) © Jeff Lemire and Vertigo Comics. All rights reserved.	224
Fig. 25	Jeff Lemire, *Sweet Tooth* #26 (2011) © Jeff Lemire and Vertigo Comics. All rights reserved.	225
Fig. 26	Having Kids, The Benefits of Smaller Families (2018) © Fair Start Movement. All rights reserved. https://havingkids.org/wp-content/uploads/2018/06/New-2100-graphic-rearranged.jpg	230
Fig. 27	Having Kids, *The Consequences of Poor Family Planning* (2018) © Fair Start Movement. All rights reserved. https://havingkids.org/wp-content/uploads/2018/07/New-2100-graphic-part-2-1.jpg	231
Fig. 28	Having Kids, Having one fewer child will save 58.6 tonnes of CO2-equivalent per year (2018) © Fair Start Movement. All rights reserved. https://havingkids.org/wp-content/uploads/2018/12/child-impact-guardian-graphic-1-358x400.jpg	251
Fig. 29	Having Kids, *The carbon legacy of just one child* (2018) © Fair Start Movement. All rights reserved. https://havingkids.org/wp-content/uploads/2018/10/27972800_1575734402545365_7111663051996650685_n-400x400.png	252
Fig. 30	Global Footprint Network, *How Many Earths?* (2019) © Global Footprint Network. All rights reserved. https://www.footprintnetwork.org/content/uploads/2019/01/Infogrpahic-Pub-Data-Circle-v3.png	253

Fig. 31	Population Matters, *Having a Smaller Family* (2021) © Shutterstock. Free to use. https://populationmatters.org/having-smaller-family	254
Fig. 32	Alberto Casetta, *Family Walking in the Rain* (2021) © Alberto Casetta. Free to use under Unsplash licence. https://populationmatters.org/sites/default/files/styles/full_width_image/public/alberto-casetta-349138-unsplash_0.jpg?itok=UzNpoVfv	254
Fig. 33	Uncrowded, *Homepage* (2021) © Fair Start Movement. All rights reserved. https://fairstartmovement.org/uncrowded-home/	255
Fig. 34	'Flights of a Bumblebee', from *Where The Animals Go* by James Cheshire and Oliver Uberti. Illustration Copyright © James Cheshire and Oliver Uberti, 2016, published by Particular Books, 2016, in North America by W.W. Norton & Co Inc, 2016, Penguin Press, 2018. Reprinted by permission of Penguin Books Limited. All rights reserved.	277
Fig. 35	'Untitled', p. 53, from *The Pillar*, a series of photographs by Stephen Gill, with words by Karl Ove Knausgård, published by Nobody Books (nobodybooks.com). *The Pillar*, 2015–2019 © Stephen Gill. All rights reserved. Any further use or distribution of the photographs is strictly prohibited. Please contact The Wylie Agency (UK) Ltd to request permission for any further use or distribution of the photographs.	279
Fig. 36	Ben De Bruyn, *Conceptual interaction network of the food web in Reservoir 13* (2021) © Ben De Bruyn. CC BY-NC.	289
Fig. 37	'Untitled', p. 113, from *The Pillar*, a series of photographs by Stephen Gill, with words by Karl Ove Knausgård, published by Nobody Books (nobodybooks.com). *The Pillar*, 2015–2019 © Stephen Gill. All rights reserved. Any further use or distribution of the photographs is strictly prohibited. Please contact The Wylie Agency (UK) Ltd to request permission for any further use or distribution of the photographs.	290
Fig. 38	Frank Stockton, 'What's a fish?', from Mark Kurlansky, *World Without Fish* illustrated by Frank Stockton, p. 142 © Frank Stockton. All rights reserved.	302
Fig. 39	Li-Vollmer and Mahato, from *Climate Changes Health*, written by Meredith Li-Vollmer and artwork by Mita Mahato, published by Public Health—Seattle & King County (2019) © Li-Vollmer and Mahato. All rights reserved.	305

Fig. 40	Image reproduced from the chapter 'Climate Change' from *Science Tales: Lies, Hoaxes and Scams* by Darryl Cunningham © Myriad Editions, UK, 2019. www.myriadeditions.com. All rights reserved.	307
Fig. 41	From *Climate Changed* by Philippe Squarzoni. *Saison Brune* by Philippe Squarzoni © Éditions Delcourt. 2012 Text and illustrations by Philippe Squarzoni. Translation by Ivanka Hahnenberger. English translation copyright © 2014 Harry N. Abrams, Inc. All rights reserved.	311
Fig. 42	Lauren Redniss, *Thunder & Lightning: Weather Past, Present, Future* (Random House, 2015), pp. 146–47 © Lauren Redniss. All rights reserved.	315
Fig. 43	From *A Fire Story* by Brian Fies. Text copyright © 2019. 2020 Brian Fies. Used by permission of Abrams ComicArts®, an imprint of ABRAMS, New York. All rights reserved.	320
Fig. 44	From *A Fire Story* by Brian Fies. Text copyright © 2019. 2020 Brian Fies. Used with permission of Abrams ComicArts®, an imprint of ABRAMS, New York. All rights reserved.	321
Fig. 45	Matthew MacKenzie, *Bears* (2018) © Alexis McKeown and Punctuate! Theatre. Environmental design by T. Erin Gruber. Pictured: Sheldon Elter. Chorus: Lara Ebata, Gianna Vacirca, Skye Demas, Alida Kendell, Zoe Glassman, Kendra Shorter, Rebecca Sadowski. Photograph by Alexis McKeown. All rights reserved.	327
Fig. 46	Miriam Buether's auditorium set for Mike Bartlett's *Earthquakes in London* at the Cottesloe, National Theatre, London (2010). Photograph by Manuel Harlan © Manuel Harlan. All rights reserved.	337
Fig. 47	Mark Rylance on the Royal Court stage set of *Jerusalem* (2009). Photograph by Tristram Kenton. © Tristram Kenton. All rights reserved.	338
Fig. 48	Extinction Rebellion 'penitent' performer in Cornwall, protesting the G7 Summit (June 2021). Photograph by Greg Martin © Greg Martin / Cornwall Live. All rights reserved.	339
Fig. 49	Red Rebel Brigade in Cornwall, protesting the G7 Summit (June 2021). Photograph by Joao Daniel Pereira © Joao Daniel Pereira / Extinction Rebellion. All rights reserved.	340
Fig. 50	Red Rebel Brigade protesting in London (2019). Photograph by Connor Newson © Connor Newson/Adams Creative Media. All rights reserved (no financial gain).	340

Fig. 51	Deke Weaver, *ELEPHANT* (Stock Pavilion, University of Illinois, 2010). Photograph by Valerie Oliviero © Deke Weaver (artist) and Valerie Oliviero (photographer). All rights reserved.	344
Fig. 52	Deke Weaver, *BEAR* (Meadowbrook Park, Urbana, Illinois, 2016). Photograph by Nathan Keay © Deke Weaver (artist) and Nathan Keay (photographer). All rights reserved.	345
Fig. 53	*EXOTE* © A Two Dogs Company. Photography © Kristof Vrancken. All rights reserved.	356
Fig. 54	*EXOTE* © A Two Dogs Company. Photography © Kristof Vrancken. All rights reserved.	357
Fig. 55	The garden party before the withdrawal of humans. *Die Welt ohne Uns*. Directed by Tobias Rausch. © Katrin Ribbe. All rights reserved.	360
Fig. 56	The view from the container: a garden without humans. *Die Welt ohne Uns*. Directed by Tobias Rausch © Katrin Ribbe. All rights reserved.	361
Fig. 57	The view from the container: the garden abandoned. *Die Welt ohne Uns*. Directed by Tobias Rausch © Katrin Ribbe. All rights reserved.	361
Fig. 58	A piece of forest. *Beaming Sahara*. Performed and directed by EdgarundAllan collective. ©edgar&allan. All rights reserved.	367
Fig. 59	Watering a piece of forest. *Beaming Sahara*. Performed and directed by EdgarundAllan collective. ©Julia von der Maur. All rights reserved.	367
Fig. 60	Crossing deserts, past and future: spectators caressing sand in *Beaming Sahara*. Performed and directed by EdgarundAllan collective. ©edgar&allan. All rights reserved.	368

Index

abjection, abject 134, 142
actant 263, 353–354, 363, 370, 375–376
actor 32, 60, 235, 243, 302–303, 325, 330, 345, 347, 353–355, 360, 362–363, 372
affect, affective 4, 14, 23, 32, 43, 46, 50–51, 53, 57–58, 65, 72, 115, 148–149, 161, 169, 174, 196–198, 232–234, 238, 240, 245, 302, 330, 341–342, 347, 365
agency, agent, agentive 11, 17, 20–21, 24–25, 27, 47, 50, 60, 62–63, 72, 74, 78–79, 97, 104, 115–116, 155, 193, 215–216, 219–221, 278, 303, 325, 327, 333, 370
AIDS 189–190
Aït-Touati, Frédérique 9, 12, 331
  Inside 9, 331–332, 355
  Moving Earths 331
Albrecht, Glenn 14
Aldiss, Brian 29, 185, 234, 236
allegory, allegorical 9, 21, 106, 149
anagnorisis 189
analogy, analogies 6–8, 49, 56, 58, 102, 105–107, 175, 189, 275, 283, 323, 325, 354
animal 25, 28–29, 31–32, 46, 55–56, 58, 72, 78, 96–98, 108–109, 111–114, 132–133, 183, 203–205, 213, 217, 221, 223, 262–264, 266, 271–275, 278–280, 282, 287, 290–291, 304, 306, 312, 323–324, 327, 337, 342–347, 355, 358, 364, 374
Anthropocene 2–3, 16–19, 22–23, 32, 60–61, 63–64, 70, 74–75, 77, 79, 102, 107, 205, 212, 220–222, 261, 263, 266, 268, 291–294, 330, 346, 355, 357, 359, 362–364, 375–376
anthropocentrism, anthropocentric 3, 20, 29, 43, 261, 263, 271–272, 325–326, 346, 348, 354, 363, 372
anthropomorphism, anthropomorphic 13, 272, 326
apocalyptic 3, 14, 29, 186, 192, 203, 221, 222, 232, 234, 235, 236, 243, 244, 326, 334. *See also* post-apocalyptic
archive 43, 63–64, 82, 188, 343, 345, 359
Aristotle 108, 189, 210
Ashman, Howard 31, 327
assemblage 72, 77–78, 90, 107, 185, 210–211, 268, 288–289, 291
astronomical 83, 85
Atwood, Margaret 236, 243, 248, 256
  MaddAddam trilogy 236, 243–248, 255
autobiographical 26, 69, 89, 308. *See also* biography, biographical

bacteria, bacterium 20, 23, 62, 85, 110, 288, 353, 357–358, 364
Bakhtin, Mikhail 24, 55
Barad, Karen 28, 210–212
Bartlett, Mike 336–337
  Earthquakes in London 336–337
B., David 26, 147–148, 155–156, 166–171, 173–175
  Epileptic 26–27, 147–150, 155–156, 166, 166–175, 169, 174–175
Bean, Richard 329
  The Heretic 329, 333
Bears

386 *Index*

Punctuate! Theatre 327
Beckett, Samuel 2, 31, 137, 328, 356
  Happy Days 328
Beer, Gillian 1, 3, 5
  Darwin's Plots 1
Bennett, Jane 70, 72, 78
Bernard Shaw, George
  Back to Methuselah: A Microbiological Pentateuch 329
biocapital, biovalue 18
biochemistry, biochemical 18, 23, 41, 51, 61, 63–65, 106
biodiversity 3, 13, 22, 30, 262, 264, 266, 268, 275, 291, 293, 344
biography, biographical 159, 308. *See also* autobiographical
biology, biological 1, 2, 3, 4, 5, 8, 10, 11, 14, 15, 16, 18, 19, 20, 23, 24, 25, 30, 31, 32, 41, 42, 43, 44, 48, 53, 63, 69, 71, 78, 80, 81, 82, 83, 84, 85, 89, 90, 93, 94, 95, 97, 98, 101, 103, 104, 106, 107, 112, 113, 125, 142, 161, 185, 187, 188, 189, 204, 205, 206, 233, 239, 240, 241, 261, 265, 266, 268, 282, 283, 286, 288, 300, 302, 304, 326, 336, 353, 354, 356, 359, 362, 366, 372, 375, 376. *See also* microbiology
  animal 304
  cellular 4, 17–18, 24
  developmental 23–24, 80–82, 84–85
  evolutionary 1
  imaginary 20, 23–24
  molecular 18, 42
  popular 4, 23, 285
  strange 10
biology, popular 4, 23
biomedical 5, 10, 15, 17, 18, 129, 149. *See also* medicine, medical
bionarratology 21, 261
biopolitical, biopolitics 2–3, 5, 10–12, 17, 19, 27, 97, 100, 108, 185, 190–191, 239
biopower 3, 10, 12, 190–191
biotechnology 12, 17–18, 48
body, bodily 5, 18–19, 24, 26, 44–45, 49–51, 53, 55–59, 61–65, 69, 71–74, 77, 79, 84, 86–87, 90, 100, 111, 123–124, 126–128, 130, 132–138, 140–143, 149, 152, 154, 166, 171, 190, 198, 200, 204, 206–207, 209–210, 215, 222, 248, 269, 307, 328–329, 331–333, 335, 338, 345, 353, 362
Boey, Bec 338
botany, botanical 32, 113, 359, 366
Bovell, Andrew 334–335
  When the Rain Stops Falling 334–335
brain 4, 7, 14–15, 26–27, 82, 84, 87, 124, 126–127, 129, 141, 147–156, 158–164, 166, 168–170, 174–175, 177–179
  brain imaging 7, 14, 26, 158–160, 162–163
Buffini, Moira 329
  Greenland 329
Burke, Edmund 23, 44–48, 50–51, 53, 62
Busch, Akiko 30, 290
Butler, Judith 103, 196, 241, 366, 371
Butterworth, Jez
  Jerusalem 337–338

cancer 6, 23, 125, 142, 358
capitalism, capitalist 11–12, 16, 20, 28, 61, 64, 90, 162, 187, 198–199, 204, 206, 213, 220, 264
care 126, 140, 190, 209, 236, 311
Carson, Rachel 13
catastrophe 5, 27, 29, 185–187, 190, 232–237, 240, 242, 245–250, 254–255, 264, 330–331, 333, 335, 344, 364, 376
categorization 25, 26, 99, 100, 101, 103, 108, 114, 141, 142, 143, 163, 169, 206. *See also* classification
cellular biology 4, 17–18, 24
Chakrabarty, Dipesh 21
chaos, chaotic 101, 125, 127, 135, 143, 175, 221, 343
character 13, 26, 28, 30, 43, 51–52, 56, 104–105, 123, 126, 132–133, 135–136, 138, 140, 142, 148–152, 155–156, 160, 176–177, 189, 192, 197, 215–216, 219–220, 235–237, 243, 248, 256, 262–263, 266, 269–272, 280–281, 283, 287–288, 293, 301, 306, 325–326, 329–330, 333–334, 341, 374

charismatic megafauna 13, 30, 95, 101, 301, 324
Charman, Matt 329
  Greenland 329
Chaudhuri, Una 323, 343–345, 347
Chekhov, Anton
  The Cherry Orchard 348
chemicals 14, 49, 61–64, 73, 84, 104, 133, 142, 148, 154, 238
Cheng, Anne 195–196
Cheshire, James 30, 275, 277
  Where the Animals Go 30, 275, 275–277, 280, 287
chronic illness 125
Chthulucene 11–12, 346–347, 355, 363, 375
Citton, Yves 23
Clarke, Gillian 3, 9, 21, 24, 69, 69–90, 70–82, 84–90, 263
  A Recipe for Water 69, 73, 73–75, 79, 79–80, 87–89, 89
  Ice 69, 72–73, 76, 79, 81, 86–87, 89
  Making the Beds for the Dead 69, 71–72, 78–79, 84–85, 87, 89
  Zoology 69, 79–80, 82, 84–87, 89–90
Clark, Timothy 17, 21–22, 43, 70, 77, 188, 242, 261
classification 46, 113, 153. *See also* categorization
climate, climatology 3–4, 12–15, 17–19, 23–24, 29–31, 43, 58, 60, 70, 78, 87–89, 96, 104, 107, 139, 221, 255, 262, 266, 268, 291, 293–294, 299–304, 306–310, 312–313, 316–319, 323–326, 328–331, 334, 336–337, 339, 346, 349, 353, 359, 363
Coccia, Emanuele 12, 364, 375
colonial 10, 12, 27, 114, 189–190, 192, 199, 213, 217, 264
comics 3, 28, 30, 31, 152, 153, 154, 155, 156, 157, 158, 160, 166, 169, 174, 175, 203, 204, 207, 213, 215, 217, 222, 223, 225, 300, 301, 302, 303, 304, 306, 308, 309, 311, 312, 313, 317, 318, 319. *See also* graphic novel, graphic narrative

community 27–28, 149, 204, 215, 223, 229, 235–236, 243, 245–247, 250, 263, 266, 304, 317, 319
consciousness 5, 7, 20, 26, 71, 147, 151, 154, 162, 166, 177, 324, 328, 349
conservation 13, 240, 280
contagion, contagious 17, 43, 198, 200. *See also* infection, infectious
cosy catastrophe 29, 185, 234, 236
COVID-19 14, 43, 185, 187, 189–191, 234, 323, 325, 348
Crichton, Michael 56, 60
  Micro 56, 60–61
crime fiction 263, 271
criticism (literary) 25, 135, 153. *See also* ecocriticism
culture 1, 3–4, 9, 14, 18–19, 25, 42, 45, 50–51, 53–54, 65, 70, 74–75, 88, 93, 95, 97, 104, 112, 114, 139–140, 152–153, 155, 161–162, 169, 177, 197, 200, 213, 216, 242, 246, 256, 262, 264, 267, 278, 308, 326, 344, 347, 365
  digital 104
  popular 140
  visual 95
Cunningham, Darryl
  Science Tales: Lies, Hoaxes, and Scams 304, 306–307
cybernetics 25, 107

Damasio, Antonio 154, 162, 166
Darwin, Charles 1, 6, 324
  On the Origin of Species 324
data 6–7, 13, 138, 154, 158–159, 177, 185, 189, 239–240, 275, 331–332
de Angelis, April 338
de Certeau, Michel 199
deep time 21, 23, 43
defamiliarization 15, 100
Defoe, Daniel 27, 187, 193, 200
demodystopia 235, 241, 249
demographic excess 232, 235, 241, 247–248
demography, demographics 3, 240
Despret, Vinciane 8, 13, 283
determinism 1, 15

diagnosis 25–26, 125, 129–131, 134, 136–137, 241
Dickinson, Adam 19, 21, 24, 44, 57, 61–64
  Anatomic 19, 24, 44, 61, 62–64, 63
Dickinson, Emily 131, 137, 141, 143
diplomat, diplomatic 32, 355, 374–376
disability 26, 123, 127, 139–143
discourse 1, 5–7, 10, 12–13, 15, 17–18, 22, 30–31, 42, 54, 58, 62, 128, 222, 238–239, 250, 300
disease 5, 17, 27, 58, 125, 126, 127, 128, 129, 132, 135, 148, 174, 189, 190, 196, 197, 198, 200, 201, 266, 288, 304. *See also* illness
displacement 221, 248, 317, 371
DNA 1, 18, 22, 41–42, 49–51, 96, 104–107, 112, 198, 217, 219
documentary 94–95, 316–317, 330, 343
dualism 127, 133, 212
Duffy, Carol Ann 328
Dumit, Joseph 147, 152, 159, 161–163, 178
dystopia, dystopic 186, 191, 193, 237, 241, 301

Earth Ensemble 31, 338–339, 341–342
earth system science 3, 9
ecocriticism 2, 4, 12, 16–17, 74, 97, 116, 205, 329, 336, 346, 374
ecological detective 16, 19, 30, 269
ecological performance 32, 353
ecology, ecological 3, 5, 8–21, 23–25, 29–32, 43, 52, 58, 60–61, 65, 78–79, 83, 93, 95–100, 102, 104, 106–110, 112, 186, 207, 213, 221, 226, 238–241, 261–263, 266–267, 269, 274–275, 282–283, 285, 287–288, 291–292, 294, 300–302, 308, 317, 323–332, 334–337, 341–343, 345–349, 358, 363–365, 368, 374
  connectedness 18, 95–96, 107, 266–267, 274, 283, 291, 301
  crisis 16, 29, 186, 262, 292, 323, 325–328, 330–331, 334–335, 364, 368
  discourse 12, 15, 30, 267
  ecological similes 8
  ecological sublime 52
  ecological theory 8
  global ecology 9
  imagination 267
ecopathography 14
ecostoicism 14, 30, 292–294
ecosystem 3, 5, 7–10, 14–15, 18, 21, 29–32, 47, 53, 56, 58–60, 63–64, 95, 100, 104, 107, 263, 267–268, 280–289, 291, 293–294, 324, 331, 333, 347, 354, 357, 364–365, 376
ecosystem modelling 3, 7, 29–30, 261, 268, 280, 283–285, 291, 294
Edgar und Allan
  Beaming Sahara 22, 32, 70, 354, 362, 362–363, 365–369, 371–372
Ehrlich, Anne and Paul 5, 240
elegy, elegiac 12–13
Emmott, Stephen 31, 330, 332
  Ten Billion 330, 332
empathy 211–212, 303, 372, 375
energy 205, 304, 372
entangled, entanglement 18, 20, 32, 78, 107, 208, 286, 324, 347, 363, 368
environmental 8, 10, 13–14, 16–21, 23–25, 29–32, 43, 47, 60–61, 63–65, 70, 74, 76, 82, 89, 100, 155, 186, 208, 238–241, 251, 256, 261, 263–264, 268–269, 273, 275, 282–283, 287–288, 291–292, 301–302, 304, 308–310, 313, 317, 324, 326, 329–330, 334, 336, 338, 347
  activism 292, 300, 308, 338
  crisis 13, 17, 31, 65, 256, 291, 301, 304, 308, 324, 326, 330, 334
  humanities 8, 16, 25
  imagination 17, 261, 268, 287, 291
  justice 32
  literary studies 238
  literature 20, 24, 64, 70, 76, 273
  networks 65
  turn 24, 70, 74
epidemic 28, 189–190, 192, 198
epidemiology 3
epigenetics 24, 87, 300
epilepsy 147–149, 156, 167, 170, 174

epistemology, epistemological, epistemic 7, 9, 14, 16, 19, 22, 26, 28, 32, 43, 45, 51, 58, 65, 155–156, 166, 169, 174, 177, 190, 212–213, 215, 347, 355
epivitality 12
ethics, ethical 7–8, 15, 22, 43, 161, 313, 347, 359
Everyman 328
evolution 1, 4, 7, 11, 21, 131, 133, 135, 359–360, 362, 366
evolutionary 1–2, 7–8, 19, 61, 328
evolutionary biology 1
experiment, experimental 4–5, 22, 42, 71, 85–86, 104, 124, 143, 149, 152, 154–155, 159, 162–165, 167, 174, 196, 211, 245, 263, 265–268, 270, 324, 354, 365–366
expert, expertise 13, 31, 124, 130, 132, 142, 154, 162, 163, 177, 178, 189, 283, 309, 331. *See also* specialization
explanatory gap 4, 26, 147, 153, 164, 166, 174
extinction 12–13, 30–31, 233, 236, 249, 300–302, 316, 324–325, 329, 339, 342
Extinction Rebellion 15, 31, 338–342
extraction, extractive 11, 19, 28, 79, 90, 203, 207, 213, 216, 221
eye 42, 44, 54, 71, 86, 87, 97, 106, 210, 211, 274, 331, 333. *See also* optical, sight, visual

fairytale 149
fantasy 20, 27, 96, 100, 147–148, 152, 156, 162, 166, 169, 174–175, 209–210, 234, 240, 244, 248, 255, 270, 327
Farinella, Matteo 26, 147, 149–152, 156–157, 165–166, 175–176
 Neurocomic 26–27, 147, 149, 149–152, 150–153, 155–158, 165, 164–166, 176, 175–177
Farndon, John 339, 341
Ferris, Joshua 25–26, 123–134, 136–143
 The Unnamed 25–26, 123, 123–143, 124, 126–143
fiction 2–5, 12–13, 15, 19–21, 23, 25, 28–29, 43, 54, 56, 95–96, 99–102, 104–106, 108, 110–111, 113, 116, 127, 137, 143, 149, 152, 175, 178, 185–187, 191, 193, 196–197, 199–200, 203, 209, 232–235, 238, 240, 242, 245–250, 254–256, 261–265, 267–268, 271, 286–289, 293, 300, 302, 326–327, 366
anticipatory fiction 14
detective fiction 19, 137, 143, 263, 264, 271. *See also* crime narrative
genetic fiction 47
metafiction 96, 104
non-fiction 21, 54, 100, 175, 282, 290, 300, 302, 326
pandemic fiction 19, 185, 186, 187, 191, 200. *See also* pandemic novel
post-apocalyptic fiction 3, 5, 29, 203, 232–234, 240, 242, 245–250, 254–255
realist fiction 245–246
science fiction 20, 29, 96, 100, 104, 111, 193, 203, 209, 233–235, 238, 247, 249, 287–288, 327, 366
speculative fiction 12, 178
utopian fiction 29, 233, 238, 246
weird fiction 20, 95, 100–102, 105–106, 110, 113, 116
Fies, Brian 317–321
 A Fire Story 317, 317–321, 318–321
figurative 2, 4–5, 7–8, 25, 42, 80, 132, 363
fire 90, 188, 199, 317–318
fish, fishing 9, 31, 90, 111, 213, 300–303, 333
flânerie, flâneur 139–140
flood 23, 168, 303, 308
focalization, focalizer 104, 135–136, 220, 223
food chain 30, 288, 301–302
forest 10, 22, 27, 32, 52, 61–62, 79, 93, 99, 107, 151, 175, 282, 326, 362, 367, 372–373, 375
form 1, 3–7, 10, 12–13, 15–16, 18–21, 25, 29, 32, 41–42, 44, 46, 52–53, 55, 58–61, 63–65, 71–72, 77–80, 82–84, 86–88, 90, 95–98, 102, 104–105, 107–108, 110–116, 127–128, 133, 149, 152–155, 159, 164, 166, 169, 175, 196, 207, 209–210, 212, 216, 224, 233, 237, 245–248, 250,

390 *Index*

261–264, 266, 268, 270–271, 274–275, 280–289, 291–292, 294, 301, 303, 313, 316, 319, 330–331, 342–343, 346, 348, 355, 363, 365, 371, 375
  aesthetic form 7, 12, 19–21, 42, 71, 98, 153, 233, 331
  bodily form 77
  cultural form 1
  formal strategies 30
  human form 59, 64, 169
  life form 1, 5, 10, 21, 32, 44, 53, 55, 60, 65, 78, 80, 95, 97, 105, 107–108, 110, 112, 116, 262, 266, 274–275, 283, 287, 289
  literary form 96, 110, 262, 282
  of the Western novel 20
  physical form 164, 169
  prose form 21, 25, 127, 149, 268
  realist form 246
  seasonal form 280, 282–283, 291–292, 294
  theatrical form 29
  visual form 152, 154, 166, 291, 316
formalism 25, 97, 116
fossil 3, 19, 28, 69, 78, 205, 317
fossilized life 3
Foucault, Michel 7, 10–11, 190–191, 366
Freud, Sigmund 157, 196
Friel, Brian
  Translations 336
fungi 3, 5, 10, 25, 93–99, 101–102, 105–110, 112–116, 189

Gaia 5, 9, 17, 353–354
Galileo 6
gender 13, 75, 230
genes, genetics 1–2, 8, 15, 18, 23, 41–44, 47–55, 62, 65, 82–83, 87, 106, 190, 219, 226
genocentrism 2
genomic 7, 8, 25. *See also* post-genomic
genre 11, 19, 21, 25, 29, 31–32, 44, 53, 58, 99, 101–102, 116, 124, 142–143, 149, 151, 153–154, 158, 169, 186, 203, 221, 232–233, 235, 237, 245–247, 249, 313, 370

geology 3, 71, 74, 81
geometry 6, 239
geontology 12
geontopower 12, 28, 204–205
Ghosh, Amitav 20, 208, 262
Gill, Stephen 2, 30, 275, 278–280, 290
  Night Procession 30, 275, 278
  The Pillar 30, 275, 278–279, 287, 290
Glaspell, Susan 31, 327
global warming 25, 30, 99–100, 221, 255, 309, 313
Gore, Al 330
  An Inconvenient Truth 330
graph 7, 252, 309, 317
graphic memoir 147
graphic narrative, graphic novel 2, 3, 15, 21, 22, 23, 26, 28, 29, 30, 150, 152, 153, 155, 162, 164, 166, 169, 175, 177, 300, 309, 317, 319. *See also* comics, graphic narratives; *See also* comics, graphic novels
grotesque 4, 15, 23–24, 42–44, 49–51, 53–59, 62, 64–65, 138, 323
guerrilla theatre 15, 31, 338

habitat 5, 342, 344
Hamann, Alexandra 4, 15, 18–19, 23–24, 27, 41, 308
  The Great Transformation—Can We Beat the Heat? 306
Haraway, Donna 11–12, 32, 212, 346–347, 355, 363–364, 366, 370, 373, 375
Hartmann, Jörg 308
  The Great Transformation—Can We Beat the Heat? 306
Hawkins, Anne 14
Hayles, N. Katherine 4, 27, 162–164
health 10, 13, 17, 26, 30, 126–127, 134–136, 138, 141–143, 187, 189, 198, 248, 302, 304, 323
hearing 85, 345
Heise, Thomas 198
Heise, Ursula 9, 13, 30, 96, 237, 241–242, 256, 275, 291, 324
Herman, David 21, 261, 264–265

heterogeneity, heterogeneous 21–22, 24, 299, 313, 346
Hickson, Ella 20, 31, 325, 334–336
  Oil 20, 31, 325, 334–336
history, historical 8, 11–12, 23, 27–28, 45, 49, 63, 78–79, 95, 97, 100, 106, 110, 113–115, 143, 150–151, 169, 174–175, 189, 197–198, 203–205, 213, 217, 233, 243, 282, 284–285, 294, 306
Holton, Gerald 6–7
Hooke, Robert 45
  Micrographia 45
horror 100–101, 103, 132, 220–222, 235, 245
Horton, Zachary 10, 16–17
Houser, Heather 13, 143
Hülsmann, Jörg 308
  The Great Transformation—Can We Beat the Heat? 306
hurricane 303, 318
Huyghe, Pierre 32, 354, 357, 368, 371
  After ALife Ahead 23, 32, 354, 357, 358–359, 362, 364–366, 371
hybridity, hybrid 28, 45, 101, 149, 153, 169, 175, 203–204, 207, 215, 219, 330
hyperactor 32, 334, 363, 374
hyperobject 6, 15, 30, 32, 208–209, 299–300, 303, 308, 313, 318–319, 324, 334, 341–342, 363

identity 28, 74–76, 125, 134–135, 140, 149, 151, 153, 156, 159, 174–175, 188, 194–198, 200, 206, 264, 273, 375
illness 14, 26, 125, 126, 127, 129, 133, 135, 142, 155, 169, 189, 357. See also disease
imagination 1–7, 10, 13, 15, 17–19, 23, 31, 42–46, 49, 53, 57, 59, 70, 74, 84, 148, 236, 242, 245, 261, 267–268, 287, 291, 301, 308, 312, 324, 346–347, 362–363, 369, 374–375
imaging 1, 2, 3, 4, 5, 6, 7, 10, 13, 15, 17, 18, 19, 23, 31, 42, 43, 44, 45, 46, 49, 53, 57, 59, 70, 74, 84, 148, 236, 242, 245, 261, 267, 268, 287, 291, 301, 308, 312, 324, 346, 347, 362, 363, 369, 374, 375. See also brain imaging

medical 7, 24–25, 86
immersion 32, 354, 363–366
immigration 185, 188, 194, 195, 200. See also migration
indigenous 3, 28, 93, 206, 213, 215–217, 301
  epistemology 28
  knowledge 93
  ontologies 3
inert 12, 70, 72, 80, 217
infection, infectious 5, 27, 185, 186, 187, 189, 191, 192, 200, 201. See also contagion, contagious
inorganic 24, 69, 71–72, 74, 76, 78, 80–81, 84–85, 89–90
insect 29, 56, 98
installation 23, 31–32, 70, 263, 282, 355, 357–360, 365–366, 369, 371–373, 375
interconnection 47, 52, 65, 77, 89, 166, 346, 347. See also interrelation
interdisciplinarity 5, 8, 19, 60–61, 209, 241
interdiscursivity, intertextuality 5
interrelation 375. See also interconnection
intertextuality 143
invisible 14, 15, 17, 23, 41, 42, 44, 46, 49, 52, 55, 56, 59, 70, 84, 86, 94, 103, 106, 109, 140, 193, 196, 323, 324, 334, 349. See also visible
ironic 141
irony 27, 52, 109, 128–130, 148, 152, 169, 174, 191, 344, 346

Jameson, Fredric 234, 241, 245

Kant, Immanuel 47, 50
  Kantian 23, 102, 240
Keats, John 88
Kim, Carole 338
  The Seed Will Search... 338
knowledge 3, 6, 7, 14, 28, 41, 42, 46, 50, 52, 53, 77, 88, 93, 116, 123, 127, 131, 150, 152, 154, 155, 156, 163, 177, 178, 189, 198, 199, 205, 212, 213, 215, 220, 240, 275, 288, 301, 306, 309, 364, 370, 375. See also indigenous knowledge, indigenous epistemology

expert 177–178
production 212
settler capitalist 220
Kurlansky, Mark 9, 300–302
World Without Fish 300, 300–302, 302

landscape 5, 22, 24, 28, 31, 44, 56–59, 62, 64–65, 70, 74, 80, 83, 86–88, 90, 103, 106, 140, 221, 223, 255, 264, 269, 274, 278, 281, 283–284, 286–287, 289, 310, 323, 328, 333, 348, 360, 362
Latour, Bruno 9, 12, 17, 31, 316, 331–332, 354–355, 371
Inside 9, 331–332, 355
Moving Earths 331
LeDoux, Joseph 26, 36, 126, 145, 154, 180
Le Guin, Ursula K. 107–108
Leinfelder, Reinhold 308
The Great Transformation—Can We Beat the Heat? 306
Lemire, Jeff 12, 22, 28, 203–204, 213–215, 218–220, 223–225
Sweet Tooth 3, 12, 14, 22, 27–28, 203, 203–226, 204–207, 213–215, 218–221, 223–225
Leopold, Aldo 282
Levine, Caroline 6, 284–285
Levine, Gabriel 327
Levine, Joseph 148
Levine, Phillip 26, 147
liberalism 8, 216
Linnaeus, Carolus 113
Li-Vollmer, Meredith 302, 304–305, 316
Climate Changes Health: How Your Health Is at Stake and What You Can Do 302, 304
Lovecraft, H. P. 100–103, 105
Lovelock, James 9, 353

Macmillan, Duncan 329–330
2071 330–332
Lungs 329, 333
macroscopic 4, 28, 43–44, 207, 209–210, 212, 220, 267, 326, 349, 363
Maeterlinck, Maurice 31, 328, 356
Les Aveugles 328

Mahato, Mita 302, 304–305, 316
Climate Changes Health: How Your Health Is at Stake and What You Can Do 302, 304–305
Malabou, Catherine 149
Ma, Ling 14, 27, 102, 183, 187–188, 191–192, 196–197, 242, 244, 249, 256
Severance 14, 27–28, 102, 183–201, 187–188, 191–193, 196, 198, 201, 242, 244, 246–249, 256
Malthus, Thomas 4, 232, 239–240, 242, 248–249
Mandel, Emily St. John 28, 187, 233, 235–237, 243, 255
Station Eleven 14, 27–29, 187, 233, 235–237, 243, 243–249, 244–248, 255, 255–256
Margulis, Lynn 9, 113, 353
masculinity 139, 140. *See also* virility
material
materiality 24, 29, 52–53, 57–58, 62, 175, 328, 355
semiotics 370, 373
materialism 116, 133, 142
matter
brain matter 147–149, 164
inorganic 24, 69, 71–72, 76, 78, 81, 84, 90
matter, materiality, material 6, 11, 16, 20, 24, 29, 33, 43, 46, 49, 52–53, 55, 57–58, 62, 64–65, 69–74, 76–81, 84–86, 90, 109, 111, 123, 126, 132, 138, 142, 147–149, 164, 174–175, 178, 206, 209, 211, 217, 222, 237, 240, 248, 256, 266, 274, 284, 319, 328, 332, 337, 346–347, 355, 362, 364–366, 369–370, 373–374, 376
Mawer, Simon 24, 48–51
Mendel's Dwarf 24, 48, 49–51, 50
McGregor, Jon 5, 9, 15, 19, 29–30, 262–272, 274–275, 278–281, 283–284, 286, 288–289, 291–294
Reservoir 13 9, 19, 29–30, 261–294, 262–275, 278, 280–294
media 2, 6–7, 13, 31, 44, 54, 94–96, 107, 109–110, 158, 185, 221, 306, 342–343

medical humanities 2
medicine, medical 2, 6, 7, 10, 13, 18, 19, 24, 25, 26, 27, 30, 43, 86, 124, 125, 127, 129, 130, 131, 132, 136, 142, 148, 159, 163, 164, 169, 174, 196, 197, 216, 256, 317. *See also* biomedical
melancholy, melancholic 28, 195–196
Menken, Alan 31, 327
mesocosm 29, 266, 268, 272, 274, 280, 284, 291
metamorphosis 24, 69, 80–82
metaphor 6–7, 25, 42, 74, 80, 107, 109, 116, 132, 157, 175, 188, 198, 221, 291, 300, 324, 329, 348, 365–366, 371
methodology 83, 159, 161, 289, 354
Meuret, Michel 8, 283
microbes 3, 22, 31, 44, 49, 51, 53–56, 58–63, 325, 327, 346
microbial environment 56–59, 62
microbiology, microbiological 2–4, 7, 17–18, 23, 27, 41, 58
microbiome, microbiota 8, 14, 18, 42, 54–55, 60–61
microcosm 77
microscopy 45, 220
migration 188. *See also* immigration
mind 47, 54, 59, 110, 123–124, 126–130, 132–135, 137, 139, 141–143, 158, 162, 166–167, 197, 243, 292, 326
Mitchell, Katie 325, 328–331, 348
modelling 3, 5, 7, 30, 268, 280, 283–286, 288, 291, 293–294
molecular environment 51, 59, 63
molecular sublime 4, 42, 44, 46–54, 62, 65
monitoring 5, 281, 283
monologue 330, 362, 375
Moore, Jason W. 212–213
more-than-human 17, 31, 261, 267, 271, 273–274, 280, 287, 291, 326, 329, 332, 348
Morizot, Baptiste 355, 364, 374
morphogenesis 82
Morrison, Toni 183, 196
Morton, Timothy 6, 13, 15, 43, 100, 109, 205, 208, 220, 263, 299–300, 318–319, 324, 335, 341–342, 346, 363, 368
multi-scalar aesthetics 21, 23

multiscale narration 4, 21, 262, 264, 266, 271
mycelial, mycelium 10, 25, 94–95, 106, 108–109
mycoaesthetics 4, 19, 25, 62, 93, 95, 97, 101, 104, 108–112, 115–116
mycology 23, 94–95, 113–114

narrative 1–2, 4–6, 13–14, 19–22, 25–31, 45, 49, 96–97, 104–105, 112, 115, 124–127, 129, 131–138, 140–141, 143, 149–150, 152–153, 155, 158, 162–164, 166, 169, 175, 177–178, 189, 191, 197, 203–205, 207, 213, 220–221, 223, 234, 236, 243, 261–269, 271–272, 274–275, 278, 280, 282–283, 285–286, 288–289, 291–294, 300–303, 308–309, 317–319, 326, 329, 375
  chaos narrative 135
  crime narrative 263–264, 269, 271–272
  eco-narrative 13
  evolutionary narrative 1, 33
  frontier captivity narrative 203
  graphic narrative 26, 29–30, 149–150, 152–153, 155, 162, 164, 166, 169, 175, 177, 300, 319
  multiple-track narrative 21
  national narrative 28
  neuronarrative 4, 10, 124
  post-apocalyptic narrative 203
  satirical narrative 26
  structure 30, 189, 236, 261, 262, 300. *See also* narrative form
  surrender 125, 136
  voice 197
narratives
  illness narratives 126–127, 155
naturalism 133, 156
naturalist 98, 105, 213
nature writing 263
neo-sublime 4
network 6–10, 25, 50, 60, 65, 94–95, 97, 107–109, 140, 148, 229, 283–284, 288–289, 372
Neufeld, Josh 302–303
  A.D. New Orleans After the Deluge 302–304

neurobiological 4, 25–26, 124, 147, 149
neurofiction 19
neurological reduction 127
neurology, neurological, neuroscience 17, 20, 123–124, 127, 130, 142, 149–150, 152, 159, 161, 163–164, 166, 174–175
neuromarketing 160–161
neuronarrative 4, 10, 124
neuronovel 25, 26, 124, 127, 138, 141, 143, 154, 178. *See also* syndrome novel
neurons 3, 148, 158, 164–166, 176
neuroscience 7, 26, 124, 127, 143, 149, 151, 153, 156–157, 178
new materialist 24, 89
Nippoldt, Robert 308
  The Great Transformation—Can We Beat the Heat? 306
Nippoldt, Studio 308
  The Great Transformation—Can We Beat the Heat? 306
noir 19, 263–264
nonfiction 175, 282, 290
nonhuman 11, 12, 24, 30, 32, 47, 61, 62, 63, 64, 70, 101, 106, 115, 116, 190, 205, 206, 209, 211, 217, 223, 226, 233, 262, 263, 265, 267, 268, 269, 271, 272, 274, 275, 280, 283, 287, 291, 303, 325, 326, 327, 332, 346, 347, 348, 354, 366, 369, 370, 373, 374, 375. *See also* other-than-human
normate 26, 140
Norris, Frank 131–133, 138
nostalgia, nostalgic 28, 80, 196–198, 201, 366
novel 2, 3, 6, 11, 12, 15, 19, 20, 21, 22, 23, 24, 25, 26, 27, 28, 29, 30, 44, 48, 49, 52, 56, 60, 95, 96, 99, 102, 103, 105, 110, 115, 116, 123, 124, 125, 126, 127, 128, 131, 133, 134, 135, 136, 137, 138, 139, 140, 141, 142, 143, 164, 186, 188, 191, 192, 193, 196, 197, 198, 201, 208, 220, 233, 235, 236, 237, 238, 241, 243, 244, 246, 247, 248, 249, 255, 261, 262, 263, 264, 265, 266, 267, 268, 269, 270, 271, 272, 273, 274, 275, 278, 280, 281, 282, 283, 285, 286, 287, 288, 289, 290, 291, 292, 293, 294, 309, 312, 317, 326. *See also* graphic novel, neuronovel, syndrome novel
contemporary novel 15
crime novel 293. *See also* crime fiction and crime narrative
diaspora novel 193
graphic novel 3, 22–23, 28, 309, 317
Naturalist novel 131, 138, 143
neuronovel 19, 25–26, 124, 127, 137–138, 141–143, 145
pandemic novel 11, 27
post-apocalyptic novel 233, 237, 256, 257. *See also* post-apocalyptic fiction
realist novel 29, 233, 238, 246, 248, 268. *See also* realist fiction
rural novel 264, 266, 291
weird novel 102
Western novel 20
objectivity, objective 10, 128, 152, 166, 197, 210, 212, 219, 273
observation, observational 23, 52, 152, 157, 240, 244, 267, 273, 318
ontogeny 24, 87, 90
ontology, ontological 16, 21, 25, 27–28, 43, 72, 78, 96–98, 108, 112–114, 204, 206, 262, 346, 371, 376
  flat ontology 262
  fungal ontology 114
  indigenous Australian ontology 206
  Inuit ontology 204
  positivist ontology 27
  settler ontology 28
  Western ontology 204
optical 45, 199. *See also* eye, sight, visual
organism, organic 2, 5, 7–10, 12, 21, 23–24, 41, 43–45, 55–57, 60–61, 63, 70–74, 76, 78–85, 89–90, 94, 103–108, 112–113, 116, 130, 162, 164–165, 197, 200, 205, 212, 265, 273, 284, 287, 332, 346, 353, 356, 358–359, 362, 364, 369, 372–373, 375
other-than-human 3, 21, 32, 373, 376. *See also* nonhuman

pandemic 3, 10–11, 13–14, 17, 19, 22, 27–28, 43, 185–187, 189–194, 196, 199–200, 203–204, 207, 233, 235–237, 239, 241–244, 249, 323, 348
Panksepp, Jaak 160, 166, 169, 178
pastoral 19, 26–27, 29, 103, 139, 235–236, 250, 264, 268, 291
perception, perceptual, perceptible 4, 6, 13–15, 17, 22, 48, 53, 70, 77, 112, 115, 133, 158, 166, 177, 207, 216, 274, 331, 363–364, 371
performance 2–6, 9, 13, 15–16, 20, 22, 29, 31–32, 248, 263, 326, 329–332, 336–339, 341–344, 346–349, 354–356, 358–359, 362–363, 365–366, 368–371, 373, 375–376
phenomenology, phenomenological 75, 149, 268
philosophy 5, 12, 15, 20, 77, 95, 100, 104, 149, 154, 178, 211, 308, 354
photograph, photography 30, 114, 275, 278–279, 290–291, 306, 319, 343, 374
physics 6, 16, 78, 89, 169, 309
physiocide 32, 364
physiology, physiological 72, 148, 150–152, 154–155, 158, 164, 166, 174–175, 178
plague 27–28, 31, 102, 187, 193, 198, 203, 206, 215–217, 219
planet 9, 14–17, 20–22, 29, 43, 45, 60, 63, 65, 70, 77–78, 88, 99, 103, 108, 232, 234, 238–241, 251, 255, 261, 266–267, 275, 287–288, 292, 294, 312–313, 316, 328, 332–333, 347–348, 356–357, 359, 365
plant 12–13, 20, 25, 27, 30–31, 51–52, 56, 79, 93, 95–98, 106, 108–109, 111–114, 166, 183, 189, 205, 262, 269–270, 272, 280–281, 284, 290, 304, 327, 330, 332, 337, 346, 348, 353, 355, 359–360, 362, 364, 372, 374
plot 1, 22, 104, 149, 189, 237, 265, 269–270, 272, 274, 281, 284, 289, 291, 306, 324, 330
  subplot 138, 262, 265, 268, 271–272, 274, 279–283, 285, 287, 291–292, 294
pluriverse 4–5, 274, 280, 291

Poe, Edgar Allan 19, 101, 269
poetry 2–3, 9, 11, 18–19, 21, 23–24, 61–63, 69–90, 137, 143, 189, 263, 266, 273, 310, 362, 373–375
population 3–4, 9–11, 14, 27–29, 190–193, 200, 232–242, 246–250, 255, 272, 275, 285, 304, 329, 358
population unconscious 11, 232, 237, 246–247, 249
post-apocalyptic 3, 14, 29, 203, 243, 244. *See also* apocalyptic
post-genomic 8. *See also* genomic
posthuman 11, 20, 31, 236
posthumanism 97
Povinelli, Elizabeth 12, 28, 204–206, 210, 216, 223
Powers, Richard 13, 51–52, 55, 292, 326
  The Overstory 51, 326
Preston, Richard 56, 60
  Micro 56, 60–61
Propp, Vladimir 5
psychology, psychological 78, 125–126, 130, 162, 164, 177, 188, 197, 200
Puchner, Martin 17
Punctuate! Theatre 327

race, racialized, racism 10–11, 28, 100, 185, 187, 190–192, 195, 200, 222, 234, 242, 292
Ramón y Cajal, Santiago 150, 164, 166, 176
Rapley, Chris 31, 330–332
  2071 330–332
Rausch, Tobias 10, 13, 32, 348, 354, 359–361, 365–366, 368, 371–373, 375
  Die Welt ohne Uns 359–361
realism 20, 27, 30, 100, 105, 166, 197, 247, 262, 265, 291, 325
Redniss, Lauren
  Thunder & Lightning: Weather Past, Present, Future 312–316, 315
Red Rebel Brigade 339–342
representation, representational 2–3, 5–10, 13–16, 18, 21–22, 25–27, 29, 31, 42–45, 48, 50–53, 55–57, 59–60, 63, 65, 71–73, 77–78, 85–86, 88, 95,

112, 115–116, 126, 132, 136, 151–158, 160–164, 166, 169, 174, 177–179, 205, 283, 291, 303, 325–326, 332, 337, 343, 346, 353–354, 372, 374–375
rhetoric, rhetorical 12–13, 45, 108, 110, 155, 159–160, 177–178, 190, 229, 249–251, 262, 310, 316
rhythm 76–77, 84, 177, 263–264, 266, 268, 282, 284, 290, 294, 359, 362
Rickson, Ian 336–337
Romantic, Romanticism 10, 48, 72, 88, 98, 107, 240
Rosebury, Theodor
  Life on Man 24, 45, 54–56, 58
Rose, Nikolas 17–18, 26, 41, 126, 130
Roš, Hana 26, 147, 149–152, 156–157, 165–166, 175–176
  Neurocomic 26–27, 147, 149, 149–153, 150–153, 155–158, 165, 164–166, 176, 175–177

satirical 7, 26, 138
scala 8–9
scale critique 16
scale, scalar 3, 9–10, 14–17, 21–23, 25, 29–30, 32, 41–44, 46, 48, 51, 53, 55, 57–58, 60, 65, 70, 76–78, 81, 88, 93–94, 97, 106, 109–110, 189–190, 199, 209, 220, 226, 229–230, 232–233, 236, 238, 243–247, 250, 255, 261–262, 264–268, 275, 288, 291–293, 299–300, 303–304, 309, 318, 323–325, 327–329, 331–334, 336, 342–344, 346, 349
  atomic scale 48
  derangements of scale 43, 66
  ecological scale 93, 329, 334
  flexibility 266
  geobiological scale 363
  geological scale 32
  global scale 199
  human scale 31, 65, 323–325, 332, 346, 349
  imperceptible scale 14, 21–23, 55, 59
  international scale 334
  large scale 10, 17, 30, 42–43, 190, 267, 275, 300, 329, 331

  literacy 17
  macro scale 309, 318, 325
  meso-scale 15–16
  microbial scale 57
  microscopic scale 15, 17, 23, 45, 57, 61, 85, 106, 209, 309, 327
  modesty 266–267
  molecular scale 43, 48
  nonhuman scale 31, 209, 267, 291
  nuclear scale 48
  planetary scale 20, 88
  poetics 266
  scale domain 232, 246
  scale effects 70
  scale of human experience 16
  scale of the climate crisis 336
  scale of the organism 81
  small scale 30, 44, 247, 250, 300
  species scale 21, 261
  timescale 21, 23–24, 29, 43, 70–71, 74, 78, 81–83, 87, 90, 209, 308, 336, 362
  translation 17
  uncertainty 23
science and literature studies 2
science communication 3, 25, 58
science fiction 20, 29, 96, 100, 104, 111, 193, 203, 209, 233–235, 238, 247, 249, 287–288, 327, 366
science, popular 4, 7, 18, 24–27, 44, 53–54, 95, 241–242
self 9, 26, 43, 46, 61–62, 64, 80, 111–112, 124–127, 132–135, 141–142, 147, 149, 151, 154, 159, 162, 166, 174, 178, 188, 196–197, 209, 212, 215, 247, 249–250, 330, 333, 369, 371
sensitivity, sensitive, sensitivities 30, 62, 268, 364–366, 368–370, 372, 374
Shaw, George Bernard 31, 327, 329
  Too True to Be Good 327
Shawn, Wallace 330
  Grasses of a Thousand Colors 330
sight 86, 160, 166, 341. *See also* eye, optical, visual
Skinner, Penelope 329
  Greenland 329

soil 22, 60, 74, 76, 94, 99, 109, 265, 294, 324, 358, 362, 367, 372, 375
solastalgia 14, 365
soul 132–133, 141–143, 210
specialization, specialist 104, 159, 326, 359. *See also* expert, expertise
species 5, 10–11, 21–22, 29–30, 52, 56, 82, 88, 97, 99, 107, 115, 189–190, 213, 226, 236, 248, 261–262, 264–266, 268, 270, 273, 275, 280–281, 284–285, 287, 291, 301–302, 316, 324–325, 329–330, 335, 342, 344, 346, 355–357, 372
speculative fiction 12, 178
Squarzoni, Philippe 309–311, 316, 318
Climate Changed: A Personal Journey through the Science 309–311
statistical 158, 163, 177
steampunk 148, 156
Steinlight, Emily 239, 247–248, 266
Stephens, Simon 330
Wastwater 330
Stevenson, Robert Louis 132–133
Stewart, Susan 24, 55–56
Stockton, Frank 300, 302, 321
stoicism, ecostoicism 14, 30, 291–294
subjectivity 20, 24, 26, 47, 72, 101, 150, 164, 209–210
sublime 4, 14–15, 23–24, 41–62, 64–65, 102, 222, 240, 248, 250
superorganism 9, 96, 106
symbiosis, symbioses, symbiotic 10, 20, 56–57, 104, 107, 109
synapses 154, 156–157
syndrome 19, 25–26, 123–124, 129, 141–143, 238
syndrome novel 19, 25, 26, 124, 141, 142. *See also* neuronovel
syntax, syntactical 129, 135, 153, 216

tactile, tactility 86, 87. *See also* touch
Talbot, Mary M. and Brian 23, 308
Rain 23, 308
technology 7–8, 10, 27, 42, 45, 53, 94, 108–109, 124, 130–131, 150, 152, 154–156, 158–164, 174, 177–178, 186, 205, 235, 237, 274, 280, 312–313

technomorphism 8–9, 108
terraforming 95, 103, 288
theatre 13, 15, 20–21, 29, 31–32, 199, 325–329, 331–332, 334–336, 338, 342–344, 347–349, 353–355, 358–359, 362–363, 366, 368–369, 372, 374, 376
themata 6
Thomson, Garland 26, 48–49, 140
Thoreau, Henry David 270, 282
Thorne, Jack 329
Greenland 329
thought experiment 154, 164, 366
touch 48, 53, 86, 89, 162, 166, 171, 174, 213, 216, 299, 367. *See also* tactile, tactility
tragedy, tragic 12–13, 132, 134, 189, 264, 328
trope 8–9, 24, 27, 56, 197, 199–200, 212, 220, 325, 354
Tsing, Anna 17, 94, 283–285

Uberti, Oliver 30, 275, 277
Where the Animals Go 30, 275, 275–277, 280, 287
uncanny 96, 100, 164, 341, 347
uncertainty 21, 23, 45, 156, 345
Urgurel, Iris
The Great Transformation—Can We Beat the Heat? 306
utopia, utopian 4, 22, 27, 29, 107, 162, 233, 238, 245–247, 249

VanderMeer, Jeff 13, 15, 19, 25, 93, 95–97, 99–100, 102–105, 108, 110, 112–113, 115–116, 221
Acceptance 95, 103, 105
Annihilation 25, 93, 95–97, 102, 102–116, 104, 108, 110, 112, 115
Authority 95
vegetal 29, 32, 108, 263, 282, 359, 364
Verdonck, Kris 22, 31, 354–355, 371–372
Exote 22, 31, 354–358, 355–356, 358, 362, 364
Vint, Sherryl 12, 18
viral, virality 14, 17, 31, 104, 204, 221
virility 140. *See also* masculinity

virology 17
virus 6, 10, 21–22, 48, 84–85, 190, 200–201, 206, 221, 224, 236, 249, 323
visible 42, 44, 46, 52, 54, 55, 65, 80, 84, 109, 139, 164, 187, 200, 212, 217, 301, 323, 341, 370, 372. *See also* invisible
visual, visualization 5, 6, 7, 9, 27, 28, 29, 83, 84, 85, 86, 93, 95, 149, 152, 153, 154, 155, 162, 166, 169, 229, 232, 249, 250, 251, 275, 287, 288, 291, 303, 309, 310, 316, 323, 324, 331, 333, 334, 336, 360, 370. *See also* eye, optical, sight
vitality 18, 24, 69–71, 73, 76, 78, 90, 210, 223, 285

walking 26, 89, 123, 126, 128, 132, 134, 136–137, 139–141, 235, 249, 254, 256, 309–310, 335, 348
Waters, Steve 31, 333
  The Contingency Plan 31, 333, 333–334, 336
waves 76, 87, 89–90, 93–95, 177, 199
Weaver, Deke 13, 31, 342–347
  The Unreliable Bestiary 31, 342–347

web 10, 25, 86, 93, 94, 95, 96, 97, 101, 104, 106, 107, 108, 109, 110, 114, 116, 274, 288, 289, 303, 346. *See also* wood wide web
weird 5, 15, 18, 20, 25, 58, 62, 95–103, 105–108, 110–111, 113–116, 268
weird ecology 96, 100, 106
weird fiction 20, 95, 100–102, 105–106, 110, 113, 116
Wolfe, Cary 266, 273–274, 283, 291
wood wide web 10, 25, 93–97, 104, 106–110, 114, 116
worm 98, 100, 111, 272, 284, 288, 294
Wynter, Sylvia 11, 217

Yong, Ed 44, 54, 58–60, 62, 94
  I Contain Multitudes 24, 44, 54, 58

Zea-Schmidt, Claudia 308
  The Great Transformation—Can We Beat the Heat? 306
zoom 9, 14–15, 17, 43–44, 52, 54, 57, 183, 293

About the Team

Alessandra Tosi was the managing editor for this book.

Lucy Barnes and Rohini Bhonsle-Allemand performed the copy-editing and proof-reading. Lucy indexed the manuscript.

Katy Saunders designed the cover. The cover was produced in InDesign using the Fontin font.

Luca Baffa typeset the book in InDesign and produced the paperback and hardback editions. The text font is Tex Gyre Pagella; the heading font is Californian FB.

Luca produced the EPUB, AZW3, PDF, HTML, and XML editions — the conversion is performed with open source software such as pandoc (https://pandoc.org/) created by John MacFarlane and other tools freely available on our GitHub page (https://github.com/OpenBookPublishers).

This book need not end here...

Share

All our books — including the one you have just read — are free to access online so that students, researchers and members of the public who can't afford a printed edition will have access to the same ideas. This title will be accessed online by hundreds of readers each month across the globe: why not share the link so that someone you know is one of them?

This book and additional content is available at:

https://doi.org/10.11647/OBP.0303

Donate

Open Book Publishers is an award-winning, scholar-led, not-for-profit press making knowledge freely available one book at a time. We don't charge authors to publish with us: instead, our work is supported by our library members and by donations from people who believe that research shouldn't be locked behind paywalls.

Why not join them in freeing knowledge by supporting us: https://www.openbookpublishers.com/support-us

Like Open Book Publishers

Follow @OpenBookPublish

Read more at the Open Book Publishers BLOG

You may also be interested in:

Right Research
Modelling Sustainable Research Practices in the Anthropocene
Chelsea Miya; Oliver Rossier; Geoffrey Rockwell (eds)

https://doi.org/10.11647/OBP.0213

Global Warming in Local Discourses
How Communities around the World Make Sense of Climate Change
Michael Brüggemann; Simone Rödder (eds)

https://doi.org/10.11647/OBP.0212

Living Earth Community
Multiple Ways of Being and Knowing
Sam Mickey; Mary Evelyn Tucker; John Grim (eds)

https://doi.org/10.11647/OBP.0186

www.ingramcontent.com/pod-product-compliance
Lightning Source LLC
Chambersburg PA
CBHW040746020526
44116CB00036B/2964